THE MIDDLE EAST

U.S. Policy, Israel, Oil and the Arabs

THIRD EDITION

**Timely Reports to Keep
Journalists, Scholars and the Public
Abreast of Developing Issues, Events and Trends**

September 1977

**CONGRESSIONAL QUARTERLY
1414 22ND STREET, N.W., WASHINGTON, D.C. 20037**

Congressional Quarterly Inc.

Congressional Quarterly Inc., an editorial research service and publishing company, serves clients in the fields of news, education, business and government. It combines specific coverage of Congress, government and politics by Congressional Quarterly with the more general subject range of an affiliated service, Editorial Research Reports.

Congressional Quarterly was founded in 1945 by Henrietta and Nelson Poynter. Its basic periodical publication was and still is the CQ *Weekly Report*, mailed to clients every Saturday. A cumulative index is published quarterly.

The CQ *Almanac*, a compendium of legislation for one session of Congress, is published every spring. *Congress and the Nation* is published every four years as a record of government for one presidential term.

Congressional Quarterly also publishes books on public affairs. These include the twice-yearly *Guide to Current American Government* and such recent titles as *President Carter* and *Congressional Ethics*.

CQ Direct Research is a consulting service which performs contract research and maintains a reference library and query desk for the convenience of clients.

Editorial Research Reports covers subjects beyond the specialized scope of Congressional Quarterly. It publishes reference material on foreign affairs, business, education, cultural affairs, national security, science and other topics of news interest. Service to clients includes a 6,000-word report four times a month bound and indexed semi-annually. Editorial Research Reports publishes paperback books in its fields of coverage. Founded in 1923, the service merged with Congressional Quarterly in 1956.

Editor: Mark A. Bruzonsky.
Copy Editor: Margaret Thompson.
Editorial Coordinator: Laura B. Weiss.
Research Staff: Robert E. Healy, Mary Neumann, W. Allan Wilbur.
Art Director: Richard Pottern. **Staff Artist:** Terry Atkinson.
Indexer: Susan Henry.
Production Manager: I. D. Fuller. **Assistant Production Manager:**
Maceo Mayo.

Book Department Editor: Robert A. Diamond.

Third Printing

Library of Congress Catalog No. 74-5252
International Standard Book No. 0-87187-113-0

Cover Photo: National Aeronautics and Space Administration.

Table of Contents

Summaries, Tables, Maps

Editor's Note

THE MIDDLE EAST: U.S. POLICY, ISRAEL, OIL AND THE ARABS (Third Edition) provides an up-to-date, in-depth analysis of the issues and disputes involving the countries of the Middle East. Emphasis is upon United States involvement in the entire region, historically and at present.

This Third Edition covers the dramatic events which have been shaping political and economic realities throughout the area. The controversial Palestinian question, the tragic events of the Lebanese civil war, the continuing debate about the future of the Israeli occupied territories, the Carter administration's evolving Middle East policies, and the issues surrounding Middle East oil are all highlighted. In addition, background information contained in the previous two editions, published in 1974 and 1975, has been incorporated.

THE MIDDLE EAST (Third Edition) is divided into three parts:

Part I contains country-by-country profiles and a chapter on Arab history.

Part II provides analyses of the major issues affecting Middle East political and economic life: Middle East diplomacy, U.S. Middle East policy, Middle East oil, the Palestinian problem, chapters on the Arab and Israeli lobbies in Washington, and a background chapter on the Arab-Israeli wars during the past three decades.

Part III is a three-section appendix. Biographies of leading figures from Middle East history during the 20th century come first. A detailed chronology covering major Middle East events from 1945 through August 1977 follows. Last is a comprehensive bibliography noting books, articles, reference sources and Congressional documents.

Introduction

Peace between Israel and the Arabs remains exceedingly difficult to accomplish. While the Arab-Israeli tragedy may have reached a turning point in the late 1970s, there appears to be little agreement on what to expect. Rarely has the future of an entire region seemed so promising and yet so dangerously uncertain.

The Changing Middle East

Unprecedented historical changes are taking place throughout the Middle East. Saudi Arabia and the four mini-states bordering the Persian Gulf are the centers of a fantastic money explosion created by the ever-rising world demand for petroleum. Egypt, with the largest population of any Arab country, desperately seeks an escape from poverty through the infusion of massive amounts of foreign assistance and the application of Western technology. Lebanon remains fractured after the recent years of chaotic destruction and may never reclaim her position as a center of commerce and the playland of Arabdom.

Israel and Iran, the two non-Arab countries in the area, are also in a period of transition. The Jewish state, still living in isolation from its Arab neighbors, will face largely altered conditions should a peace settlement be achieved or should another war erupt. Iran, already the regional superpower, is attempting to make the leap to a European level of industrialization.

The other countries covered in this volume are also undergoing rapid change. King Hussein's Jordan has prospered despite the loss of territories on the West Bank of the Jordan River. Yet the future for the three million Jordanians continues to be uncertain, especially with Palestinian nationalism now asserting a claim in the area. Syria's regime, deeply involved in turbulent Lebanon, faces internal challenges and remains the least willing of the four confrontation states bordering Israel to consider normalization of relations with Israel.

Iraq and Libya nurture and finance what has come to be termed "the rejection front," a coalition of Palestinian factions, leftist intellectuals and Moslem fanatics unwilling to accept any form of Jewish state in their midst.

Some of these changes occuring throughout the region have made a compromise peace between Israel and the Arabs possible, maybe for the first time since the conflict flared as British forces withdrew from Palestine three decades ago. Yet, despite the enormous diplomatic efforts underway, such a peace continues to be, at best, elusive.

Even if achieved in principle, a compromise settlement along the lines the Carter administration began advocating early in 1977, would be only a gamble. Still, to simply allow the massive arms build-up to continue within such a volatile political environment would be even more risky.

American Involvement

Politically and economically, there has been a significant turn toward the West by the Arab Middle East in the mid-1970s. Egypt and Saudi Arabia have led the way. The United States has become, indisputably, the key outside participant in the affairs of the entire region. Whether the measure be political involvement, arms sales, economic and technological aid, cultural interchange, or efforts to

---

U.S. Policy Statement

"What of the future? Is it a future in which Israel's three million people try by force of arms alone to hold out against the hostility and growing power of the Arab world? Or can a process of reconciliation be started—a process in which peace protects Israel's security, a peace in which the urge for revenge and recrimination is replaced by mutual recognition and respect?

"America has a special responsibility and a special opportunity to help bring about this kind of peace. This comes about first of all because of our unique and profound relationship with the state of Israel since its creation more than a generation ago. Our sense of shared values and purposes means that, for Americans, the question of Israel's survival is not a political question but rather stands as a moral imperative of our foreign policy....

"It is precisely because of our close ties with both Israel and her Arab neighbors that we are uniquely placed to promote the search for peace, to work for an improved understanding of each side's legitimate concerns, and to help them work out what we hope will be a basis for negotiation leading to a final peace in the Middle East."

Vice President Walter F. Mondale
June 17, 1977

resolve Arab-Israeli tensions, the United States plays a pivotal role in today's Middle East.

The Carter administration's efforts to prevent another Arab-Israeli explosion point up crucial American national interests which would be seriously jeopardized should a new war occur. *(Carter administration policy, p. 80)* These interests include access to the region's petroleum resources and trade markets, the survival and welfare of Israel which enjoys a unique relationship with the United States, continuing friendly relations with most states in the area, and minimization of the Soviet Union's political and military influence.

Economically, Saudi Arabia and Iran dominate the region with a combined Gross National Product approximately half that of the entire area. American involvement with both countries is considerable and still growing. Each is engaged in a rather frantic rush to modernize, and each is experiencing changes from traditional ways which are as unavoidable as they are unpredictable.

Iran's preoccupation is with internal diversification and industrialization in anticipation of the expected depletion of her oil reserves by the turn of the century. The Shah is building a gigantic military machine to protect his country from external challengers, as well as from internal subversion. American arms, technological transfers and political support are crucial to the Shah's plans.

Saudi Arabia's preoccupation is with regional stability and a healthy international economy. Without both, there would be a potential for serious tension in the still growing American-Saudi friendship—the cornerstone of all Saudi policies. U.S. support is a bulwark for the Saudi political order against potentially hostile neighbors or possible internal disorder.

Since 1974 the American-Egyptian friendship has blossomed. President Anwar Sadat has twice visited the United States and the American economic aid program to Egypt now surpasses all aid being channeled to the rest of Africa and Latin America combined. Sadat's Egypt has made a basic decision to align itself with American and Saudi interests.

Though American-Israeli differences have become more pronounced since leadership changes brought Jimmy Carter and Menahem Begin to office, the American-Israeli partnership remains strongly rooted and sure to weather the current storms. Israeli vitality has attracted the American spirit and Jewish history has nurtured both sympathy and admiration. Though the U.S. may seem to some to be pushing Israel into an undesirable, unfair and dangerous settlement, others believe Israel is tenaciously clinging to positions the U.S. cannot accept. *(U.S. policy, p. 80; Middle East diplomacy, p. 65)*

Culturally, the modernization taking place throughout the Arab Middle East and Iran, made possible by the industrial world's thirst for petrochemical products and by the Organization of Petroleum Exporting Countries' (OPEC) price rises, involves a curious blending of traditional Islamic ways with modern Western life-styles. London, for example, is being partially recreated as a modern-world Arab capital and meeting ground. While the Britishers seek the summer sun, many Arabs prefer to vacation in the English fog and mist. Meanwhile, students from throughout the region flock to American and European universities where they experience new ways of life in addition to academic instruction. And Western businessmen and tourists bring to the Middle East more than raw technology and Western standards.

U.S. Palestinian Policy

As this book went to press the administration took steps to give meaning to President Carter's advocacy of a "Palestinian homeland," first made in a statement March 16, 1977.

At a news conference July 28, 1977, President Carter indicated that "The major stumbling block" to the Geneva Conference and an attempt to work out a comprehensive settlement "is the participation of the Palestinian representative." The President followed this statement with an offer to begin formal and direct discussions with the Palestine Liberation Organization and to endorse a role for the PLO at peace negotiations once a goal of coexistence is accepted.

This basic alteration in American Middle East policy did not begin with the Carter Administration. Henry Kissinger, in many ways the architect who made possible the contemporary quest for a comprehensive settlement, considered the need for a new approach to the Palestinian predicament during the 1975 "reassessment" of U.S. Middle East policy. But he chose to pursue more limited goals. Questioned by journalist Edward Sheehan as to why he did not devise a new American Palestinian policy, Kissinger responded, "Mr. Sheehan, do you want to start a revolution in the United States?"*

But President Ford and Kissinger took a major initiative in November 1975 when State Department official Harold Saunders testified before Congress. "In many ways, the Palestinian dimension of the Arab-Israeli conflict is the heart of the conflict," said Saunders. Because the PLO "has not stated its readiness to negotiate peace with Israel; and Israel does not recognize the PLO or the idea of a separate Palestinian entity," Saunders continued, "we do not at this point have the framework for a negotiation involving the PLO." In his conclusion, Saunders noted that "It is obvious that thinking on the Palestinian aspects of the problem must evolve on all sides. As it does, what is not possible today may become possible."

Less than two months after becoming President, Carter decided to formally endorse the idea of a "Palestinian homeland." In doing so, he chose a phrase which brought back memories of the 1917 Balfour Declaration which first gave international recognition to Zionist aspirations with a call for a "national home for the Jewish people."

Carter's statement came at the very time the Palestine National Council, the supreme policy-making body of the PLO, was holding its thirteenth meeting in Cairo. The President's initiative was clearly designed to encourage Palestinian policies emphasizing goals of negotiations and a settlement rather than unending struggle.

President Carter's July and August positions gave further signs that the United States had decided to continue efforts to achieve new PLO policies that would justify direct diplomatic contacts, Palestinian participation at peace negotiations, and eventual creation of a Palestinian state in some form.

*The Arabs, Israelis, and Kissinger (Reader's Digest Press: 1976), p. 167.

Psychologically, the Arabs are finding it imperative to adjust to a world where they must take account of other views. The reality of Israel coupled with the sheer strength and determination of Jewish nationalism have recently begun to penetrate Arab thinking, raising the previously unspoken possibility of peaceful coexistence with Israel in the context of an overall settlement. For the first time, Egyptian President Anwar Sadat and other Arab leaders have begun to speak of a complete and permanent peace with Israel, including diplomatic relations, within a few years of a political settlement. Pursuing such a settlement has become the dominant political effort the U.S. is making in the region. Having willingly accepted a role as intermediary and conciliator, the U.S. now finds itself being thrust into an arbitrator's capacity as well.

The Arab-Israeli Conflict

The deeply entrenched conflict between Israel and the Arabs is the central focus of this book. The confrontation is broadly affected by the historical transformations taking place in the countries of the region as well as by the mounting influence of oil and petrodollars. Yet, the basic clash of aspirations remains much as it has been since the early decades of this century. Many years before the actual outbreak of international warfare between Jews and Arabs in 1948, it had become evident that Palestine would be a major arena of conflict. Zionism and Palestinian nationalism had developed in near-complete isolation from each other. Competing Western imperial interests further exacerbated differences. At the end of World War II, the Cold War grafted Great Power politics onto the local and regional clashes.

Since the Yom Kippur War in October 1973 the conflict has partly returned to its original form—a struggle between competing national movements. One, Zionism, inextricably rooted in Jewish history and religion, has already achieved phenomenal results. The other, Palestinian nationalism, arising from a feudalistic, colonial past, has been a mounting response to the Zionist challenge.

The Palestinian national movement has made tremendous strides and suffered serious setbacks during the past decade of its rebirth.

Almost universally considered homeless refugees just a few years ago—as in the famous 1967 reference in Security Council Resolution #242 to "a just settlement of the refugee problem"—the Palestinians are now widely recognized to be stateless, even though some of them have found temporary homes. President Carter's advocacy of a "Palestinian homeland" represents a historic shift in American Middle East policy. With Carter's recognition of the legitimacy of the movement for Palestinian self-determination has come an increasingly widespread awareness of Palestinian aspirations.

It was Israeli occupation of the West Bank and Gaza Strip during the 1967 Six-day War which caused resurgence of the idea of a separate Palestinian state. From 1948 through 1967 the West Bank had entered a process of Jordanization, and the Gaza Strip existed as a largely forgotten, mostly destitute, Egyptian-administered ghetto. With Israeli occupation came the rebirth of a dormant Palestinian identity, one for which the Palestine Liberation Organization has achieved a large measure of international legitimacy.

Still, the Arab-Israeli conflict has been considerably more than a competition between Zionism and what can now be termed Palestinianism. Ever since Israel's creation, Arab states have championed the Palestinian cause. And this cause, until recently, has been the elimination of a separate Jewish national existence in historic *Eretz Israel* (the Biblical Land of Israel). As long as that had been the goal for which the most powerful Arab leaders were willing to sacrifice, there was no possibility for a negotiated settlement.

Zionism, the belief in Jewish peoplehood and the right to a national existence in *Eretz Israel*, is a widely shared aspiration of Jews throughout the world. Moreover, America's commitment to Israel is a unique one, transcending more transient foreign policy concerns. As English author Henry Fairlie has stated, "If America does not ensure the survival of Israel, the American people will endure a despondency of spirit beside which their defeat in Vietnam will appear as one restless night."

Today, the possibility of a Palestinian state coexisting with the Jewish state—an approach debated and approved by the United Nations after World War II—has again returned to the agenda of international diplomacy. If Jewish and Palestinian aspirations can each be given partial fulfillment, possibly the hatreds and phobias of past decades can be contained.

The Historic Opportunity

In the introduction to the second edition of this book, it was noted that during Henry Kissinger's "shuttle diplomacy" in 1974 and 1975 that "such fundamental questions as permanent boundaries, international guarantees, the future of the Palestinians and the status of Jerusalem have deliberately been shelved while the Egyptian president and the Israeli premier have haggled over a few kilometers of desert in the Sinai Peninsula."

The Carter Administration entered office aware that a historic opportunity had presented itself for reversing the bitter course of Arab-Israeli relations through seeking a comprehensive political settlement. Turning the parties away from a fifth war has become a major challenge for Washington.

What is envisioned is not good neighborliness within a period of months, but a commitment by all the parties to a coexistence which can then be molded into a firm, lasting peace within a period of years. Israeli scholar Shimon Shamir has said: "A comprehensive settlement is not necessarily identical with a historical termination of the Arab-Israeli dispute and a complete normalization of relations between the two societies—for those can be achieved only through a protracted process which a diplomatic act is more apt to initiate than to conclude."

Surely the time has come in the Middle East when reason should control emotions, when intercultural empathy should transcend chauvinistic impulses, and when a vision of a productive future should finally triumph over memories of the fiery past.

Mark A. Bruzonsky
August 1977

COUNTRY PROFILES

LONG-STANDING STRUGGLE FOR NATIONAL SURVIVAL

In mid-1977, Israel was a troubled, anxious country. Increasingly politically isolated, economically strained to meet military expenditures, diplomatically feuding with its single patron, the United States, the Jewish state was in the midst of considering various paths toward the future.

The war of October 1973 and its aftermath brought hardships of the kind Israelis have overcome repeatedly in their short, remarkable history. But it also posed new questions that were difficult to formulate, let alone answer. Settling a new land, recovering from the Nazi holocaust, establishing a new state, building a modern economy...these tangible needs absorbed the best of Israeli energy and thought.

Of all the needs, none was more clear than defense against the surrounding Arabs, whose constant goal was to end the Zionist dream of a Jewish state on land the settlers considered their own. Israelis felt they were waging a struggle for survival, which greatly simplified the question of relations with the Arabs. They could be viewed above all as enemies and dealt with primarily through military force. Since the 1973 war, however, Israel's relations with its Arab neighbors can no longer be viewed in such simple terms.

A number of Arab states, including Egypt, Jordan and Saudi Arabia, have suggested they might be willing to coexist with a Jewish state of Israel if Israel would withdraw from the territories occupied in the 1967 war and allow the creation of a Palestinian state. By 1977, some Palestinians gave guarded support to this possibility, though the Palestine Liberation Organization (PLO) officially remained glued to its Palestine National Charter which calls for the end of the Jewish state.

Furthermore, the Carter administration soon after coming into office in January 1977, made statements which appeared to support the idea of Israeli withdrawal from the occupied territories and creation of a "Palestinian homeland" in exchange for real peace leading to normalized relations between Israel and her Arab neighbors.

By mid-1977 Israel was involved in a diplomatic battle with her major supporter, the United States, over the outline of a possible settlement that might be discussed at a resumed Geneva conference; a propaganda battle with the Arab states over which party was being intransigent—the Arabs about living in peace with Israel or Israel about withdrawing from Arab territories; and an economic battle at home where taxes, inflation and unemployment (new to Israel) were combining to cause an increase in the rate of emigration and social unrest.

On May 17, 1977, all these problems combined to result in a stunning defeat for Israel's Labor Party at the polls. After 29 years of continuous rule, ever since Israel was founded, Labor was replaced by the right-wing, nationalist opposition headed by the former underground leader in pre-state days, Menahem Begin.[1]

---

1 Begin's history in the underground Irgun, which often resorted to terrorist tactics, is detailed in J. Bowyer Bell's *Terror Out of Zion* (St. Martin's Press: 1977) and also in his own book, *The Revolt*, written in 1948 and republished by Nash Publishing in Los Angeles in 1972. Begin was once asked how it felt to be the father of terrorism in the Middle East. His response: "In the Middle East? In the whole world!" (noted in "Book World," *The Washington Post*, June 26, 1977, p. E5).

Israeli Prime Minister Menahem Begin

Impact of 1973 War

The most obvious shock of the war that began on Yom Kippur, the Day of Atonement, was the improved performance of the Arab armies. Although Israeli forces were poised on the brink of a smashing victory when the United States and the Soviet Union imposed a cease-fire, the first days of the war had revealed a dramatic change from the Arab collapse in the Six-day War of June 1967.

In a carefully coordinated surprise attack, Egyptian forces crossed the Suez Canal, and Syrian armor attacked in the Golan Heights. The spectacle of Israeli soldiers in retreat and even captivity was the reverse image of what Israelis had learned to expect from previous wars. More troubling still was the unfamiliar success of the Arabs in using surprise and in achieving coordination. As the battle dragged on, it became clear that the Arab forces were employing sophisticated modern weapons with deadly skill; Soviet-supplied SAM-6 missiles and anti-tank weapons cost the Israelis heavily in planes and tanks. When the cease-fire ended the fighting after 16 days, about 2,500 Israelis had been killed. For a population of only 3.2 million, the loss was severe. Had the United States suffered proportional casualties in a war, it would have lost 150,000 men (50,000 Americans died in eight years in Vietnam).

The cost in human life was felt in most Israeli homes, and the cost of future wars seemed likely to prove even higher. Most disturbing of all, perhaps, was the realization by Israelis that even their overwhelming victory in 1967 had not forced the Arabs to sue for lasting peace. If military triumph could not ensure Israel's safety, what could?

By the end of the war it was clear that Arab forces were still no match for Israel's army, air force and navy, but it was not at all clear how Israel could match a new weapon used by the Arabs with awesome results. By placing an embargo on oil shipments to the United States, selectively cutting supplies to other countries and raising oil prices, the Arabs had shaken the industrial world.

The impact in the United States revealed a conflict between this country's special commitment to Israel and its need for assured supplies of oil from the Middle East. The embargo also created strains between the United States and its European allies, who, being much more dependent on Arab oil, showed vexation at what was considered excessive U.S. support for Israel.

A new uncertainty in the future of U.S.-Israeli relations thus appeared at the very moment when Israel's dependence on U.S. support had been demonstrated by an airlift of vitally needed military equipment in the midst of war and an emergency aid package totaling about $2.2-billion. Then, during peace negotiations in the spring of 1975, the United States suspended consideration of Israel's new $2.5-billion aid request pending "reassessment" of its Middle East policies, exerting thinly veiled pressure on the Israeli government to be more forthcoming with concessions to Egypt. And though Vice President Walter Mondale publicly assured Israel in June 1977 that "we do not intend to use our military aid as pressure on Israel," doubts remained about what forms of pressure the United States might again resort to if Israel refused to make the concessions decided necessary by the new Carter administration in Washington to achieve a peace settlement.

European nations and Japan, heavily dependent on Arab oil, have grown increasingly impatient with Israeli occupation of Arab territories taken in the 1967 war. In June 1977, the European Economic Community formally went on record supporting the creation of a "Palestinian homeland" and repeating its insistence that Israel withdraw from the territories occupied 10 years earlier. Third World countries have shown growing hostility in the United Nations. Only Bolivia, the Dominican Republic and the United States voted with Israel in unsuccessfully trying to prevent Yasir Arafat, head of the Palestine Liberation Organization, from being the first non-governmental representative to address the United Nations General Assembly.

Israel's characteristic reponse to its growing isolation has been a reaffirmation of its tradition of stubborn self-reliance. Digging in for what then Prime Minister Yitzhak Rabin foresaw as "seven lean years," the government devalued the Israeli pound by 43 per cent in November 1974, sharply curtailed the market for consumer goods and maintained even higher levels of military mobilization than before the war. Numerous small devaluations have followed—often on a monthly basis—while Israel's military strength has continued to increase.

And then in May 1977, Israelis ejected at the polls the socialist Labor Party—the party of Israel's founding father David Ben-Gurion and of former Prime Minister Golda Meir—and replaced it with the anti-socialist, right-wing Likud coalition headed by the perrenial opposition figure Menahem Begin.

Many felt the Israeli people, by choosing the uncompromising Begin as their leader, were preparing for a difficult time with the United States. Others saw Begin's victory as less of a positive vote for him and more of a protest vote against the Labor Party for its indecisiveness on both domestic issues and international matters. Still others

saw Labor's defeat as caused by Israel's increasingly difficult economic problems and the series of scandals which just a month before the election brought about Prime Minister Rabin's resignation after revelation of his wife's illegal bank account in Washington, D.C., where he had served many years as ambassador. Veteran foreign affairs writer Tad Szulc suggested that the emerging outline of a Carter administration plan for a Middle East settlement that might eventually be imposed on Israel contributed to Begin's democratic coup d'etat.[2]

Signs of Stress

The strain showed in demographic figures. In 1974, the number of Israelis emigrating to other countries rose to 18,000, almost double the average annual exodus since the founding of the state. The number of immigrants in 1974 dropped to 32,000, compared with 55,000 in 1973. But the birth rate, indicating perhaps an instinct to preserve the nation, rose from 66,000 births in 1973 to 84,000 in 1974.

By 1977, official figures indicated that as many people were leaving the country as were coming. Unofficial figures showed the emigration rate in excess of the immigration rate. In January 1977, *The New York Times* reported that "a quarter of a century after the Zionist dream of a Jewish state was fulfilled, one out of every 10 Israelis [350,000] has left home" and about 300,000 reside in the United States. Though many of these Israelis insist they are only living abroad temporarily, most have probably left Israel for good, the *Times* reported. According to the Israeli Consul General in New York, "They are living here but their soul is in Israel."[3]

Reflecting on the "earthquake" of the 1973 war, Israeli novelist Hanoch Bartov felt that its powerful impact derived from the previous Six-day War, "in whose glowing aftermath we lost all sense of reality, all sense of what it still means to be Jewish in a hostile world." He continued:

> That "normalization of the Jewish people" for which we had so long yearned, that Zionist promise which had for so long eluded us, seemed finally within our grasp. Insecurity gave way to a sense of ease, a feeling that the sky was the limit. A nation which had always taken pride in achievements of the spirit felt entitled, for the first time in its history, to a share in the good things which come with material and military success.[4]

Psychologists and psychiatrists meeting in Tel Aviv in January 1975 found a high level of tension among Israelis after the war. "I find a restlessness among the population," psychiatrist Haim Dasberg reported, "quick switches in values and an increased willingness to take crazy risks." But the challenges posed by the war also stimulated morale. "The threat to survival is as natural to an Israeli as the air he breathes," according to Israel Stockman of Hebrew University. "I am so used to it that it gives a kind of exhilaration. I don't know if I'd want it any other way. It makes us a great people."[5]

Psychiatrist and foreign experts meeting in the United States also noted the war's centrality in Israeli life. At a State Department conference, Dr. Rita Rogers

2 See "How Carter Fouled the Israeli Election," *New York Magazine*, 6 June 1977. A review of the situation leading up to Begin's victory is in Bernard Avishai, "A New Israel," *The New York Review*, 23 June 1977.

3 James Feron, "The Israelis of New York," *New York Times Magazine*, 16 January 1977. Also see Amnon Kapeliouk, "Israeli Emigrants: Deserters or Realists?" *The Guardian*, 15 May 1977.

4 Hanoch Bartov, "Back to Abnormal," *Commentary*, March 1974.

5 *Newsweek*, Jan. 20, 1975, report on the International Conference on Psychological Stress and Adjustment in Time of War and Peace.

reported on interviews conducted in Israel. "All the subjects interviewed," Dr. Rogers noted, "clearly exhibited the existence of a new emotional calendar in Israeli society." "Even a simple remark to an Israeli such as 'That's a nice coat,' is replied to with 'I bought it before the war.' "[6]

Possibilities for Peace

The stalemate at the end of the 1973 war provided, many thought, the best opportunity yet to try to end the 25 years of hostilities between Israel and its Arab neighbors. But what would Israel's place be in a peaceful Middle East? Would it remain a tiny outpost of Western culture in the heart of the Moslem world? Or would it slowly be absorbed, like the medieval crusader states, in the civilization of the Middle East? Would the pre-Israeli pattern of relations between oriental Jews and Arab societies ultimately reassert itself? Or would Israelis play a leading role in the transformation of Arab societies? These questions perplexed Israelis as all signs in mid-1977 led to the realization that historical developments, sure to determine many answers to these questions, would be taking place within the coming few years.

Instead of attempting an overall settlement to the Arab-Israeli conflict after the 1973 war, Secretary of State Henry A. Kissinger embarked on a step-by-step approach involving what came to be termed "shuttle diplomacy." He eventually achieved two disengagement agreements in the Sinai between Egypt and Israel and a separation of forces agreement between Israel and Syria. But he was severely criticized for not capitalizing on the opportunity to attempt a comprehensive settlement, including a solution for the nagging Palestinian problem.

Former Under Secretary of State George Ball wrote in February 1976 of his "deep regret as I watched the United States turn its back on a serious effort to solve the problem in favor of a tactical maneuver that bought time at the expense of ultimate peace."[7] Journalist Edward Sheehan wrote that Kissinger achieved his successes by avoiding "the very essence" of the Middle East problem, the Palestinians.[8]

By the time the Carter administration took office in January 1977, there was general agreement that step-by-step diplomacy should be replaced with an effort for an overall settlement. But the questions concerning the form of such a settlement and how to reach agreement on it remained illusive and highly controversial.

Constant wars in the past enabled Israelis to concentrate on urgent, less complicated questions. Indeed, the militancy of the Arab world in some ways helped Israelis avoid the ambiguities that would accompany any peace process. By mid-1977, the new wealth of the Arab states, the oil weapon, pressures on the United States for "even-handedness" and Europe's waning support for Israel combined to feed the anxieties of a people that had brought much of the West with them in the search for their oriental roots. The peace process posed new challenges and new dangers.

For a short time, Israeli spirits were lifted by the July 4, 1976, raid on Entebbe Airport in Uganda by Israeli soldiers who freed 103 hostages being held by Palestinian terrorists. But such psychological lifts did not alter the basic political and economic realities of Israel's predicament.

By the summer of 1977, the search for a comprehensive settlement was underway again, and the new government of Menahem Begin appeared to be considerably apprehensive about what concessions and what risks would be demanded of Israel.

Zionist Background

Like the United States, Israel is a country founded and developed by immigrants from many different ethnic and cultural backgrounds. The Jewish population of Israel today is almost as hybrid as that of the United States. Since the founding of the state in 1948, more than 1.5 million immigrants have come to Israel from 101 countries. Israel encourages this "ingathering" of Jews from all parts of the world, counting on thir common Jewish heritage to help cement their union with other Israelis—a task that presents serious difficulties. This devotion to "ingathering" is the essence of Zionism, which might be called the founding "religion" of the Israeli nation. But even Zionism is coming under question in the changing temper of modern Israel.

Zionism is rooted in the reformist zeal that swept over eastern Europe in the latter part of the 19th century. It emerged from a ferment of nationalist, socialist, populist and utopian ideas that were inflaming the youth of that time and place. As nationalists, the Jews were not unlike other minority groups within the Russian and Austro-Hungarian empires that chafed under foreign rule. The Jews had a special impetus, however, because of anti-Semitic persecutions.

Zionism began with a small group that in 1882 formed an organization called Lovers of Zion.[9] These young Jews conceived the idea of sending groups of colonists to Palestine, then a neglected backwater of the Ottoman Empire, to establish Jewish communities in the land of their forbearers. A pamphlet distributed at that time projected the rationale of the new movement: "Everywhere we are rejected," it stated. "We are pushed out from everywhere. We are considered aliens.... [But] Judea shall rise again. Let our own lives be an example to our people. Let us forsake our lives in foreign lands and stand on firm ground on the land of our forefathers. Let us reach for shovels and plows...."[10]

Palestine at that time had a Jewish population of around 25,000, mostly descendants of refugees from the Spanish Inquisition and pious pilgrims who had come to the Holy Land to pray and die. They were very poor, very religious, and lived closed in on themselves in the largely Moslem cities. The resident Jews looked with hostility on the new arrivals, whom they considered dangerous radicals and religious renegades, while the newcomers were equally repelled by a life-style that seemed to contain the worst aspects of the ghetto life from which they had fled.

Earliest Settlements

The new arrivals moved beyond the squalid cities to the sparsely populated coastal plains, "a silent, mournful expanse, ravaged by centuries of warfare, fever, piracy and neglect."[11] This was the ancient land of the Philistines. Ruins of antiquity lay under shifting sand dunes. Much of the area was malarial swampland, the result of clogged

6 "The Psychological Impact of the Seventeen-Day War," Department of State, 29 March, 1974.
7 George Ball, "How not to handle the Middle East," *Atlantic Monthly*, February 1976.
8 See Edward F. R. Sheehan, *The Arabs, Israelis, and Kissinger* (Reader's Digest Press: 1976).

9 From Zion, the name of a hill in Jerusalem on which King David's palace stood.
10 Amos Elon, *The Israelis: Founders and Sons* (Bantam edition: 1972), p. 92. The formal founding of the Zionist movement took place at an international congress held at Basel, Switzerland, in 1897, under the leadership of Theodore Herzl, an Austrian journalist whose book, *The Jewish State*, published in 1896, had a powerful influence in marshaling support for the Zionist idea.
11 Elon, *ibid.*, p. 109.

waterways. Despite the unpromising scene and the difficult climate, a succession of immigrants succeeded over the next few decades in founding several dozen communities. Land was purchased from absentee owners with funds that came largely from philanthropy, including large sums given by Baron Edmond de Rothschild in Paris. Many could not stand the rigors and left, but those who stuck it out managed to establish a viable agriculture, although they came to depend increasingly on Arab farm labor.

The settlement of Jewish Palestine, later Israel, was marked by a series of waves of immigration, known by the Hebrew word *aliyah*, meaning ascension (to Zion).[12] The first *aliyah*, 1882-1903, brought some 20,000-30,000 Jews to Palestine. The second *aliyah*, 1905-1914, which brought 35,000-40,000, was the formative one that set the tone for the nation-to-come, the one that produced the first leaders of independent Israel.[13] These immigrants came as unattached young men and women, in their late teens or early 20s, burning with zeal to create a utopia. They believed that only through socialism could a society be created free of the evils of selfishness, materialism, exploitation and the aberrations that produced anti-Semitism. "What had so dismally failed in Russia (in the abortive revolution of 1905) some now hoped would succeed in one of the more destitute corners of the Ottoman empire—a safe haven for Jews, and a new paradise to boot. A kingdom of saints, a new world purged of suffering and sin."[14]

Although not religious in the traditional sense, they were driven by an intense, near-mystical devotion to their cause. Working the soil for them was not merely a pioneering necessity, but a sacred mission. They scorned all luxury, wearing the plainest clothes and eating the simplest food.

Founding of Kibbutzim

"The immigrants of the second *aliyah* brought with them to Palestine not only their powerful ties to Jewish history and traditions as well as to contemporary political and social movements...in their countries of origin, but also ideologies and principles concerning the nature and institutions of the Jewish community and society they intended to create." This wave of immigrants "did in fact later become the political, social, economic, and ideological backbone of the Jewish community in Palestine, and large sectors of life in Israel today are organized around institutions created by immigrants arriving in the Second *Aliyah*.[15]

These were the founders of the kibbutz, which author Amos Elon has described as "the single most powerful cultural force of the entire Zionist enterprise." The first kibbutz was founded in 1910 on swampland near Lake Tiberias in the Jordan Valley; it was the cheapest land available. The new collective community was named Deganiah ("God's Wheat"). Life was hell in this burning pesthole, and yet it was marvelously exciting. The average age of the group [of founders] was 20. They were like an extended family. They took their meals together in the communal dining hall and...talked and danced the hora and sang sad Russian songs until dawn."[16]

The kibbutz idea appealed to the most ardent idealists among the new pioneers, and soon other agricultural collectives of similar mold were established. Although the kibbutzim never held at any time more than 10 to 12 per cent of the Jewish population, they created the national ideal of the tough, vital, selflessly dedicated, sun-bronzed farmer-soldier and patriot. The kibbutz provided the nation's governing elite and some of its best soldiers. Tales of their heroism in defending Israel's borders and in its wars are legend; although only 4 per cent of the Israeli population has lived on kibbutzim in recent years, Kibbutzniks accounted for one-fourth of the fatalities in the Six-day War.[17]

Immigration

World War I brought Jewish immigration to a halt. Furthermore the Turkish regime, which was on the side of the Central Powers, expelled many Jews who had come from Allied nations. Other Jews left voluntarily because of deteriorating economic conditions. During the war the Jewish population of Palestine dropped from 85,000 to 56,000. But a third *aliyah* soon got under way, encouraged by the Balfour Declaration of Nov. 2, 1917, in which the British government, by then the occupying power in Palestine, expressed sympathy for the Jewish dream of a homeland. Between 1919 and 1923, the third *aliyah* brought 35,000 Jews to Palestine. They included mainly new contingents from Russia, similar in their drives and conditioning to the prewar pioneers.

A new wave—the fourth *aliyah*, 1924-1931, brought some 82,000 immigrants, mainly middle-class Jews from Poland. "An economic depression, combined evidently with anti-Semitism, touched off widespread economic, social and political sanctions and discrimination against the Jews in Poland and subjected them to increasing pressure toward emigration."[18] Another factor was newly enacted U.S. legislation restricting immigration from eastern Europe.[19] The new quotas enacted by Congress reduced the potential flow to the United States. This was also a period when Palestine's immigration from Asian and African countries rose from less than 5 per cent of the total during the third wave to 12 per cent during the fourth.

The fifth *aliyah*, 1932-1938, was massive; in all, 217,000 Jews came to Palestine in those years. The rise of Nazism in Germany and its expansionist moves in central Europe occasioned the first sizable influx of immigration to Palestine from Germany and Austria, as well as from Czechoslovakia, Hungary and Greece. The Nazi threat also accounted in large part for a renewed flow, totaling 91,000, from Poland. Unlike the pioneers, who were young, unattached and fired up about working on the soil, the *aliyah* of the 1930s included a number of settled middle-class families, headed by mature men who had made their mark in business and in the professions. Some of them came with appreciable amounts of capital. This group established a number of industries, commercial enterprises, financial organizations and cultural institutions that are still important in Israeli society.

Throughout the entire period of Jewish settlement, large numbers had left Palestine. Nevertheless by the end of

---

12 Detailed accounts of the various *aliyah* movements can be found in *Encyclopedia of Zionism and Israel* (Herzl Press, McGraw Hill: 1971).

13 Among them were David Ben Gurion and Ben Zvi, later prime minister and president.

14 Elon, *op. cit.*, p. 137.

15 Judah Matras, "The Jewish Population: Growth, Expansion of Settlement and Changing Composition," *Integration and Development in Israel* (1970), S. N. Eisenstadt, Rivkah Bar Yosef, Chaim Adler, editors, p. 310.

16 Herbert Russcol and Margalit Banal, *The First Million Sabras* (1970), p. 57.

17 The kibbutznik's good performance as soldiers is attributed in part to the hardiness engendered by kibbutz life and in part to the communal methods of child rearing which emphasized love of the land and intense group-loyalty to one's peers.

18 Matras, *op. cit.*, p. 311.

19 The Quota Act of 1920, revised and continued by the Immigration Act of 1924, imposed country-by-country quotas beyond the western hemisphere, greatly favoring emigrants from Britain, Ireland and Germany at the expense of most of the other.

1938 the Jewish population had risen to 413,000. There followed a slowdown of immigration due to Arab demonstrations and a general strike in protest against the large influx of Jews. The British government, ruling Palestine under terms of a 1920 mandate from the League of Nations, responded to that protest by ordering a reduction in immigration quotas and restricting Jewish land purchases.

This policy, set forth in a government White Paper on May 17, 1939, remained in force throughout World War II. Nevertheless, 75,000 Jews entered Palestine during the war years, 29,000 of them coming in illegally. They were mainly Polish, German, Romanian and Czech, but they also came from the Soviet Union, Bulgaria, Hungary, Austria, Yugoslavia and Italy. Jewish immigration from Middle Eastern countries also rose. Menahem Begin, who became Israel's Prime Minister in 1977, reached Palestine in 1942 while in the Polish army. By the end of the war, the Jewish population of Palestine stood at 564,000.

The next wave drew mainly from the 200,000 homeless Jews—Russian, Polish and German—who were living in so-called "displaced persons" camps after World War II. In the years 1946-48, 61,000 came to Palestine, nearly one-half of them slipping through a blockade imposed by the British to halt further immigration.[20] The ban on immigration drew protests throughout a world haunted by the disclosure of Hitler's death camps, and provided an additional ingredient to fuel a three-sided civil war developing in Palestine among the Jews, British and Arabs. A hard-pressed Britain notified the United Nations in 1947 it could no longer continue its role in Palestine and withdrew its forces the following year. With the ending of the British mandate, Jewish Palestine declared its independence, took the name Israel and abolished all restrictions on Jewish immigration.[21] A census of November 1948 recorded a Jewish population of 716,678, of whom 35 per cent were native-born.

A mass immigration then began that lasted until 1951. Within three years the population doubled to 1.4 million, 75 per cent of it foreign-born. The Jews came not only from the displaced persons camps but from other locations in Poland and Romania, which had lifted restrictions on Jewish emigration. And for the first time there was a large flow from non-Western countries. "This immigration changed the character of the Jewish population of Israel considerably," Judah Matras has written. In 1977, more than half of Israel's population was from North Africa.

Oriental Jews

Altogether in the 13-year period, 1948-1960, nearly one million Jews from 40 countries entered Israel to stay, and nearly one-half of them were from Asia and Africa. Together with a high birth rate, this flow accounted for a population growth in that period from 649,000 to 1.9 million. "The number of children under 10 years of age almost quadrupled, the Jewish population of Asian birth increased fivefold, and that of African birth increased 15-fold. Substantial previously unsettled areas of Israel were populated, the number of settlements more than doubled, and the rural population almost quadrupled."[22]

There has not since then been an immigration flow of this magnitude, although the stream continues, mainly from the Soviet Union. Despite the great emotional and financial support for Israel among American Jews, the exodus of Americans for settlement in Israel has always been modest, although it did pick up somewhat after the 1967 war. From 1949 through 1972, a total of 25,797 American Jews emigrated to Israel, according to figures provided by the Israeli embassy in Washington. Those who go there to settle may be given immigrant status, which may lead to Israeli citizenship, or some form of resident status.

In the years 1965 through 1968, 2,066 American Jews were given immigrant status; from 1969 through 1971, another 1,583; and in 1972, another 502. However, in 1969-71, a total of 12,920 American Jews were classified as potential immigrants—that is, they were given some sort of residence status after going to Israel. The figure for that category in 1972 was 3,361.[23]

Israel is now looking hopefully to the Soviet Union's three million Jews as the remaining large pool for supplying it with immigrants. A lowering of Soviet barriers to Jewish emigration accounted for an influx to Israel of more than 90,000 by the end of 1974. "We must prepare to absorb 250,000-500,000 newcomers in the next five years," said Minister of Immigration Nathan Peled in January 1973. "The final figure," he added, "depends largely on the policy of the Soviet Union." But the number coming from the Soviet Union has become much smaller in recent years and a campaign to "Free Soviet Jewry" has gained considerable support among American Jews.

Unifying and Divisive Forces

The heavy Jewish immigration was a prime necessity for realization of the Zionist dream, but it has presented many problems and it accounts for many of the tensions in Israeli society today. Its basic contributions were in producing the desired Jewish majority and providing manpower for defense, for economic expansion and for colonizing empty areas of the country. The post-1948 immigration alone accounted for the founding of 500 villages. Recent emigres from the United States and the Soviet Union have been a particular boon because they have tended to be well-educated and to possess needed skills; 40 per cent of them are in the professions. The Americans also brought in moderate amounts of capital; in 1971 alone, according to the immigration minister, new arrivals accounted for an investment of $150-million in various enterprises.

Israel's efforts to court immigrants of this caliber have caused resentment among earlier settlers and native-born Israelis who do not enjoy the same advantages. Housing is a particular sore point because of its scarcity. Couples who have been on a waiting list for an apartment for many months or years find it unfair that the new arrivals are so promptly settled into new quarters. The oriental Jews, many of whom arrived penniless a decade or more earlier, recall that they were first quartered in tents, and even now

---

20 The 1939 White Paper stated that after five years "no further Jewish immigration will be permitted unless the Arabs of Palestine are prepared to acquiesce in it." See "Palestine Crisis," *Editorial Research Reports,* 1948 Vol. I, p. 168.

21 Under the 1950 Law of Return, every Jew has the right to emigrate to Israel and to become a citizen of that country.

22 Matras, *op. cit.,* pp. 320-321.

23 The two organizations chiefly responsible for raising funds for Israel in the United States are the United Jewish Appeal and the Israel Bond Organization. UJA headquarters in New York is a tax-exempt organization under United States laws. It raises money for Jewish causes within the United States as well as contributions on a "people-to-people" basis for rescue, rehabilitation and recovery programs within Israel. Accordingly, UJA made direct contributions of $250-million in 1973 and $500-million in 1974 within Israel. The bond organization sells bonds and securities for the state of Israel, and funds are used for development programs, a spokesman for Israel Bond headquarters in New York said. These bond sales provide significant sums toward the overall development budget—for agriculture, industry, roads, civil aviation, harbors, refineries, pipelines, etc. Recent figures showed bond sales of $502-million in 1973 and $281-million in 1974.

many of them live with their families in crowded slums. There were also complaints that immigrant students were pushing native-born Israelis out of the universities. Immigrants' duty-free privileges for bringing in household goods added another grievance.

For the newly arrived Americans and Russians, on the other hand, the favored quarters given them were usually below the standard to which they were accustomed. There have been occasions when new arrivals from Russia have refused assignment to quarters in new towns or villages in development areas distant from the pleasures and excitement of the big cities. Israel apparently has its share of bored ex-urban housewives in fine, new, well-equipped homes, located in culturally barren "new towns."

Efforts to Equalize Schooling

Two institutions serve as the main vessels of Israel's melting pot—the schools and the army. The country provides 10 years of free, compulsory education, from kindergarten for five-year-olds through grade nine. There are two types of public schools—general and religious. The anomaly of tax-supported religious schools in the public system of a secular state is a tribute to the power of a small but high-pressure ultra-orthodox religious minority that has always held the balance of power in a coalition government. In 1977 this religious minority gained more power than ever as part of the Begin coalition government. Zevulun Hammer, of the National Religious Party, became Minister of Education.

Basic curricula are the same in all schools, but the religious schools naturally stress the religious aspects, and they keep up the ancient rituals that sustained Jews over the many years of wandering and persecution in the Diaspora. Such schools serve to counteract the assimilationist effect of public schooling by perpetuating the religious, and often political, conservatism of their founders.

Parents may send their children to either type of public school—or to private school, some of which are run by various Christian denominations. There is little coeducation between Israeli Arabs and Jews. By 1972, the Arabs had 225 kindergartens, 277 elementary, 66 secondary and 25 postsecondary schools of their own, and there were 500 Arabs attending Israeli universities. Educational attainment of the Arab population is much lower than the norm, and the university drop-out rate is high. Centralization of school administration has necessarily prevailed in the early years of nationhood, but efforts to decentralize are under way. Kibbutzim are free to operate schools in their own manner; these naturally make an effort to inculcate the young with the values of collectivization.

As in the United States, the schools have been set the task of attacking poverty and cultural alienation by equalizing educational opportunity for children from widely diverse home backgrounds. In the early years of the nation, the problem of the "educationally disadvantaged" was overlooked in the scramble to provide the physical facilities and staff to handle the sudden accretions of school population. About 14 years ago, the schools began developing special programs for children known as *te'unei tipuah*, meaning "requiring care." These were chiefly children of poorly educated or illiterate parents who had lived in poverty in Moslem countries before emigrating to Israel. Many suffer a language barrier in the schools. Hebrew is the official language of the country and therefore the language of its schools.

"We cannot yet claim a breakthrough," Yigal Allon, deputy prime minister and minister of education and culture, said in a report on education issued in January 1973. An indication of the failure is the decline of school attendance among children from Moslem countries as they reach higher grades. This group constitutes 63 per cent of kindergarten attendance, 60 per cent primary, 43 per cent secondary and 13 per cent university attendance. Tuition is lowered according to family income, even dropped, to encourage this group to stay in school beyond the compulsory-attendance years. University tuitions have gone up considerably for everyone, however, in recent years.

But the problem is not only economic but cultural. In some of the smaller towns and villages, where nearly the entire population is of Moslem-country origin, there is no opportunity for integrating children of Ashkenazi (European) and Sephardic (oriental) backgrounds in the classroom. Experienced teachers are reluctant to serve in these cultural backwaters. The sense of alienation of these people from the Israeli mainstream therefore tends to deepen. Educational television, recently introduced, is being directed toward ameliorating this situation.

Army Life, Role of Women

Army service is a welding influence because it is a near-universal experience and because the army itself, as an institution, stands so high in the regard of the people. All Israeli youth, of both sexes, come up for military service at age 18. Physical or educational handicaps are not necessarily a barrier to acceptance. Men serve three years and women for two years. The only exceptions are for religious scruples or, for a woman, marriage.

One of the conditions Prime Minister Begin accepted in order to gain the support of one of the religious parties that joined his government was that a mere declaration of religious opposition would be sufficient to keep a dissenting young woman from military service.

The Israeli army is unusual in that it depends more on the individual motivation of the soldier than on ritualistic forms to maintain discipline. It is egalitarian in spirit; there is little saluting and no hazing of green recruits. Everyone enters as a private. Officers come up from the ranks and are often addressed by their first names. They wear minimal insignia, and all uniforms are cut from the same cloth, though in recent years more medals and awards have been given out.

Army service does not have as alienating an effect on recruits as it might in other nations. Israel is a small country. Soldiers thus serve close to home and get frequent passes for home visits. The hitchhiking soldier is a familiar weekend sight. Furthermore, the army continues to be an important part of his life pattern well into middle age. He remains a member of the reserves until he is 55; as such he is back in the army for about 40 days every year and is subject to call-ups for emergencies at any time.

The generation gap in Israel has taken the odd form of a cooling off rather than a heating up of passionate rebellion in the young. The women's liberation movement, as it is known in the United States, has made little headway. Equal rights for women are written in the law and are regarded as a truism in Israel. But it was the pioneer woman of the past who shucked off the cares of homemaking and child rearing on the kibbutz to toil as an equal beside pioneer man. Young women in Israel today feel equal, there is an easy camaraderie between the sexes and career opportunities are open. But most of them opt for marriage and babies in their early 20s, and their employment tends to be in service jobs traditionally reserved for females. Even in the army their function is to do work that releases the male soldier for combat.[24] Although one of the nation's best known Prime Ministers, Gold Meir, was a woman—a pioneer woman—there have been few others of her sex in government.

For many years, the old guard has complained of a decline in idealism among Israeli youth. The complaint is that they are too concerned about "making good" in their personal lives. Actually the young are impatient with the oft-told tales of their elders about the heroic deeds of the pioneer past, and especially with sentimental nostalgia for the good old days. And many do not want to hear more about the holocaust, when so many Jews went passively to their deaths. "Why didn't our army save them?" an Israeli child asked when told of the Jewish tragedy.

This is a question from the new Israeli, confident, secure in his nationhood and no longer marked by the emotional hang-ups of the minority Jew. It has taken three generations, David Schoenbrun wrote, to produce the true Israeli. The early settlers, though they rebelled against the ghetto mentality, were nevertheless marked by it and inevitably passed on some of the malaise to their children. But the grandchild of the early settlers cannot conceive of anything but a free Israeli living on his own land. He is patriotic, but his attitude is "Why the fuss?" He is not interested in the old philosophic issues over which the pioneers wrangled—questions such as "What is a Jew?"

The change in outlook may spell the end of a distinctive style of Jewish humor, just as the adoption of Hebrew as the national language tossed aside the highly expressive Yiddish as a language of the ghetto Jew. "Sabra humor hardly excels in the kind of biting yet humane self-irony that is the hallmark of traditional [Yiddish] humor," wrote Elon. "It is cooler, a bit distant, or abstract, in the shaggy-dog style. In his humor, the sabra is critical, not of himself, but of the pathos of the older generation."[25]

Strains in the Social Order

There are strains in Israeli society that spell far more serious trouble ahead than the so-called generation gap, which is actually more a matter of manner than a clash of values. Three situations are potentially explosive. One is the social-cultural-economic gap between the oriental Jews and the so-called Ashkenazi, whose background and outlook is Western. Another is the tension between the ultra-religious and the secular Israelis, which has not abated over the years of their testy coexistence. The third has to do with Israel's future relations with the Arabs, both those within Israeli borders and the Arab world outside which has sworn eternal enmity to the very existence of a Jewish national state. This situation has become more pressing because of the issue of what to do about the "administered"—occupied—territories.

The tension between the oriental and Western Jew is sometimes compared with that of the black and white people in the United States. This is partly because the oriental Jew is darker skinned than the European Jew and falls low on every modern index of social well-being

---

24 Army chieftains told journalist David Schoenbrun that the army did not press induction on women who did not want to serve or whose families objected, as many religious families do. Although the army takes in low I.Q. or near-illiterate males and gives them basic education, it does not have enough teachers to give intensive education to all, "so we don't take women if they're a problem."—David Schoenbrun, *The New Israelis* (1973), pp. 144, 153.

25 Elon, *op. cit.,* p. 321.

Statistics on Israel

Area: 7,993 square miles.

Capital: Jerusalem; Tel Aviv is the diplomatic capital recognized by the United States.

Population: 3,591,000 (1977 est.).

Religion: Predominantly Judaism; Arab minority is largely Moslem; also Christian Arabs who are chiefly Greek Catholic and Greek Orthodox.

Official Language: Hebrew. Arabic is spoken by 15 per cent of the population. English is widely used.

GNP: $11.4-billion, $3,380 per capita (1975 est.).

—income, education, occupational status, quality of housing, etc. The black-white analogy was stressed when a group of dissident oriental Jews took the name of Black Panther and began to demonstrate in the classic manner of civil rights activists. Like American blacks, Sephardic Jews claim the system is stacked against them.

Concessions to religious orthodoxy have put a curiously theocratic overlay on secular Israel. Civil marriage is forbidden, in effect banning mixed marriage. Divorce laws are strict. Blue laws close down many facilities, including public transit, on the Sabbath. But the effect of such laws is hardly more than a nuisance compared with the pressures of the religious minority on major policies such as those pertaining to the occupied territories. The religious orthodox are chauvinist about this question and unbudging in their righteousness. They claim a Talmudic justification for holding on to the acquired lands, a belief also held by Prime Minister Begin. Religious influence in Israel substantially increased when the Begin government took power in June 1977. Begin's Likud Party relied on two smaller religious parties to form a government. The religious parties demanded certain cabinet seats and social reforms. For the first time in Israel's history the National Religious Party was assigned the Ministry of Education. This meant that Israel's political system allowed a minority of 15 to 20 per cent to determine many ways of life for Israel's secular majority of 70 to 80 per cent.

All issues and all tensions come back to the Arab question. Despite the terrorist attacks by Palestinian guerrillas and the retaliatory attacks on bordering Arab countries, Israel has maintained a relatively peaceful relationship with the Arabs within its borders. The Israeli policy is to try to integrate the Arabs into the Israeli economy, winning their cooperation by the demonstrated benefits to be gained by a rising standard of living while allowing them to maintain a separate cultural identity as a minority within the Jewish state. As it has worked out, the booming Israeli economy has come to depend more and more on Arab labor. To some Israelis, the dependence on Arabs to do the kind of manual labor the pioneers once gloried in is a negation of the Israeli ethos and the beginning of decadence.

But the relationship with "Israel's Arabs" has been questioned in recent years. Demonstrations in March 1976 spread from the occupied West Bank into Israel proper. There were six deaths of Israeli Arabs that month.

Increasingly, it seems, the once docile Arabs of Israel are identifying as Palestinians, seeing themselves aligned on

Ambassador Dinitz on Peace

Just before the Israeli election in May 1977, Ambassador Simcha Dinitz spoke with *The Middle East* editor, Mark Bruzonsky, about the possibilities of peace in the Middle East.

"For true peace," Ambassador Dinitz insisted, "major portions of the territories now held by Israel could be returned." But he stressed that so far the Arab leaders were only ambiguously speaking of an "end to the state of belligerency," not a real, lasting peace.

"Peace as defined by President Carter," the Ambassador stated, "is not only a declaration—definitely not merely a cessation of a state of war—but rather peace with components of realism in it; of open borders, of exchange of trade, of cultural exchange, of exchange of people, of exchange of tourists, of diplomatic exchange, etc., etc."

Such a peace is what Israel demands, Dinitz continued, because "we are living with a very transitory Arab world, an Arab world that can have a policy of accommodation today and a policy of confrontation tomorrow. Therefore, if we are to assure that the State of Israel's permanence, not its fact, but its permanence, is acquiesced to by the Arab people, by the Arab world, then something realistic has to happen, something that the man in the street in Cairo and Damascus and Amman will feel has happened in the Middle East. Only then will it be, not impossible, but difficult, for any subsequent Arab leader to change this reality by reneging on a commitment he took."

The entire interview with Ambassador Dinitz—along with a similar interview with Egyptian Ambassador Ashraf Ghorbal—appears in Worldview magazine, July-August 1977.

political and national matters with other Arabs, and especially with the West Bank Palestinians.[26]

"Deep mistrust underlies the entire relationship between Israeli Jews and the country's Arab minority," CBS News reported at the time of the March demonstrations. "Most Israeli Arabs, as a result of Israel's policy of unofficial apartheid, live apart in their decaying and neglected villages, ignored by the Jewish minority.... Every Arab citizen of Israel knows that at some point, no matter how qualified he may be, his path will be blocked by a closed door, that there's a line he cannot cross."[27]

This situation with the Arabs has not helped Israel's world image as a thriving democracy. Chaim Weizmann, Israel's first president, was quoted as saying in November 1947, the day after the U.N. vote to establish a Jewish state, "I am certain that the world will judge the Jewish state by what it shall do with the Arabs." Beginning in 1967, Israel found itself not just having a small Arab minority to deal with, but in control of nearly 1 million Arabs in the occupied territories. Israel's occupation of these Arabs remains a central element in the continuing Arab-Israeli conflict.

West Bank Occupation

A decade after taking over the area known as the West Bank (and also the Gaza Strip located along the coast some 35 miles south of Tel Aviv), Israel was still undecided about the future of the occupied territories. Dozens of Israeli settlements now dot the area and the Likud Party—which has come into power—ran on a platform of annexation. Yet soon after coming into power, Begin's government became ambiguous about the territories which the United States has repeatedly stressed must be considered both negotiable and returnable in exchange for a real peace. "The territory has not in any real sense been absorbed into Israel," the London *Economist* reported in April 1977. "Even east Jerusalem, except for the currency, the goods on sale (and the ring of half-empty Jewish housing estates), is almost as Arab as ever it was."[28]

But there is increasing tension in the areas and much apprehension that the Israelis will keep "creating facts," by building settlements, until at least some of the areas become part of Israel, even if only *de facto*. A *Sunday Times* story in June 1977 alleged that "Torture of Arab prisoners is so widespread and systematic" in the occupied territories "that it cannot be dismissed as 'rogue cops' exceeding orders. It appears to be sanctioned as deliberate policy."[29] Israel vehemently denied the charges in a detailed, case-by-case report. And the *Sunday Times* did acknowledge that "Very few if any of the Arab countries would emerge unstained from a comparable inquiry into methods used there by police and prison authorities." Still, there remained reports of brutal treatment and suppression over Israeli occupation policies.

The controversy over the future of the West Bank is intense in Israel. It involves considerations of what would happen to the very character of the Jewish state if it attempted to absorb an additional million-plus Palestinian Arabs. And it involves Biblical claims as well as security considerations. The Begin government appeared, upon election, to be strongly committed to retaining the areas under Israeli control if not sovereignty, leaving the Golan Heights and the Sinai to be negotiated over with the Arabs. Others in the opposition have disagreed. Yehoshafat Harkabi, for instance, one of Israel's leading Palestinian experts, stated in March 1977, "I personally would like to see Israel get rid of the West Bank. It's corrupting our soul."[30]

The PLO

But even if Israel decided to rid itself of the West Bank and maybe the Gaza Strip, who would it give it to? Jordan was relieved by the Arab Summit Conference in Rabat in 1974 of responsibility for the West Bank. Responsibility was given to the PLO. But Israel has refused to deal with the PLO. And though secret meetings in Paris had begun during 1976 between members of the PLO and a party of Israeli leaders formerly associated with the Labor Party, little had come of these attempts at reconciliation by mid-1977. In March, the *Economist* reported that "This motionless Israeli-Palestinian deadlock, as of two intertwined wrestlers each afraid to move for fear the other will hurl him to the ground, can be broken only inch by gingerly inch."[31] But by summer 1977, the Begin government had refused even to contemplate recognition of the PLO and the PLO remained committed to the Palestine National Covenant, which refuses to accept coexistence with a Jewish state in Palestine no matter what its boundaries. *(See Palestinian Question chapter, p. 109)*

26 For background on Israel's Arab minority see Sabri Jiryis, *The Arabs in Israel* (Monthly Review Press: 1976).

27 Tom Fenton reporting from Israel, CBS Evening News, 29 March 1976.

28 *The Economist*, 23 April 1977, p. 78.

29 "Israel and Torture," *Sunday Times* (London), 19 June 1977.

30 *Christian Science Monitor*, 8 March 1977.

31 The *Economist*, 12 March 1977.

The Economy and Preparations for War

While talk of peace and possible resumption of the Geneva Conference continued throughout 1977, Israel continued to plow about 35 per cent of its GNP and nearly half of its budget (if interest on the spiraling national debt mainly incurred for previous wars is included) into preparations for war. Former Defense Minister Shimon Peres had pushed Israel toward self-sufficiency in the production of many arms during the last years of the Labor government. But the attempt was proving extremely expensive.

"Israel has become an arms producer of major proportions," the *Jerusalem Post* reported in May 1977, listing fighter planes, machine guns, tanks, missile boats, mortars, and all kinds of ammunition.[32] It is also suspected that Israel has produced, or is capable of producing within a very short time, nuclear weapons.

While the costs of arms productions on such a scale are tremendous, it is felt a measure of political independence is gained and by producing at home, much of the money that would have gone into foreign procurement goes back into the Israeli economy. Concern has also been expressed that the huge amounts of American aid flowing to Israel since the Yom Kippur War could not be counted on indefinitely. "Any Israeli who expects the U.S. to continue pumping $2-billion per year into Israel—mostly for arms—is living in a dream world," the *Jerusalem Post* further noted quoting one source who said "The bubble is going to burst, and we have to ensure that our security is not totally jeopardized when it happens."[33]

Israel has also become an arms supplier nation. More than 20 nations, including South Africa, bought Israeli military products in 1977. Income to Israel was estimated as much as $500-million in precious foreign exchange in 1976 and could reach $1-billion by 1978.

Still, the arms burden is so heavy for a small country that there has been talk of having to rely on a nuclear force now or at some point in the future. "I think we have reached the limit...of the military burden the country can carry," former Defense Minister and then Begin's Foreign Minister, Moshe Dayan, stated in April 1977. *The Washington Post* went on to report that "It is not possible for Israel, with its limited resources, to contemplate forever matching the Arab world, tank for tank, and plane for plane. Dayan has advocated a 'two layer' military capability. 'Conventional weapons of a limited number' and a 'nuclear capability.' "[34]

U.S.-Israel Relations

In large part, Israel's fate will be determined by its relationship with the United States. And while the United States has repeatedly stressed the special and unbreakable bond of friendship that exists between the two countries,[35] significant strains over policies began to develop after the Yom Kippur War and grew through the Nixon, Ford and Carter administrations.

By mid-1977 the Carter government appeared to be moving toward support for at least two policies the Israelis have bitterly opposed: return to approximately the 1967 boundaries and creation of a "Palestinian homeland" in the West Bank and Gaza Strip. In late June 1977, just weeks before the first visit of Menahem Begin to Washington as prime minister, the State Department released a blunt statement reaffirming these views. It was "an act of rudeness unprecedented in the relationship between Israel and the United States," according to former Prime Minister Rabin, and it in a sense symbolized the unprecedented differences between the two countries.

By the summer of 1977 it appeared the lines were being drawn for a major political contest between the Carter administration and the Begin government; not over continuing American friendship or economic and military aid, but over what concessions the Israelis would be pressured into making.

After generally pro-Israeli *New Republic* weekly, soon after Begin's victory, summed things up in this way: "We cannot rule out a monumental political crisis between Israel and the Carter administration, and between the administration and Israel's friends in Congress, with Americans forced to choose up sides in a bitter fight."[36]

But just a few months before, the same journal stressed, in an article by English author Henry Fairlie, that "if America does not ensure the survival of Israel, the American people will endure a despondency of spirit beside which the defeat in Vietnam will appear as one restless night; not merely because America helped create the nation, but because the invisible Jew would once again be on their backs. The Jew has to have a nation, in order for us not to see him where he is not."[37]

The Future

Though Israel's problems have multiplied enormously since the pre-Yom Kippur War days, the Zionist vision of a Jewish state remains alive and strong. In April 1977 Labor Party leader Shimon Peres told a visiting delegation from the United Jewish Appeal, "We shall march forward together with you so as to attain, in the year 2000, a country of six million Jews, with a firm and sound economic infrastructure and with a political and social future guaranteed for our coming generations."[38]

But no such visions could mask the uneasiness felt by Israelis and American Jews throughout 1977. It was this sense of confusion and anxiety, above all else, which caused Israelis to elevate a new party and a new leader. Menahem Begin's appearance on the political scene caused considerable reevaluation about the peace process at first, but within a short time it seemed that that process as outlined by President Carter would have to continue and that Begin would have to cooperate.

Nevertheless, Israelis had clearly voted for a change and the change they voted for was a considerable one of style, policies and beliefs. For decades, in fact, Labor Party leaders had condemned Begin, for both his personal history and his policies. David Ben-Gurion, in fact, was said never to have spoken directly to Begin in the Knesset, but only to have referred to him with loathing as the leader of the opposition. Raanan Lurie, the Israeli who is an internationally syndicated political cartoonist, noted immediately after Begin's victory that "30 years ago [he] was the commander of the Israeli PLO, the exact equivalent."[39]

32 Hirsh Goodman, "The Army Imbroglio," *Jerusalem Post International Edition;* 31 May 1977, p. 12.

33 *Ibid.*

34 *Washington Post,* 30 April 1977, p. A16.

35 In his speech before a joint session of Congress on 28 January 1977, Prime Minister Rabin discussed the historical bonds between the two countries stressing "our common heritage" and our shared democratic values.

36 *The New Republic,* 28 May 1977.

37 *The New Republic,* 5 February 1977.

38 Jewish Telegraph Agency Report, 19 April 1977.

39 Public Broadcasting System, MacNeil/Lehrer Report, 19 May 1977.

But others saw Begin as a determined Zionist with an uncompromising background who would nevertheless turn statesman now that he had finally achieved power. And when he accepted territorial and political compromises, many pointed out, there would not be any major right-wing opposition to challenge him as would have been the case had Labor retained power.

Begin's flexibility, or lack of, to deal with the United States and the moderating Arab states is what was most in question, in fact, in mid-1977. Syndicated columnist Joseph Kraft wrote that Begin "is a true believer, about as flexible,

as I know from personal conversations, as Gordon Liddy"[40]—a reference to the Watergate figure who absolutely refused to cooperate with investigators. But Begin began showing signs of what appeared to be flexibility almost as soon as he took office. The consensus seemed to arise both among the Israelis and in the United States that Menahem Begin should be given a chance to lead Israel toward the peace and acceptance she has for so long struggled to achieve.

40 *The Washington Post,* 21 June 1977.

BECOMING A MIDDLE EAST SUPERPOWER

In October 1971, the Shah of Iran, Mohammed Reza Pahlavi, invited foreign dignitaries to celebrate the 2,500th birthday of his country's monarchy in an air-conditioned tent city beside the ruins of Persepolis. While kings and presidents, sheikhs and prime ministers enjoyed the creations of French decorators, French chefs and French costume designers, Western journalists snickered at these trappings of power and raised their eyebrows at the bill (more than $100-million, according to some accounts; $16.6-million, by official Iranian estimate).

Now the Shah wields the substance of power—oil—and the bill he has presented his erstwhile guests is staggering. The five-fold increase in the price of Persian Gulf oil since October 1973 has been partially inspired by the Shah's leadership of the oil exporting countries.

Iran's annual revenues from oil shot up from an annual rate of about $2.5-billion before the increase to more than $20-billion in recent years. The country's gross national product, growing for a decade at an average rate of about 10 per cent, vaulted 34 per cent in 1973 and a further 41 per cent in 1974. But a drop in demand for Iranian crude oil slowed the growth rate in 1975 by as much as half.

Western businessmen now flock to Teheran like flies to the honeypot, in the words of one Iranian entrepreneur. And heads of powerful industrial states shudder when his imperial majesty, king of kings, light of the Aryans, threatens further increases in the price of the fuel that drives their economies.

The Shah has used his new wealth to buy into those economies. Some of his government's transactions dramatized Iran's new status among the industrial giants of the world. Germany's legendary Krupp empire, in a deal announced July 17, 1974, sold to Iran, for an estimated $100-million, a 25 per cent interest in the cartel's steel and engineering division and gave the Iranian government the right to nominate a member of the supervisory board of the parent holding company. A similar interest has been purchased in a major West German manufacturer of power-generating machinery—Deutsche Babcock and Wilcox, A.G. Less publicized ventures have brought Iranian money and participation to other countries in both Western and Eastern Europe, Asia and the Middle East.

"Iran...[must] assure the security of the Gulf—which is also the security of your oil supply—don't ever forget it."

—Shah Mohammed Reza Pahlavi

In 1975, Iran proposed, then dropped, a $300-million loan-purchase investment in Pan American World Airways, ailing partly as a result of high oil prices. The rescue financing was to be part of Iran's planned $10- $12-billion long-range U.S. investments. An Iranian official cited Pan Am's "internal affairs" and a new order of Iranian priorities for development projects as the reasons for dropping the plan. Also, Iran's oil production was running $3-billion short of 1974 levels by mid-1975. In need of funds for the increasing number of development projects and investment undertakings, the Shah has continually been in the forefront of the Organization of Petroleum Exporting Countries' (OPEC) efforts to keep oil prices continually rising.

Iran has negotiated large economic agreements with and has provided Italy, in exchange for technological aid, with $5-billion in credits to help reduce its large balance-of-payments deficits. Iran also has promised more than $10-billion in aid to developing countries, of which $3-billion had been allocated by mid-1975.

In 1974 alone, Iran undertook credit and loan commitments totaling $8-billion, divided equally between developing and industrialized countries. These commitments obligated about 6 per cent of Iran's GNP for "development assistance."

Iran and France have entered into an agreement for collaboration in energy development and industrialization worth about $6-billion. France also planned to aid Iran in the construction of at least five nuclear reactors.

But it is with the United States that Iran has the greatest ties. In May, 1975, the two countries signed a trade agreement designating $15-billion over a five-year period for American products—equally divided in arms, nonmilitary trade products and disbursements for development of Iran's industries, food supplies, housing and public services. An additional $7-billion was agreed to for six to eight nuclear power plants to be provided by the United States. During 1976 a 4-year trade package involving about $50-billion was being negotiated.

Iran's massive Fifth National Development Plan (1973-1978) calls for fixed investments totaling $70-billion from both the public and private sectors. Iran's GNP in 1975 was $50-billion, with the oil sector accounting for nearly half and industrial exports expanding about 30 per cent yearly. For a nation which in 1972 had a $15-billion GNP, Iran is indisputably a boom country.

Broadening the Base

The Shah is impatient to transform his own country, much of which remains an impoverished agricultural land, by establishing an industrial base that will preserve new living standards after the oil reserves are depleted. If existing production levels are maintained, known reserves will be used up by 1990.

Lower demands for oil from depressed industrial countries and inflation in the cost of imports are cutting into

Iran's financial power. The Shah, meanwhile, has decreed that Iran's income from other sources should equal its oil income within 10 years. His government predicts that by 1988 Iran will have a gross national product as large as France's in 1975, and larger than Britain's, although some economists view such projections with skepticism.

Western governments, faced with huge new payments deficits to oil-producing countries, are eager to cash in on Iran's development programs. Nonmilitary imports to Iran more than trebled during 1974 to $10-billion and since then each year has brought more imports. Secretary of State Henry A. Kissinger called the 1975 trade accord "the largest agreement of this kind that has ever been signed by any two countries."

Arms For Petrodollars

Massive sales of weapons to Iran have also eased the balance-of-payments burdens imposed on the United States by higher oil prices. The Shah's huge military budget rose 44 per cent in 1974 to a total of nearly $8-billion. Cash sales of military goods from the United States reached $2.1-billion in fiscal 1973, jumped to $4.8-billion the next year and are continuing at a high rate.

Between 1971 and 1977, Iran bought $15-billion in military equipment just from the United States. This makes Iran the single largest purchaser of U.S. arms in the world.

In 1974, Iran purchased six American destroyers, and in 1975 80 Grumman F-14 jets at a cost of more than $1-billion. This purchase included a $75-million high-interest loan from Iran's Bank Melli, which helped save Grumman from bankruptcy. Some 2,000 American ground and service personnel were sent by Grumman to Iran in conjunction with this sale. In 1977, many of the 31,000 Americans in Iran were there as a result of Iran's American arms purchases.

The weapons the Shah is purchasing include the most sophisticated available. Rockwell International has been involved in building an intelligence base from which Iran can monitor military and civilian electronic communications throughout the Persian Gulf region. Iran is paying billions for 160 F-16s (in addition to its F-14s) and is seeking 140 more. The Shah also wants F-18s and wide-bodied transport aircraft still under development. In April 1977, Iran ordered seven Boeing 707s worth $1.1-billion for an early warning system. The system, called AWACS (Airborne Warning and Control System), partially substituted for a stupendous $32-billion mountaintop radar system. AWACS was a bargain. Instead of 41 mountaintop stations, only 12 to 16 would be needed to compliment the AWACS. But due to congressional anxiety, the sale became questionable in August.

Strategic Concerns

Why does Iran want such an arsenal? It is not enough to worry Russia, and more than enough for defense against Iran's other neighbors. Pakistan, wounded by the loss of Bangladesh, is too concerned about India to make trouble for Iran, with whom in any case it has friendly relations and common interests. Turkey has no recent or outstanding dispute with Iran and is preoccupied with Cyprus. Afghanistan is too underdeveloped to challenge Iran; Iraq has only a third of Iran's population; the other Persian Gulf states are much smaller. In the longer perspective of Iranian history, the Shah's fascination with modern weapons may find some justification. The country is full of minorities—Arabs, Turks, Kurds, Qashqais, Baluchis, Turcomans, Lurs—with turbulent pasts and no deep love for the government in Teheran. Some of these groups have brethren in neighboring countries, and the great powers have not hesitated to encourage local or tribal rebelliousness in order to bring pressure on the central government. Other groups in Iran—religious, indeological, socio-economic (e.g., the bazaar merchants in Teheran)—have risen against the government in the past and might do so again at the first sign of weakness at the top.

The size and modernity of the Shah's army inspires awe among would-be rebels at home and their supporters next door, persuading them of the hopelessness of opposing his rule. As long as his armed forces seem overwhelming, rebellions will not even begin, Iranians will appear to acquiesce in his autocracy and foreign governments will see few opportunities to meddle.

From the beginning of the 19th century until after World War II, Iran was squeezed between the Russians to the north and the British in India. By balancing their rival imperialisms, Iran preserved at least nominal independence, although British influence was felt by Iranians to be pervasive in their own country. The withdrawal of Britain from India in 1947 and the subsequent waning of British power east of Suez forced Iran to seek another balance to Russian pressures, and the United States gradually assumed that role.

No matter how many weapons he buys, the Shah obviously could not withstand a Soviet attack—he must rely on the Americans to deter that. But Soviet influence with Iran's other neighbors troubles him. He expressed this concern in an interview in October 1973:

> The Cold War is over. But the question with Soviet Russia will always be the same and, when negotiating with the Russians, Iran must remember the chief dilemma: to become Communist or not?... There exists what I call the USSR pincer movement. There exists their dream of reaching the Indian Ocean through the Persian Gulf.[1]

The Shah's fears have historical validity. Soviet attempts to absorb parts of his country after World War II can never be far from the Shah's mind. That Soviet move, which the United States prevented, continues to provide validation for the assertion by Soviet Foreign Minister Vyacheslav Molotov to Nazi Foreign Minister Joachim von Ribbentrop, in November 1940, that "The territorial aspirations of the Soviet Union lie south of Soviet territory in the direction of the Persian Gulf."

A 1973 coup in Afghanistan has proved less pro-Russian than the Shah feared, and his March 1975 agreement with Iraq may have reduced the reliance on Moscow of that country's radical government. In exchange for the end of Iranian support of the Kurdish rebellion, Iraq probably gave Iran private assurances that it would cut back its support of potential rebels in Iran and on the Arab side of the gulf. (p. 34)

The conservative government in Oman, at the southern tip of the gulf, enjoys the support of Iranian troops and planes in its miniature war in Dhofar Province against guerrillas backed by the avowedly Marxist regime in South Yemen. It is a minor skirmish in a primitive society inhabiting some of the most desolate territory on earth, but it serves the Shah as a symbol of hegemony.

His solicitude for the stability of the gulf makes the Shah, in the eyes of some Western strategists, the natural ally of

---

1 Interview with Italian journalist Oriana Fallaci, quoted in *The Economist*, May 17-23, 1975, Survey, p. 74.

the West in the region. In its May 1975 survey of the Persian Gulf region, *The Economist* noted the Shah's effort to make Iran a "countervailing power against Soviet influence" in the region, and particularly his interest in helping to patrol the sea routes through the northwestern Indian Ocean. "There are excellent reasons why he should be helped to do this," *The Economist* said.

> These sea routes are the most unprotected parts of the Western economic system, and so long as inflation continues it is unlikely that the American or British or French navies will be able to afford to give them adequate supervision. The Shah's interest in the Indian Ocean complements the West's very neatly, and it should be the aim of the governments in Washington and London and Paris to coordinate the two as closely as possible.[2]

In the same survey, *The Economist* outlined possible Western military action in the Persian Gulf to seize oil fields if either intolerable oil price increases, or a new embargo sparked by the Arab-Israeli conflict, left the West no choice. This scenario and similar ones in American magazines located the hypothetical invasion in Saudi Arabia, not Iran. But the Shah has made clear that he wants protection of the gulf to be left entirely to the states of the region with no interference from *any* outside powers.

His efforts to establish regional cooperation at least raise the question whether the arms Iran is purchasing in the West might someday be used to protect Persian Gulf oil from the West. The mere possibility could make Western powers more hesitant to try a military response to future oil price increases and thus, in effect, raise the price level, which they would consider tolerable. Should Iran use its forthcoming nuclear power potential for weapons development, the cost of interference in the region by either the Russians or the West would rise substantially.

The Shah's long-standing ties to the United States and his continuing reliance on this country to counter Soviet pressures make scenarios of Iranian-U.S. conflict highly speculative as long as his regime is in power. His government has a remarkable record of stability in a region notorious for instability. The ongoing economic and social transformation of Iran, however, has the proportions of an earthquake. For all his eagerness to accelerate technological change, the Shah has gone to extremes to prevent political change.

He has divided and intimidated all opposition to his regime, and this policy has been so consistent that any liberalization could be interpreted by Iranians as a sign of weakness and an opportunity to rebel. The result has been to prevent national debate on the most fundamental issues. But fundamental questions are being posed by changes in Iranian civilization, and the silence only hides historical forces whose power, sooner or later, will be felt. As long as the economy is booming, dissent will find few champions and few followers. But if the great expectations raised by the boom are later frustrated by slackening economic growth, the situation could become explosive.

In the traditional absence of formal political institutions entrusted with the advancement of the national interest, Iran's shahs have often been the sole embodiment of central authority and the principal architects of national policy. Since the turn of the century they have adamantly opposed any encroachment on their sovereignty by representative government. Because the present shah is no

exception, he provokes vociferous criticism from the left and arouses mild irritation among American policymakers who have been largely responsible for underwriting his ascendance. "It's not democratic leadership, but it's decisive leadership," a State Department official has said.

In the Shah's opinion, democracy itself is an irritation, a poor substitute for divine inspiration. His public candor before his people is possible, he explained to an American television audience, because he doesn't "fear and tremble before them" or "look at the polls to see if they're up or down two degrees."[3] Absolute monarchy is apparently easier to defend in a country that has experienced it for over a thousand years. At 57, the Shah himself is an amalgam of technocracy and royalty who is often portrayed as being ruthless yet religiously inspired. In an interview with the prominent Italian journalist Oriana Fallaci, the Shah confided to her: "I believe in God, and that I've been chosen by God to perform a task."[4]

U.S.-Iranian Friendship

If the Shah has been ordained by God, he has also received able assistance from the secular powers of the United States. U.S. friendship with Iran dates from 1946, when official U.S. protests to the Soviet Union led to Soviet troop withdrawals from northern Iran.

As part of a World War II agreement with Britain, the Soviet Union and Britain were to station troops in Iran until the end of the war and then withdraw. The Soviet troops remained beyond the agreed date for their departure, and they provided the cover for a Communist-inspired separatist movement in the northern Iranian province of Azerbaijan. Following the Soviet troop withdrawal, the Iranian government put down the rebellion.

In 1947, a U.S. military mission was established, and a grant aid program was initiated in 1950. And in 1953, it has been alleged, CIA support restored the Shah to the preeminent position he currently enjoys. But his gratitude for American support has by no means diminished his desire for an independent foreign policy, even when that policy conflicts with Washington's. He amply demonstrated that independence, and celebrated a national holiday, by nationalizing the Western-held Iran Oil Participants Ltd. on March 21, 1973, the Iranian new year. Iran did not join the oil embargo imposed by the Arab states during their 1973 war against Israel, whose government Iran recognizes. But the Shah enthusiastically championed the increases in oil prices that began with the embargo.

Iran cannot be labeled an American "client" state. But its interests seem to coincide with America's own interests in the Middle East insofar as Iran's strength, rather than Russia's, tends to fill the "vacuum" left by the withdrawal of British forces "east of Suez" in 1971. Defense Department official James H. Noyes told a House Foreign Affairs subcommittee in 1973 that the United States has definite "security interests" in the Persian Gulf. The bulk of U.S. arms support goes to Iran because it is perceived to be the most stable and responsible guarantor of those interests, he said. Reliance on this concept of regional defense as a substitute for deterioration of the Central Treaty Organization[5]

---

2 *Ibid*, p. 11.

3 On "Sixty Minutes" (CBS-TV), Feb. 24, 1974.

4 *The New Republic*, Dec. 1, 1973, p. 16.

5 The organization had its origin in a pact of mutual cooperation signed at Baghdad initially by Iraq and Turkey and soon afterward by Britain, Pakistan and Iran, in 1955 at U.S. urging, to cover an area between the reaches of the North Atlantic Treaty Organization and the Southeast Asia Treaty Organization. Iraq withdrew in 1959. The United States is an associate member.

as a credible deterrent against potential Soviet aggression has logically contributed to the present U.S. Iranian policy.

25 Centuries of History

The Shah's festivities in 1971 to commemorate his country's 25th century as a continuous civilization served to remind the world of Iran's antiquity and cultural heritage. Ancient Persia, or Iran *(see box)*, extends back even farther than that time span, as Old Testament chronicles attest. However, the sixth century B.C. was when Cyrus the Great established the greatness of the Persian Empire which, with Darius' subsequent conquest of Babylonia and Egypt, extended from the Nile Valley almost to Asia Minor.

The empire shrank under the weight of Greek and Roman conquests and internal disintegration. By the seventh century A.D. it was beset by Arab invaders who brought with them the religion of Islam and foreign rule. The Arabs were gradually deposed, but the religion remained. Today more than 90 per cent of Iran's people belong to the Shia sect of Islam, the state religion. Together with the Arabs and the Turks, the Iranians played a major role in shaping Islamic civilization. Within the world of Islam, however, the Shia sect is a minority, retaining a formal priesthood and influential clerical establishment and differing in many ways from the orthodox Sunni majority.

Modern Iranian history begins with a nationalist uprising in 1905, the granting of a limited constitution in 1906 and the discovery of oil in 1908, which intensified a British-Russian rivalry that had already (in 1907) divided Iran into spheres of influence. During World War I the country underwent actual occupation by British and Russian soldiers, although it officially remained neutral in the war. The war interrupted the sporadic growth of constitutionalism. In 1919 Iran made a trade agreement with Britain that formally affirmed Iranian independence but in fact established a British protectorate over the country. After Iran's recognition of the new Communist government, Moscow renounced the imperialistic policies of the czars toward Iran and withdrew any remaining Russian troops.

A second revolutionary movement, directed largely by foreign influence in Iranian affairs, was initiated in 1921 by Reza Khan, father of the present Shah and founder of the current Pahlavi dynasty. In 1925 he was placed on the throne and made a start toward modernizing his country. But his flirtation with the Nazis led to another British and Russian occupation in 1941 and forced him to abdicate in favor of his son. Separatist Azerbaijani and Kurdish regimes, established under Soviet patronage in northern Iran, crumbled after Soviet forces were withdrawn in 1946 under pressure from the United States and United Nations.

Nationalism, Mossadegh and Oil

With the Soviet crisis over, Iran was left with the subtle but disturbing presence of the British. The most onerous symbol of British influence in Iran's domestic affairs, and one which served to bolster Iranian nationalism, was the Anglo-Iranian Oil Company. This consortium of British and American interests antagonized both the political right and left in Iran. The domestic political climate was further aggravated by deteriorating economic conditions resulting from the allied occupation during World War II. By 1951, the moderate center had rallied around the Shah, who had then been on the throne for 10 years without real power.

Hopeful of avoiding civil war, the Shah sought an accommodation with the oil industry. But the more radical demands prevailed. The Communist-backed Tudeh Party, largely middle class with strong support in the army, and the rightist National Front, a heterogeneous collection of groups led by the "opposition" nationalist leader in parliament, Dr. Mohammed Mossadegh, demanded nationalization of the oil consortium. Finally, on March 15, 1951, the Iranian Parliament voted unanimously for nationalization, and on April 27 named the 73-year-old Mossadegh as premier. Given the prevailing atmosphere, the Shah had little choice but to accede.

Britain and the United States took a dim view of those events, while the Soviet Union did curiously little to exploit the unrest. The British strenuously rejected Iran's claim of the sovereign right to nationalize. Secretary of State Dean Acheson later recalled that the United States feared that "Britain might drive Iran to a Communist *coup d'etat,* or Iran might drive Britain out of the country."[6] Although Iran did neither, what did happen was unsatisfactory to all parties. The oil industry refused to cooperate with the new national ownership lest it set an example in the Middle East. Consequently, oil production virtually stopped for the next several years.

New evidence brought to light by Senate Foreign Relations Committee hearings on multinational companies in March 1974 indicated the extent of oil company and U.S. government collusion in reacting to Iran's unsettling decision to nationalize and to the growing instability of the Mossadegh regime. Recently declassified State Department files made public at the hearings revealed concern as early as 1951 over monopolistic practices of American oil companies in the Middle East, including Iran, and urged that such practices be ended. The Justice Department responded with an antitrust suit. The suit was quashed, however, when in 1953 the National Security Council invoked claims of "national security." According to testimony from a former Exxon executive before the same subcommittee in April 1974, Secretary of State John Foster Dulles encouraged the industry's presence in Iran to prevent the country from falling under Soviet influence.

Internal Stability

By 1953, stability was restored in Iran. Mossadegh was deposed on Aug. 19 by royalist sympathizers with the apparent backing of the CIA. Whether by chance or by design, the outcome was that the Shah gained power and the foreign-owned oil industry returned. With Mossadegh imprisoned and his National Front allies in parliament reduced to marginal effectiveness, the Shah's retribution was quick and decisive. The Tudeh Party was smashed and hundreds of its members in the army's lower ranks were rooted out, as were the sympathizers in all significant interest groups and associations. It remained for the Shah to consolidate his strength by rewarding his supporters—chiefly in the officer corps—and make peace with the nationalist-minded clerical authorities. The oil companies were free to renegotiate more favorable terms, and Iran could expect continued financial assistance from the United States in addition to $45-million it had received immediately upon Mossadegh's ouster. For the remainder of the decade, the government applied itself to economic development.

But reform programs were not introduced until after 1961, when the results of two contrived national elections incited virulent criticism, bringing about the resurgence of the National Front and threatening the stability of the

6 Dean Acheson, *Present at the Creation* (1969).

Iran or Persia?

Iran is the ancient name for the country and its people. The word Persia gradually evolved from European usage of the word Fars, an ancient province in which the Iranian people were first known to have lived. In a spirit of nationalism in 1935, the Iranian government asked all foreign governments to use the name Iran rather than Persia.

government. In January 1963, the Shah announced the "white revolution," an ambitious plan to give social growth equal priority with production. The plan promoted women's suffrage, literacy and health, nationalization of natural resources, sale of state-owned factories and profit-sharing for workers. With the addition of eight new principles by 1975 to the original 1963 six-point declaration, the final charter for reform included 14 basic objectives.

The cornerstone of the "white revolution" was land reform. In most developing countries the concentration of land in the hands of a few wealthy owners is a serious obstacle to social reform, and Iran was no exception. In some regions a single landowner might own 50 villages—as many as 100 in parts of Azerbaijan. Predictably, the landed classes, with allies among the Shiite clergy, incited violent demonstrations in June 1963. Predictably, too, the Shah crushed them instantly. After that, reforms mollified those with moderate demands and repression silenced the others.

Regional Foreign Policy[7]

In foreign affairs, Iran's traditional preoccupation with the Soviet Union gradually gave way to a greater concern for what it perceived to be the more menacing threat of Arab socialism. But in 1955, Arab socialism was still in its infancy while the Soviet Union was foremost in mind. Without an adequate defense capability of its own, Iran that year joined the Baghdad Pact. Up to 1955, postwar relations with Russia had been officially cordial as both countries chose to ignore the Soviet-inspired insurgency in the northern provinces in 1945-46.

But Iran's participation in the Central Treaty Organization provoked a furious propaganda attack from Moscow. Above all, the Russians feared Iran would permit the United States to station nuclear missiles on Iranian soil, as Turkey had done when it joined the organization. Iran assured the Kremlin in 1959 that it would not do so. Russia ceased its propaganda activity against Iran, and relations between the two countries steadily improved thereafter. Beginning in 1962, the year Iran assured the Kremlin that no foreign missiles would be allowed on Iranian soil, Iran and the Soviet Union entered into a series of extensive commercial ventures, the most significant being the 1966 agreement whereby Russia built Iran's first steel mill, at Isfahan, in exchange for large quantities of natural gas piped from southwestern Iran to the Soviet interior.

Ironically, in strengthening its ties to the Soviet Union during this period, Iran also managed to improve its relations with the United States. After it joined the Baghdad Pact in 1955, Iran was dissatisfied with the low level of arms shipments it received and soon wanted a stronger defense commitment than the United States was

willing to provide. But the American attitude changed considerably as Iran's relations with the Soviet Union began to thaw, especially in 1958, when Iran announced its intention to seek a nonaggression pact with the Russians. Instead, the result was a friendship treaty with the United States in 1959. By skillfully playing off both sides to its advantage, Iran had substantially improved its position with the great powers by 1960.

But by that time Iran had calmed its fears of Soviet belligerence only to replace them with equally intense suspicions of Arab socialism in the Persian Gulf. When, in July 1958, a military group overthrew the Hashemite monarchy in Iraq, relations between Baghdad and Teheran quickly deteriorated. Up to that point, Iran had maintained reasonably good relations with radical Arab regimes in Syria and Egypt, and with conservative monarchies and sheikdoms in Saudi Arabia and Oman. But the cumulative effect of the Iraqi coup was to nurture fear that other friendly monarchies would fall.

After witnessing the U.S. reluctance to back its CENTO partner, Pakistan, against India over Kashmir in 1965, Iran began to entertain serious doubts about the value of its defense alignment with the United States. It drew the conclusion that it would have to rely on its own strength to protect Iranian interests in the Persian Gulf. Unlike Iraq, Kuwait and Saudi Arabia at the head of the gulf, which have pipelines to transport their oil to the Mediterranean Sea and the West, Iran must export its oil entirely by water through the gulf and the Indian Ocean. The measure of Iran's heavy reliance on uninterrupted sea transport is the measure of its vulnerability. Since the British departure from the gulf, this very dependence on the Persian Gulf for access to the Indian Ocean seemed to dictate that Iran would become its *de facto* guardian.

Problems of Gulf Stability

The evolution of Iranian military and economic power has changed Iran's regional policy in the past decade from a reactive one to an assertive one. "Iran has gone out of her way to impress both local and outside powers with her ambitions and her potential for realizing them," stated the London-based International Institute for Strategic Studies.[8] These impressions were forcefully conveyed by a series of successful military ventures conducted during the four years between the announcement and implementation of Britain's withdrawal east of Suez. The Labor government of Prime Minister Harold Wilson announced in July 1967 that Britain must abandon its historical role east of Suez and the following January set a timetable for withdrawal of military forces from Singapore, Malaysia and the Persian Gulf by the end of 1971. Before 1968, British troops had left Aden, a seaport fortress vital to British interests in the Indian Ocean since the 19th century.

Iran's first move came in April 1969 and grew out of a long-standing dispute with Iraq over navigational rights in the Shatt al-Arab, a broad estuary which is of strategic importance to both countries, especially for Iran as an access route to its oil facilities.[9] Iran contested Iraq's nominal control of the waterway when it sent a freighter with jet fighter

7 See "Iran: The Making of a Regional Superpower," pp. 470, 471 in A. L. Udovitch (ed.), *The Middle East Oil, Conflict and Hope* (Lexington, N.Y., 1976).

8 "The Soviet Dilemma in the Middle East, Part II: Oil and the Persian Gulf," by Robert E. Hunter, *Adelphi Papers* of the International Institute for Strategic Studies, October 1969.

9 Iran has since moved the bulk of its refineries to a safer site near the desert-edged port on Kharg Island.

escort through the mouth of the estuary and successfully defied Iraqi threats of retaliation, which never materialized.

Finally in March 1975, in return for an end to Iranian support of the Kurdish rebellion, Iraq accepted the Iranian claim that the deep-water line in the Shatt al-Arab defined the boundary.

Another dispute, also resolved by gunboat diplomacy, was Iran's subsequent seizure of Abu Musa and the two Tumbs, three gulf islands near the Strait of Hormuz. After protracted and futile negotiations with the British over its claims to the islands, Iran landed troops there on Nov. 30, 1971, despite the protests of the Arab littoral states. To placate the Arabs, Iran dropped its ancient territorial claim to the large gulf island of Bahrain after a U.S.-sponsored referendum revealed that the Bahrainians preferred their independence to Iranian dominion.

It has been suspected that when the Shah pressed his claims to Bahrain, he did so with the sole purpose of relinquishing them with apparent magnanimity, as he did in May 1970, to obtain his real objective, the three islands in the strait.[10] Hence, by the end of 1971, with the passing of the last vestige of Pax Britannica, Iran had consolidated its position in the Persian Gulf. Success, however, was bought at the price of encouraging speculation about Iranian expansion. But both the United States and the Soviet Union and the littoral states generally seem content with Iran's frequent denials of territorial ambition. Nevertheless, in view of the rapid arms build-up in the gulf, weaker littoral states cannot cling to that belief with certainty.

Consequences of Modernization

If stability—however temporary—characterizes the Persian Gulf, it also characterizes the gulf's chief guardian.

The government of Iran appears to be stable and, despite the threat of assassination, the Shah's reign is as free of usurpation as it has been since he assumed power. There were serious mass demonstrations in 1976 protesting the Shah's changing from the Islamic calendar to one based on the coronation of Cyrus the Great. But there had not been such public disorder in over a decade. Only remnants of the Tudeh Party have survived; the several hundred remaining members are badly splintered among Soviet and Maoist factions. The Kurdish tribesmen in northwest Iran, unlike their Iraqi brethren, have been comparatively docile since 1946.

More important, the Shah continues to enjoy the enthusiastic loyalty of the armed forces. Aside from the close historical bonds that have prevailed between the Persian military and the monarch, the Shah has been directly responsible for the army's rising pay and stature, to the extent that some fear that one day a military elite will, like Frankenstein's monster, arise against its creator. But that is not regarded as likely in the forseeable future.

Furthermore, the Shah has neutralized the opposition of the religious hierarchy to the process of modernization by denying effective political power to its spokesmen and by imprisoning some of its leaders. And the urban middle class, the instrument of development programs, is unwilling to risk losing its prosperity by translating criticism into action. If the Shah has not earned their enthusiastic devotion, he at least has bought their diffident cooperation.

Yet it is sometimes said that the Shah has paid too high a price. Just as his methods of preserving gulf stability draw

criticism, so do his methods of dealing with domestic affairs. An Iranian economist at the International Monetary Fund told Editorial Research Reports that in his country "development is always linked to torture." A U.N. panel has documented gross violations of human rights, and *U.S. News & World Report* reported on Aug. 6, 1973, that "about 70 persons have been executed for subversion in the last several months." The secret police (SAVAK) has become a ubiquitous and efficient means of intimidation, according to Iranian political dissidents. It is reported that students have suspected demonstration leaders of being paid police informers, because only they could afford to be so daring.

In March 1975, the Shah dropped even the pretense of a two-party system. The majority Iran Novin Party and the opposition Mardom Party were combined into a new "National Resurgence" movement. The new and only party is called Rastahiz. The regime organized a campaign to turn out votes in the June 20, 1975, elections in which the new party put up 700 candidates for 268 seats in the national

> *"Iran is one of the most important allies the United States has. When you realize that 50 per cent of the world's oil comes through the Straits of Hormuz and the only armed forces to protect it are Iran's, to refuse him [arms] would be sheer stupidity on the part of the United States."*
>
> Senator Abraham Ribicoff
> (D Conn.), July 1977

assembly and 30 in the senate. Although the Majles (Parliament) functions as a rubber stamp for the Shah's programs, his government was anxious to give voters the impression of participating in politics. Premier Amir Abbas Hoveida, head of the National Resurgence, has told Iranians that those who failed to vote would have to explain their failure to the party, which would regard it as a breach of discipline.[11]

In his 1961 autobiography, the Shah wrote: "If I were a dictator rather than a constitutional monarch, then I might be tempted to sponsor a single dominant party such as Hitler organized or such as you find in Communist countries. But as a constitutional monarch, I can afford to encourage large-scale party activity free from the straitjacket of one-party rule or the one-party state." But the change to a single party is not granted deep significance by traditionally cynical Iranians, who hitherto had jokingly boasted that Iran had not one but two puppet parties.

While this cynicism has prevented the Shah's reforms from dashing expectations among the upper classes—since they are practiced at political indifference[12]—some intellectuals and critics have labeled the Shah's reforms cosmetic substitutes for real changes. They say the real purpose of land reform, for example, was never to redistribute land but to break the back of the aristocracy. When modernization is

---

10 "The Persian Peninsula, Iran and the Gulf States: New Wealth, New Power" (1973), a summary of papers presented at the 27th annual conference of the Middle East Institute.

11 *New York Times*, June 16, 1975.

12 See Marvin Zonis' *The Political Elite of Iran* (1971), and J. A. Bill's "Plasticity of Informal Politics," *Middle East Journal*, spring 1973.

Statistics on Iran

Area: 636,293 square miles.
Capital: Teheran.
Population: 34,424,000 (1977 est.).
Religion: 98 per cent Moslem, with great majority members of the Shia sect; 5% of Sunni sect. Minority groups include Jews, Bahais, Zoroastrians and Christian Armenians and Assyrians.
Official Language: Persian or Farsi. Kurdish, forms of Turkic, Arabic spoken by leading minorities.
GNP: $50-billion (1975 est.), $1,600 per capita (1976 est.).

conducted from the top, it invariably thwarts political participation from below. Although the prospects of a coup appear remote, the rapid growth of the middle class may increase pressures for republican government.

Great Power Intervention

Iran's success in its new role as caretaker of the Persian Gulf ultimately hinges not on its own strength but on its ability to convince the United States and the Soviet Union that the continued assertion of its power is necessary to preserve regional stability. It has done so thus far, wrote Professor Sepehr Zabih in stating the consensus viewpoint in the February 1974 issue of *Current History*. The Persian Gulf is of such vital strategic importance as an energy source that the Soviet Union and the United States have been unwilling to risk any provocation that might impair the oil lifeline.

Iran has been the chief beneficiary of this understanding, because in serving its interests in the Persian Gulf, it has convinced the great powers that it serves their interests as well. Whether this tacit understanding can endure if the tenuous Soviet-American detente begins to unravel is a matter of concern for gulf states. A renowned Iranian political scientist, Professor Roulhollah K. Ramazani, is less than sanguine about the future of great-power cooperation. "The gap between the rhetoric of the Soviet-American rivalry will probably widen in the near future," he observed, "largely because both superpowers, in fact, perceive their interests in the Persian Gulf to be too important to allow them to pursue a policy of complete non-involvement."[13]

Iran still faces gigantic problems. A majority of the people remain deeply conservative and traditional and many cultural forces oppose the secular industrialization the Shah is demanding.[14] There continues to be much apathy and cynicism throughout the country.[15] Many of the nation's 61,000 small villages are backward, impoverished areas.

And in international affairs, there has been increasing criticism of the Shah's repressive rule. The important relationship with the United States, still strong, has undergone strains. President Carter's new emphasis on human rights and arms sale reductions has caused anxiety for the Shah. Already, in May, 1977, Secretary of State Vance met with the Shah and a Middle East newsweekly reported that "It is understood in Teheran that Iran will lose her virtual carte blanche with U.S. arms merchants—given to the Shah by former President Nixon in May 1972."[16] By mid-1977 it appeared that the $4-billion sale of 250 Northrop F-18s to Iran might not be approved, and Congress was debating the desirability of selling Iran the seven Boeing 707 AWACS planes for which a tentative deal was made in April.

13 Rouhollah K. Ramazani, *The Persian Gulf: Iran's Role* (1972), p. ix.
14 See the six-part *Washington Post* series on Iran by Richard T. Sale, 8 May through 13 May, 1977.
15 See "Iran's single party rule shaken by apathy," *Events*, 3 June 1977, p. 25.
16 "U.S. takes a second look at arms deals with Iran," *Events*, 3 June 1977, p. 34.

BIRTH OF ISLAM TO SEARCH FOR NEW DESTINY

The greatest historical achievement of the Arabs was the development and spread of Islam. It was an astonishing achievement.

In the Arabian peninsula on the fringe of the civilized world, a lone prophet in the seventh century converted the pagan Arabs to a faith which they carried, within a century, to distant Spain and the steppes of central Asia. This was not merely the propagation of a religious doctrine; the compartmentalization of religion as a separate aspect of life has always been unthinkable in Islam. Respected Arabist William R. Polk notes that "Islam must be thought of as at once a religion and a social order.... Islam, the religion of the overwhelming majority of the Arabs, permeates every aspect of society."[1]

As it conquered Persia and pushed back the Byzantine Empire, Islam absorbed much of both cultures but transformed what it borrowed into a new and distinctive world view and way of life. At its height in the ninth and 10th centuries under the Abbasid caliphate in Baghdad, this flourishing culture included Christians, Jews and Zoroastrians among its creators, Greek philosophy among its intellectual stimulants and Persian administration and statecraft among its political principles. But it was dominated by the religion and language of the Arabs, whom Mohammed led out of Arabia. "Mohammed," English scholar Peter Mansfield writes, "was undeniably one of the few men who have permanently changed the world.... Through him this small race of lean and hungry camel-riding nomads, who provoked a blend of contempt and fear in the contemporary civilized world, became one of the most potent forces in the history of mankind."[2]

Even in the decadence that followed, the Crusaders that began to arrive in 1096 found a culture far more sophisticated than the Europe they had left behind. Later waves of Moslem expansion were to carry the message of Mohammed to India and even Indonesia and the Philippines; under the Ottoman Empire the armies of the prophet reached, in 1529 and again in 1683, the gates of Vienna.

Arabs' Decline

Historians trace the beginnings of Arab decline as far back as the 10th century, although the fall of the Abbasid caliphate to invading Mongols in 1258 is the customary benchmark for the end of Islam's classical period.

Turks migrating from central Asia converted to Islam and soon dominated its armies. They gained *de facto* political power in Baghdad in the 11th century, and after the fall of the caliphate, Turkish minorities increasingly dominated the governments of the Middle East. The major creative cultural impulse within Islam, meanwhile, passed from the Arabs to the Persians, whose Indo-European language, written now in Arabic script and enriched by Arabic vocabulary, provided the dominant literary medium for this next phase of Moslem history. The Arabs were eclipsed in the civilization they had launched long before

Who Is An Arab?

It is not easy to define accurately the term "Arab." The British geographer W. B. Fisher, in his book *The Middle East: A Physical, Social and Regional Geography,* stated: "From the point of view of the anthropologist, it is impossible to speak with accuracy either of an Arab or of a Semitic people. Both terms connote a mixed population varying widely in physical character and in racial origin, and are best used purely as cultural and linguistic terms respectively." Thus the so-called "Arab" countries are those in which Arabic is the primary language and which share a common culture.

As Islam spread from the Arabian peninsula, what took place has been described as the "twin processes of arabization and islamization"—closely linked but not identical. As Peter Mansfield writes, "Arabization began some two centuries before the Prophet Mohammed, with the overflow of Arabian tribes into Syria and Iraq, and reached its greatest impulse during the first decades of the Arab Empire. Islamization lasted much longer and still continues today, especially in Africa. The consequence is that although Arabic language and culture retain a special and predominant place in the world of Islam, only about one-fifth of the one-sixth of mankind who are Moslems are Arabic-speaking."[4]

Currently the Arab countries are Egypt, Syria, Jordan, Lebanon, Iraq, Saudi Arabia, Yemen Arab Republic, South Yemen, Kuwait, a series of tiny sheikdoms along the Persian Gulf, and the North African states known collectively as the Mahgreb—Algeria, Morocco, Tunisia and Libya. Islam is the predominant religion in all of these countries. But that in itself does not define an Arab country. Turkey, Iran, Afghanistan and Pakistan are Islamic, but not Arab, nations. The restive Moslem Kurds, who inhabit areas of Iraq, Iran and Turkey, have their own culture and religion and are also not Arabs. And there are Christian minorities in Arab countries—some with ancestral roots antedating Moslem conquests and others converted by missionaries—who speak Arabic and consider themselves Arabs.

the Ottoman Turks in the 16th century brought almost the entire Arabic-speaking world under their control.

"Through a millenium of Turkish hegemony," historian Bernard Lewis writes, "it came to be generally accepted that Turks commanded while others obeyed, and a non-Turk in authority was regarded as an oddity."[3] It would be a mistake to read back into the past, however, an Arab resentment of foreign rule such as the 20th century takes for

1 William R. Polk, *The United States and the Arab World* (Third Edition), (Harvard University Press: 1975), p. xi.

2 Peter Mansfield, *The Arabs: A Comprehensive History,* (Crowell: 1977), p. 22.
3 Bernard Lewis, *The Middle East and the West,* (Harper Torchbooks: 1966), p. 20.
4 Mansfield, *op. cit.,* pp. 45-46.

granted. Modern nationalism is a recent import from the West into the "Arab world," where religious, local and family identities had more political significance than ties to the diverse and scattered peoples who spoke the Arab language. In its heyday, the Ottoman Empire brought glory to the community of the faithful, and the fact that the sultans were Turks did not make their state illegitimate in the eyes of the Arabs.

By the 18th century, however, the Ottoman Empire had decayed from within and no longer could protect Moslems from the advancing civilization of the "infidel" West. The ease with which Napoleon occupied Egypt in 1798 shocked Moslems' faith in their superiority over the West, and the next century saw Arab peoples increasingly dominated by European power and wealth.

The early history the Arabs had ill prepared them for this fate. In its formative century, their religion had swept all before it, and even conquerors of the Arabs had bowed before the revelation of their prophet. Islam, like the Judaism and Christianity it recognized as predecessors, was distinguished by the ultimate importance it attached to history. But unlike those older Semitic faiths, which had early been forced to cope with defeat and persecution, Islam took shape amidst triumph after triumph, a historical experience that seemed to confirm that the Koran was, indeed, God's final revelation.

What is the modern Arab, then, to make of his own recent history? Even after political liberation, Bernard Lewis observes, "the intelligent and sensitive Arab cannot but be aware of the continued subordination of his culture to that of the West." Lewis continues:

> His richest resource is oil—but it is found and extracted by Western processes and machines, to serve the needs of Western inventions. His greatest pride is his new army—but it uses Western arms, wears Western-style uniforms, and marches to Western tunes. His ideas and ideologies, even of anti-Western revolt, derive ultimately from Western thought. His knowledge even of his own history and culture owes much to Western scholarship. His writers, his artists, his architects, his technicians, even his tailors, testify by their work to the continued supremacy of Western civilization—the ancient rival, once the pupil, now the model, of the Muslim. Even the gadgets and garments, the tools and amenities of his everyday life are symbols of bondage to an alien and dominant culture, which he hates and admires, imitates but cannot share. It is a deeply wounding, deeply humiliating experience.[5]

Two World Wars

The faint beginnings of Arab nationalism can be traced back into the 19th century, when it was strongest among Christian Arabs, who did not identify with the larger Islamic community and who were more susceptible to Western ideas. The British occupation of Egypt in 1882 sparked development of a local nationalism there, but until the First World War, the Moslem faith still supplied the predominant bulwark against the encroaching West.

As a popular movement, Arab nationalism began during World War I. The defeat of the Ottomans provided the Arabs with an opportunity to pursue political ambitions which they had habitually left to Turks, and the presence of

5 Lewis, op. cit., p. 135.

victorious European armies stirred new fears and aspirations.

At the beginning of the war, most Moslem Arabs favored the Turks against the Allied powers. But in 1916, the British organized the Arab revolt in the desert, immortalized by the adventures and writings of T. E. Lawrence. Bedouin troops then supported the British forces advancing through Palestine and Syria. Their leaders had been promised Arab independence after the war, but once the war ended the imperialist powers were reluctant to give up their advantages in the area. Instead of the hoped-for independence, the Arabs of the former Ottoman Empire found themselves divided into a series of states under British or French mandates. The mandates, as formalized by the League of Nations between 1922 and 1924, provided that the British and French would administer and develop the territories until they were ready of independence.

Under the terms of the mandates, the British were given control of Palestine (which then included today's Jordan and Iraq. Through British efforts both Transjordan (the area to the east of the Jordan River) and Iraq became ruled by Arab kings—under the constant supervision of British advisers and troops. Palestine was run by a British commissioner who, under the League of Nation's mandate, was to allow development in the area of a national home for the Jews. Syria, which then included what is now Lebanon, was administered by the French, who took over by forcing out the Arab king and partitioned the area. In Egypt, the British in 1923 recognized Egyptian independence but retained advisers and the right to station troops to oversee the Suez Canal. Iraq's independence came with the end of the British mandate in 1932.

The situation in the Arabian Peninsula, now Saudi Arabia, was different. The strength of the Ottoman Turks had never penetrated deeply there. There was a major rivalry for power between Hussein, king of the Hejaz, and Ibn Saud, king of Riyadh. The French and British were content to let them fight it out, and in 1927 Ibn Saud became sovereign over the whole area.

In the period between the wars, Saudi Arabia, while independent, was too inward-looking to lead the move for Arab unity that had begun during the war. The other states, under their tutelary rulers, were concerned with achieving a greater degree of independence from occupying powers rather than with working for pan-Arab nationalism.

The unification movement was reawakened by World War II. The Arab countries in 1939 were more than half-way between the complete servitude of 1914 and the total independence they would come to achieve. The war removed the French from Syria and Lebanon and the Italians from Libya, leaving Britain the only colonial power in the Middle East. Arab nationalism revived and intensified, and the eventual end of the British role appeared inescapable.

Near the end of the war, the Hashemite Arab leaders of Iraq and Syria proposed plans to unite several Arab countries under their leadership. The plans were opposed by both the non-Hashemite Arabs and the British. They instead backed the formation of a loose federation of the Arab states which would safeguard national sovereignties but at the same time enable them to work for the common interest. For the British, the plan appeared to have the advantage of giving them a mechanism through which they could pursue their interests in the Middle East. The loose federation grew out of two conferences among Egypt, Saudi Arabia, Yemen, Transjordan, Syria, Lebanon and Iraq and became known as the Arab League.

The Arab League

The birth of the first pan-Arab organization in 1944 stirred high hopes among many Arabs, but its capacity for action turned out to be considerably smaller than many had hoped. The League had largely a negative function in its early years. The seven original states were unequal in wealth and prestige and had differing political goals. None wanted to sacrifice its own sovereignty to a federal ideal, and there were destructive personal rivalries among the rulers of Egypt, Saudi Arabia, Jordan and Iraq.

The one area in which members of the League were in complete agreement in the early years was opposition to growing Jewish claims in Palestine. After the United Nations voted to partition Palestine in 1947, the Arab League made a joint declaration of war against the new state of Israel. But then, instead of acting in a truly coordinated fashion, each Arab state tried to help a particular client group among the Palestinian Arab population which they believed would help them to pursue their own goals. This uncoordinated and conflicting effort led to the defeat of the Arabs in the first Arab-Israeli war, bitter feuds among the Arab governments and the influx of hundreds of thousands of unwanted Palestinian Arab refugees into other Arab countries. *(Arab-Israeli Wars, p. 73)*

The move against Israel led to the end of any British influence in the Arab League, and the Arab defeat nearly led to the end of the League itself. It was resuscitated by the signing of a mutual security pact designed to protect Syria from the ambitions of the Hashemite kings.

Neither the mutual security pact nor other collective actions by the League were of great practical value. Nevertheless, the League did perform some useful functions for the Arab world. In his book *From War to War,* Middle East authority Nadav Safran speculated that the League "gave the idea of pan-Arabism an institutionalized expression that made it part of Arab daily life." During the 1950s it also was responsible for useful administrative and cultural initiatives.

The Arab League has remained an important Arab forum, though one where rival Arab countries often display more disunity than unity. In 1977, the League had 21 members including the newest addition, the Palestine Liberation Organization (PLO). Notably, when Syrian forces were used, beginning in mid-1976, to quell the destructive Lebanese civil war, they did so under the colors of the Arab League. *(See box p. 114).*

Postwar Developments

The factionalism and failure of the Arab League to meet its primary goals were reflected in the larger history of the Arab world during the post-World-War-II period.

As the British systematically gave up their remaining control, the political histories of the newly independent nations were characterized by dictatorships, coups, assassinations and abdications. Moreover, the Arab countries were continually interfering in each others' affairs. Egypt alone attempted to instigate or support revolutions in Syria, Lebanon, Iraq, Jordan, Saudi Arabia and Yemen. Egypt became involved in a full-scale war with Yemen and at one point had as many as 70,000 troops there. Almost all of the Arab countries at one time or another were involved in machinations intended to bolster one state against another.

Arab unity was a byword. Until 1973, however, attempts to achieve fruitful pan-Arab cooperation were failures.

For much of the postwar period, Egypt, the most populous state, sought to be the dominant country in the Arab world. The monarchy was overthrown by a military dictatorship in 1952. The dictatorship was headed by Gamal Abdel Nasser, a vigorous and charismatic figure, between 1954 and 1970. Nasser preached Arab nationalism, but his feuds with other leaders often made him a divisive force. His successor, Anwar Sadat, at first seemed a weak figure by comparison; but after the 1973 Arab-Israeli war, Sadat began to emerge as a more constructive worker for Arab unity.

Britain relinquished its role in what was then called Transjordan in 1946. Emir Abdullah of the Hashemite clan became king. He was assassinated in 1951 by a Palestinian because of the fear that he was attempting to achieve peace with Israel. He was succeeded briefly by his son and then, in 1953, by his grandson, Hussein, who took over at the age of 18. Hussein has remained on the Jordanian throne since then, though his kingdom has faced a number of serious crises. Generally, Hussein has been more pro-Western and moderate than other Arab leaders and it is usually said that Jordan will be the second country to make peace with Israel—its position within the Arab world not allowing her to go first. Hussein's regime has faced sharp conflicts with other Arab nations in the past because of the problems posed to the state by the existence of a large number of Palestinian refugees who fled in 1948 and in 1967 from areas controlled by Israel. A civil war broke out in Jordan in 1970 when the PLO attempted to take over the country but was defeated by the King's largely Bedouin army and the threat of Israeli military intervention supported by the United States.

Saudi Arabia was ruled from 1964 until his assassination in 1975 by King Faisal, son of King Ibn Saud, who had unified the country in 1927. The world's richest oil reserves had been discovered there in 1933, but the country remained an economically backward land of desert tribes. Fearful of challenges to its power, the royal family did not even develop an army. In recent years Saudi Arabia has become the financial capital of the Arab Middle East. Saudi political influence has multiplied enormously since the Yom Kippur War and the quadrupling of oil prices, making the U.S.-Saudi alliance one of the central political facts of world politics.

The postwar political history of other major Arab countries has been more complex. Syria and Iraq are cases in point. By some counts there had been 20 coups d'etat in Syria before the current ruler, Gen. Hafez al-Assad, seized power in 1970. Assad's position was unsure, and he did not have the freedom of action of his Egyptian counterpart, Sadat, during negotiations following the 1973 Arab-Israeli war. Iraq was also bedeviled by a series of coups after the Hashemite king was murdered in 1958. *(For more detailed political histories of the major Middle East nations, see individual country profiles starting on p. 1)*

During the postwar period, there were several unsuccessful attempts to form federations of Arab states. Syria and Egypt in 1958 united in a common government to form the United Arab Republic. The federation was ended by a Syrian coup in 1961. Nasser threatened but was powerless to act; Syria's nearest point was 130 miles from Egypt. Subsequent efforts to unite Syria and Iraq while they were both controlled by the Baath Party also fell apart. Syria, Egypt and Libya in 1971 formed a loose Federation of Arab

Republics. Subsequent plans for an actual merger of neighboring Libya and Egypt collapsed in 1973, when Sadat turned from Libya to Saudi Arabia for support.

Most recently Jordan and Syria have begun steps toward a form of limited unification. In addition, Syria's post-civil war control over Lebanon may mark a revived Syrian quest for regional hegemony. There is even discussion of a possible federal link between Syria, Lebanon, Jordan and a new state of Palestine which some hope will be created out of territories now occupied by Israel. Egypt and Syria, after more than a year of at times bitter hostility, agreed in December 1976, to a joint political command. They were joined a few months later by Sudan, Egypt's southern neighbor.

Palestinian Issue

Since 1948, the existence of Israel has been a constant irritant to the Arab world and has led four times to all-out warfare. *(See Palestinian Question chapter, p. 109).*

Ironically, Israel has been at once both an aid and a detriment to Arab unity. Whenever inter-Arab relations were seriously strained, common hostility to Israel was a unifying force. But the existence of Israel, physically separating Egypt from Syria and Iraq, impeded Arab attempts at unity.

A major offshoot of the wars was what came to be known as the Palestinian refugee problem. The war that followed the partition of Palestine led to the displacement of more than 700,000 Palestinian Arabs who had lived in the area. Since then, controversy has continued over whether the Arabs fled of their own accord or were forced out by Israeli design. The Israelis contend that the Arab leaders first fled the country and that the populace followed after receiving broadcast orders from their absent leaders to leave. The Arabs point to acts of Israeli terrorism, such as the massacre of 200 Arabs by the dissident Irgun group at Dir Yassin in April 1948, as the explanation for the mass exodus.

Whatever the cause, the refugees left, and the Israelis, not wanting to saddle the new state with a large dissident minority, did not want them back. The Palestinians became refugees in Jordan (about 50 per cent with most going to the West Bank area), the Gaza Strip (about 25 per cent) and Syria and Lebanon (most of the remainder).

Proposals for settling the problem have been put forward since 1948. The basic document still is a paragraph of the U.N. General Assembly's Resolution 194, adopted Dec. 11, 1948, which provided: "That the refugees wishing to return to their homes and live at peace with their neighbors should be permitted to do so at the very earliest practicable date, and that compensation should be paid for the property of those choosing not to return and for loss of or damage to property which, under principles of international law or in equity, should be made good by the governments or authorities responsible." A Conciliation Commission for Palestine attempted to work out a settlement in 1949 and 1951, but negotiations broke down, because the Israelis wanted to receive only a limited number of the refugees and the Arabs opposed a refugee settlement which recognized a Jewish state.

The Arab states also did not want to absorb the Palestinians. Many were settled in dreary United Nations refugee camps which grew into permanent Jordanian, Syrian and Lebanese towns; there were 63 such camps in 1974. The Arab leaders refused to absorb them, on the grounds that their lands lacked resources and were already overpopulated and that resettlement would imply the permanence of Israel. In addition, the compensation issue was never settled, and the Palestinians received nothing for the homes, orange groves, farms and businesses which were taken over by the Israelis. The festering situation led to the birth of numerous Palestinian terrorist groups, most of which now belong to the Palestine Liberation Organization, which since 1974 has become widely recognized as representing the Palestinians.

The refugee problem increased after the 1967 war when Israel occupied the Gaza Strip and the West Bank where the majority of refugees lived. While most of the Palestinians remained under Israeli occupation, others fled, some for the second time in one generation, across the Jordan River or into Lebanon.

In 1977 there were an estimated 3 million Palestinians in the world, 35 per cent living in the occupied West Bank and Gaza Strip, 30 per cent in Jordan, 15 per cent in pre-1967 Israel, 15 per cent in Syria and Lebanon and 5 per cent scattered throughout the world. Just over half of the Palestinians, about 1.6 million, are still registered as refugees with the United Nations Relief and World Agency (UNRWA), though this figure is probably substantially inflated. UNRWA-registered refugees are distributed as follows:

Jordan	550,000	Gaza	325,000
Israeli-occupied		Lebanon	185,000
West Bank	280,000	Syria	170,000

Soviet Involvement

The end of World War II and colonialism found many Arab nations economically underdeveloped but with strong anti-Western feelings. As a result, a number of Arab leaders turned to the Soviet Union for help in modernization. The arrangement was a two-way street; the U.S.S.R., which had been unable to gain a foothold in neighboring Turkey or Iran, was eager to develop clients in the strategic southern Middle East.

Russian postwar involvement in the area grew to major proportions after 1955, when the U.S.S.R. took over the role formerly held by Britain and the United States as munitions supplier and financier to Egypt. It also became heavily involved in the less stable nations of Syria and Iraq. Russian aid to Arab nations ranged from building the Aswan Dam in Egypt to port construction to supplying large numbers of weapons and economic and military advisers. In return, Russia received military facilities in Egypt, Syria, Iraq, Algeria and the Yemen. It aided both revolutionary and reactionary states and achieved good relations with most Arab countries; a major exception was anti-Communist Saudi Arabia.

The dependence of the Arab countries on Soviet aid increased substantially after the disastrous Arab defeat in the 1967 Arab-Israeli war. The Egyptian arsenal that was rebuilt was more sophisticated and extensive than the one destroyed in 1967. Then, in an unexpected move on July 18, 1972, President Sadat ordered the bulk of Soviet military personnel to leave the country. The decision was partly motivated by the "no war, no peace" policies of the Soviet Union and partly by Egyptian chaffing at Soviet attempts at dominance. ("No war, no peace" is a phrase used frequently to describe Soviet Middle East policy. It means that the Soviet Union preferred Arab-Israeli tension but was unwilling to help the Arabs force Israel to withdraw from Arab territories occupied in 1967.)

Egypt, however, was careful to avoid a complete break with Russia, and the value of the aid it had received was clearly indicated by the results of the 1973 war; Arab armies had profited greatly from new sophisticated training, and new Russian weapons left Israel far more vulnerable than it had anticipated.

The Ramadan War* and the "Oil Weapon"

While the Arabs were united in a common opposition to Israel during the period 1948-73, they were rarely able to translate this unity into effective human action. An Arab League boycott of Israeli goods was one of the only fruitful, though largely symbolic, joint actions taken until 1973. That year, the Arab countries were finally able to unite in a coordinated surprise attack on Israel and to use the "oil weapon" against Israel's supporters.

An oil embargo against pro-Israeli nations had been briefly tried after the 1967 war but had never really worked. It had been undertaken by King Faisal of Saudi Arabia under pressure from Nasser. Faisal was not a Nasser sympathizer; oil was allowed to "leak" to the West, and then the embargo was quickly ended.

By 1973, the situation was different. Nasser had died in September 1970 and his successor, Sadat, quietly set about a policy of "de-Nasserization." He eased restrictions on foreign capital, ended nationalization of industry and asked help from Saudi Arabia in economic development. In September 1973, plans for a political union with Saudi Arabia's enemy, Libya, foundered. As a result, Faisal, apparently at some point in mid-1973, indicated to Sadat a willingness to use his country's vast oil resources to gain leverage against Israel. By that time, with Westerners predicting an "energy crisis," oil was a much more potentially powerful weapon than it had been in 1967.

On Oct. 6, Egypt and Syria went to war and fought with conspicuously more success than in 1967. The earlier defeats had been immensely humiliating to Arab pride, and the comparative success in 1973 enabled the Arabs to recoup much stature in their own eyes. Sadat called the war "the first Arab victory in 500 years." The emotional impact of the success apparently led the Arabs to close ranks as they never had before. At a November 1973 meeting in Algiers, an oil embargo was proclaimed in retaliation for massive American military and financial aid to Israel. Since then, the fear of a future embargo coupled with the increased financial power of the Arab world due to ever-spiralling oil costs has acted as a stimulus to proposals to settle the Arab-Israeli conflict along lines more favorable to Arab interests. *(See Middle East Oil chapter, p. 123)*

A Senate Foreign Relations Committee staff report, "The Middle East Between War and Peace," released March 10, 1974, gave much of the credit for newly achieved Arab unity to Sadat. It said, "That unity has been made possible largely by the personality of President Sadat. Nasser was a charismatic leader, whom other Arab leaders—especially one supposes, King Faisal—would not have wanted to win a war. The fact the President Sadat is

moderate and uncharismatic, without pretensions to the leadership of the Arab world, makes him a safe ally."

Another theory for the new unity was expressed by Nadar Safran in the January 1974 issue of *Foreign Affairs*. He postulated that the "second circle" Arab countries which did not have common borders with Israel had become concerned by the vast growth of Israeli power and territory after 1967, which now had the potential to affect them directly. "Their support for countries of the 'first circle,'" Safran said, "became an investment in their own security."

The solidarity, achieved as a result of the 1973 war, brought the likelihood of continued Arab cooperation, but the outlook was unclear. A possibly major divisive factor was Libya, whose revolutionary head, Col. Muammar el-Qaddafi, was the one leader of an Arab state to hold to the old position that the Arabs should not come to terms with Israel under any condition but should continue to fight for its destruction. Iraq was also a question mark, although in 1975 it began to seek rapprochement with its neighbors.

Moreover, even if the Arab countries did continue to operate cooperatively, it was unclear how many initiatives they would take outside of the area of oil. The Senate Foreign Relations Committee report commented: "There has been one common characteristic of just about every discussion of the peace conference we have had, with both officials and private citizens, in Egypt, Saudi Arabia, Kuwait, Lebanon, and now Syria: a pervading sense on the Arabs' part of it all being somebody else's responsibility and that somebody else is the United States. They themselves are passive, skeptical and pessimistic."

The picture of the Arab as "passive" is also part of the Arabs' self-image, an image fostered by colonialism and deepened by humiliation in wars with Israel. Determination to break free of that sense of impotence has been one of the most powerful drives behind Arab terrorism. Ironically, that same self-image of passivity has long plagued the Jews, especially after the extermination of six million of them in Hitler's holocaust, compelling Israelis to compensate with demonstrations of prowess. The attitudes of both sides thus reinforce each other in escalating conflicts.

The stalemate at the end of the last, largest war, however, may provide an opportunity to break out of that vicious circle. Arab armies took the initiative in the war and fought with anything but passivity. The Israelis, by the end of the war, were again demonstrating their military prowess, but with results that showed that military power could not bring the peace they need. Both sides in the war paid a terrible price in human lives and are still paying the heavy economic price of continuing confrontation.

Since that bitter stalemate, the leaders of Egypt and Jordan have said they will recognize Israel's existence in exchange for a return to its pre-1967 borders. Jordan's King Hussein calls this a "historic change" in the Arabs position.[6] But that recognition of Israel also depends, Hussein and other Arab leaders say, on Israel's willingness to "recognize the legitimate rights of the Palestinians in their homeland." The Palestinians meanwhile are divided about the possibility of living peacefully with Israel, no matter what its boundaries. It appears that a pragmatic majority may be willing to coexist with Israel should they be allowed to create a Palestinian state in the area of the West Bank and Gaza Strip. "This development," English scholar Peter Mansfield writes, "is the most significant in the history of the Arab-Israeli conflict."[7]

---

* The war that began on October 6, 1973 has different names. In Israel and the U.S. it is best known as the Yom Kippur War since it began on the Jewish Day of Atonement. The Arabs call it the War of Ramadan, since it began during their month-long period of daytime fasting. A "neutral" term is the October War. Throughout this book Yom Kippur War is used in deference to popular usage in the U.S.

6 *The New York Times*, May 7, 1975.
7 Peter Mansfield, op. cit., p. 518.

The Arab World Today

In the aftermath of what has been described as "a kind of Arab civil war"[8] in Lebanon, a new sense of unity and power seems to have come over much of the Arab world.

Lebanon erupted in late 1975 and was partially destroyed during the next year. Tens of thousands were killed and up to one-quarter of the population fled with hundreds of thousands taking up residence in Syria. Syrian-Egyptian relations reached a new low and Arab-Israeli peace negotiations were stalled. Finally at two summit meetings late in 1976, the first in Riyadh and the second in Cairo, the decision was made that the Lebanese conflict must be brought to an end. Syrian troops already in Lebanon were converted into an Arab League peace-keeping force. The Palestinians who had allied themselves with leftist and Moslem forces were battered and forced to yield to the will of the Arab states, including Syria which had once been their most outspoken ally.

The conflict in Lebanon continued to smolder in 1977—and there was something of a *de facto* partition of the country between Maronite Christian and Moslem

8 Norman F. Howard, "Tragedy in Lebanon," *Current History*, January 1977, p. 1.

areas—but the war appeared to have ended. Many believe, however, that only a general Middle East settlement which makes provision for the Palestinians can solve the danger of another eruption in what was once called the Switzerland of the Middle East.

Especially since 1973, depleting oil and accumulating petrodollars have brought great changes to the Arab Middle East. Some countries, such as Egypt, Jordan and Syria, still have serious economic problems. Others, especially Saudi Arabia and Kuwait, are troubled by the difficult task of preserving their Arab way of life while modernizing at a rate dependent only on how fast the West can sell them the required goods and technology.

Politically a Riyadh-Cairo axis, supported by the United States, could become that upon which a settlement to the Arab-Israeli conflict can be built. At least that has been the hope in some quarters as talk of resumption of the Geneva Conference grew in early 1977. The roles to be played by Damascus, Baghdad and the Palestinians seemed less certain. There were forces of both stability and revolution in the Arab world of 1977. And though the West, primarily the United States, had the dominant involvement at the moment, it appeared certain that the Soviet Union would continue to demand a role in Middle East developments.

Much will be determined by whether there is a settlement to the festering Arab-Israeli conflict and by the nature of such a settlement.

← No leadership development under Turks, French, some under British.

Jews in Europe who did succeed did so by competition → leaders, but rest remained passive. Sabra, selfreliance, etc, → they are leaders.

SADAT'S GAMBLE: PEACE AND MONEY

On June 5, 1975, Egyptian President Anwar Sadat donned a white admiral's uniform, boarded an Egyptian destroyer and led a ceremonial parade of ships south through the newly cleared Suez Canal. Two hours later, the commercial ships entered the canal passage, the first to do so since its closing on June 6, 1967, during the Six-Day War that humiliated Egypt.

There was much symbolism in the Sadat-staged spectacle. The lead warship had been renamed "the Sixth of October," commemorating the day in 1973 when the Egyptian army crossed the canal, broke through the Israeli defense lines and established the first Egyptian presence in the Sinai since it was overrun by the Israelis in 1967. The convoy included the cruiser "Little Rock," flagship of the U.S. Sixth Fleet, not only a tribute to the U.S. Navy for its help in clearing the Suez of mines and debris, but a sign that Sadat had turned Egypt around from dependence on the Soviet Union to a new reliance on the United States—Israel's benefactor and the only nation in the world Sadat considers capable of pressuring the Jewish state into concessions which might lead to an over-all Middle East settlement. Simultaneous with the canal reopening, Israel voluntarily thinned out its troops and military equipment along the Sinai armistice line, a move cautiously hailed by Sadat as a peace gesture.

Sadat's naval uniform was a reminder to his people that they had cheered him as "the hero of the [Oct. 6] crossing." Yet, in January and March 1975, Egyptian workers, demonstrating in Cairo and Alexandria against the country's grave economic ills, had chanted, "Hero of the crossing, where is our breakfast?"

Two years later—in January 1977—79 people were killed in rioting that was sparked by the government's attempt to remove food subsidies in an effort at essential economic reforms. This followed Sadat's re-election in September 1976 with what then appeared to be broad popular support. The rioting in early 1977 was the worst internal upheaval in decades and it cast doubt on Egypt's ability to carry out measures being demanded by international lending agencies to pull Egypt's economy from potential bankruptcy. The rioting also called into question Sadat's longevity. In an effort to strengthen his hand, in February 1977 Sadat had the Egyptian people ratify a law-and-order decree outlawing demonstrations, strikes and sit-ins along with membership in "organizations that are opposed to the regime."

A Political Moderate

"To a large extent," Professor Alvin Z. Rubinstein at the University of Pennsylvania writes, "Sadat's foreign policy has been shaped by economic considerations."[1] It is Egypt's desperate economic circumstances requiring billions of dollars of aid annually, coupled with the need to promote the newly resumed Egyptian-American friendship, which jointly act as powerful incentives for the political moderation Sadat has exhibited over the past six years.

1 "The Egypt of Anwar Sadat," *Current History*, January 1977, p. 19.

Egyptian President Anwar Sadat

During a U.S. visit in April 1977, Sadat went beyond talk of "ending the state of belligerency" with Israel. He spoke of a "normalization" of relations over a period of years once Israel has withdrawn from all occupied territories and the Palestinians have created a Palestinian state linked to Jordan.

On April 6th, at the conclusion of his American visit, Sadat made a statement to a small group of senior journalists elaborating on his peaceful intentions. "I didn't say at all that peace will be postponed for the next generation," he noted in response to the widespread interpretation made of earlier remarks. "I am for full peace, permanent peace, and then everything will be normalized. For instance, the issue of boycott automatically will be finished because whenever we sign this peace agreement everything is going to be normal. For instance, now the Israeli cargo passes the Suez Canal, but after the peace agreement, sure, the Israeli ships can pass the Suez Canal because we have solved the whole problem." Sadat added another example of what normal relations would lead to by indicating that Arab and Israeli journalists would be able to visit each other's countries "whenever we sign this agreement."

Dependence on the United States

Sadat's hopes are pinned on a combination of funding from oil-producing neighbors, especially Saudi Arabia, and investments plus aid from Europe, Japan and especially the United States. In March 1975, Sadat's regime ended what was to have been a fifteen year Soviet-Egyptian Treaty of Friendship and Cooperation. Taking this step completed the reversal of policy from the years of President Gamal

Abdel Nasser's rule when Kremlin-Cairo bonds were robust and Egypt was leading the Arab world toward what became known as "Arab socialism."

Looking back just a few years Sadat may have underestimated the difficulties in returning Egypt to a Western economic as well as political orientation. "The problem is that we declared an open door policy before we provided basic infrastructural prerequisites,"[2] Dr. Ahmed El Ghandour, Deputy Minister of Economy and Economic Cooperation said in 1976. Reporting on the situation, Andrew Lycett noted that soon after Sadat's reversals in 1974, "enthusiastic capitalists were flying in from all corners of the globe." "But it did not take long for them to become disillusioned," Lycett added. "Egypt was in a bigger mess after her years of tutelage to the Soviet Union than even President Sadat must have imagined. Her debts were huge. Her bureaucrats were not geared to dealing with enterprising businessmen and talking in terms of growth."[3]

By 1977 the economic situation had begun to improve—at least on paper. Phones, public transportation, office space all were scheduled to be substantially improved. Politically as well, Sadat has clearly placed his fate in American hands. "America holds 99 per cent of the cards," Sadat has grown fond of repeating, stressing his belief that only the United States can bring about a political settlement along the lines he is advocating. In a bicentennial "Message to America"[4] Sadat focused on his "policy of dialogue and openness between us and the world." "America, as the leader of the Western world, is expected by us all to play a pivotal role and pioneer the Western world in this dialogue," he stated. "Together we can face this challenge through effective peace and establish a world system based on real harmony."

It was back in 1971 actually that Sadat became the first Arab leader to speak of signing a peace treaty with Israel. He remains today the most pragmatic Arab leader, the one most willing to find a path toward coexistence with Israel and the development of Egypt. Still, Sadat's future remains uncertain and the entire political framework in the Middle East is consequently uncertain. A 1977 visitor to Cairo, *The New Republic's* Morton Kondracke, returned with a rather gloomy assessment: "Nearly everything seems to be going sour. His economic 'open door policy' has brought in no massive productive investment to benefit Egypt's wretched millions, only playthings for the rich that mock the poor. The army is starved for weapons. Sadat is welching on his promises of democracy. If his great gamble of all falls through—his reliance on the U.S. to wrest Sinai from Israel and get him peace—Sadat will have failed utterly, and will not even be in a good position to recoup by starting a war.... This year there is real doubt in Cairo whether Sadat will stay in office into next year. One admirer said, 'He may find himself hanging from a lamppost.' "[5]

Sadat has told his country and his financial and political backers that around 1980 he expects to have pulled Egypt out of the economic doldrums. Revenues from the enlarged Suez Canal and from oil discoveries are being counted on as is capital investment from the West, once Egypt's basic infrastructure is improved and peace is on the horizon. It is a rather ambitious undertaking Sadat has accepted, and the continuing population boom threatens to make it even more difficult. Sadat will be keeping one eye on those whose might provoke further rioting and disorder. Looking to the future C. L. Sulzberger asks, "Today Egypt is flat broke and if Mr. Sadat succeeds in all he forecasts, he can retire early with a clear conscience. But the program outlined is herculean. Is Sadat Hercules?"[6]

Terrain

Egypt is at the geographical center of the Arab world. Situated at the northeast corner of Africa, it lies at the crossroads of Europe and the Orient, of north Africa and southwest Asia, a location which accounts for its prior history of foreign domination. Its location on Israel's western border and its central role in the Arab world enhance its importance to U.S. policy in the Middle East.

Occupying an area slightly larger than California, Nevada and Arizona combined, Egypt stretches northward to the Mediterranean Sea. To the east is Israel and the 1,200-mile coast of the Red Sea. The Sudan lies to the south and Libya to the west.

Egypt includes the Israeli-occupied Sinai Peninsula, seized by Israel during the 1967 war. The 1949 armistice agreements between Israel and the Arab states also granted Egypt administrative jurisdiction over the 28-mile-zone Gaza Strip, another area that Israel subsequently occupied.

Since ancient times, Egypt's lifeline has been the Nile River, which flows from Uganda and Ethiopia through upper Egypt, northward to the Mediterranean. Despite a 2,140-mile coastline and 2,100 miles of inland waterways, much of Egypt is an arid wasteland. The 100-mile-long Suez Canal, opened in 1869, links the Red Sea to the Mediterranean. Egypt's major ports are Alexandria, founded in 332 B.C. by Alexander the Great, and Port Said and Suez, both on the Suez Canal.

The ribbon-like Nile Valley, the Nile delta and a few oases provide Egypt's only arable lands. Less than 3 per cent of the country is cultivated, and 0.7 per cent is covered with inland water. The Aswan high dam, however, is expected eventually to increase cultivable land to 13,000 square miles.

The Nile Valley divides two other regions of Egypt: the Arabian Desert to the east and the Libyan Desert to the west. The eastern desert is distinguished by wadies, trenchlike formations that were once important means of communication between the north and the Red Sea. Mountains along the Red Sea in this region rise to above 7,000 feet. The Libyan Desert in Egypt's western area is broken at three places by oases, created by seepage from the Nile.

The Nile Valley nonetheless dominates the country. While Egypt has an overall population density of 93.2 persons per square mile, the figure leaps to 2,631 per square mile in the pencil-thin valley. The largest cities are concentrated there and in the Suez Canal Zone.

Cairo, Eygpt's capital and Africa's largest city with more than seven million people, is located inland along the northern end of the Nile. Although it is still partly a city of Arabic grandeur, it suffers the problems of overcrowding, urban rot and pollution. In parts of Cairo there are a quarter million people per square mile, one of the highest densities in the world. Alexandria's population is more than two-million.

People

The population was estimated at 38,791,000 in 1977, an increase of 2 million in two years. Egypt's high birth rate is a

---

2 "Egypt Economic Survey," *African Development*, November 1976, p. 7.

3 *Ibid.*

4 *Time Magazine*, 28 February, 1977, p. 35.

5 "Sadat's Sorry State," *The New Republic*, 19 March 1977, p. 14.

6 *New York Times*, 15 January 1977.

factor contributing to its ongoing serious economic difficulties. If population continues to grow at the 1977 rate Egypt will be a land stuffed with more than 70 million in less than 30 years. Even today Egypt has by far the largest population of any Arab country and is second only to Turkey in the Middle East. Unlike several of its Arab neighbors, Egypt does not have a significant nomadic population. However, a rural exodus in recent years can be attributed to the lure of cities in the wake of industrialization.

Ninety per cent of Egyptians are of eastern Hamitic stock, resembling their dark, stocky ancestors, who have inhabited the area since the beginning of historic times. Greeks, Italians and Syro-Lebanese, living primarily in the north, constitute the other 10 per cent of the population.

Illiteracy remains high in Egypt with about 50 per cent of the population functionally literate but only 25 per cent actually able to read and write. Most Egyptians—92 per cent—are Sunni Moslems. Another seven per cent are Coptics, continuing the Christian creed retained since the early days of Christianity when Egypt was among the first countries to adopt the religion. Islam was introduced into Egypt in the sixth century.

Economy

Egypt has long relied on the Nile for its life support. With few other natural resources, the Egyptian economy continues, as it has for centuries, to be based on agriculture. Nearly half the population is directly engaged in farming, and many other people work in agricultural processing and trade. Egypt's agricultural products are the net earners of foreign exchange. Although Egypt was called the granary of Rome in ancient times, its main cash crop today is cotton, which accounts for nearly half of export earnings.

Egypt's mild climate allows for multiple cropping, thereby doubling the country's yearly agricultural yield. Farming practices are inefficient by modern standards, but production continues to grow. Produce and commodities are raised primarily for the market, and there is little subsistence farming.

Faced with a rapid population growth rate and limited resources, the Egyptian government has funneled its major efforts into increasing industrialization as a means of raising productivity and the standards of living. In the early 1960s, the government nationalized all major foreign and domestic industries. Today many of these industries and all public utilities are run by organs of the central government. Most agriculture and trade, however, remain in the hands of private owners. Significant gains have been made in textile, chemical, cement, food processing, petroleum refining and construction industries.

In the mid-1970s, however, Sadat reversed the state socialism trend initiated by his predecessor in the presidency, Nasser. In April 1974 Sadat issued a statement on Egypt's long-term economic and social objectives. It was called "The October Paper" to symbolize Sadat's belief that the October War had ushered in a new era for Egypt. The paper candidly criticized some of Egypt's problems and failings and promoted decentralization and regionalism, not further centralization. Concretely, Sadat has offered incentives for foreign investment and a free-trade zone at Port Said. Also in 1974 the Egyptian State Council ruled illegal the confiscation of private property under Nasser, and certain lands and properties were returned to their former owners.

De-Nasserization of the Egyptian economy complemented Sadat's political initiatives aimed at the West, especially the United States. There is also the major objective of luring Western capital, essential to build Egypt's industrial base. An estimated 2.5 million Egyptians still subsist on incomes of about $200 a year. The inflation rate is still about 25 per cent. And Egypt's balance-of-payments remains seriously in deficit, while paying interest and principle on outstanding loans of about $15-billion. Chronically short of cash, Egypt has been forced to spend one-fourth of its national budget on the military. The director of Egypt's Central Bank has been quoted as saying the country has spent or borrowed at least $25-billion for military equipment and operations since 1967 and has lost as much again in war damage and through loss of Suez Canal revenues.

The discovery of oil resources in Egypt in the mid-1960s thrust the country into a field of growing importance. Oil concessions had been granted to Phillips Petroleum and American International Oil Company in 1963. Since 1966, oil production has increased with the development of fields in the Gulf of Suez and the Western Desert. The Sinai oil fields at Abu Rudeis were returned to Egypt after the 1975 second Israeli-Egyptian disengagement agreement. Together with other oil, by 1980 Sadat is counting on production of a million barrels daily.

A critical food shortage is a major contributor to Egypt's recurrent trade deficit. The country requires $1-billion a year in foreign currency to feed its population in spite of government efforts to boost wheat output. Capital goods constitute another sizable portion of Egyptian imports. Cotton is the main export item, although rice, petroleum and manufactured goods are assuming increasing importance.

Soviet Bloc Trade

Communist countries became Egypt's main trading partners after 1967. That arrangement had gained its initial footing in 1956 when West European countries boycotted Egyptian trade as a result of the Suez crisis. Communist countries since have purchased as much as 50 per cent of Egypt's exports. Thirty-five per cent of Egypt's imports come from Communist countries, primarily the Soviet Union, East Germany and Czechoslovakia until recently.

U.S. Trade

Before Egyptian-U.S. relations were broken off during the 1967 war, Egypt imported more goods from the United States than from any other country. Egypt received nearly $700-million in American Food for Peace program wheat in 1949-67, largely meeting Egypt's demands for that commodity. The country received no U.S. assistance of any kind from 1968 through 1971, then began to receive modest aid sums again in 1972. From World War II through 1974, Egypt received a total of $977,400,000 in American economic loans and grants.

The resumption of diplomatic ties between the two countries on February 28, 1974, opened the way for renewed and expanded trading relations. Sadat's turn toward the West in the aftermath of the 1973 war quickly resulted in Congress voting $250-million in Egyptian aid for fiscal year 1975. During fiscal 1977, Sadat's Egypt received over $700-million in economic aid plus an extra amount sought by the new Carter administration after the January 1977 rioting highlighted Sadat's serious problems.

Statistics on Egypt

Area: 386,659 square miles, including 22,500 square miles occupied by Israel.
Capital: Cairo.
Population: 38,791,000 (1977 est.).
Religion: 94 per cent Moslem, 6 per cent Coptic Christian and others.
Official Language: Arabic; English and French widely known by educated Egyptians.
GNP: $11.2-billion, $300 per capita (1975).

In fact, the United States aid program to Egypt has grown from a standing start in 1974 to the largest assistance program in the world. More American assistance went to Egypt in 1977 than to the rest of Africa and Latin American combined. John J. Gilligan, administrator of the Agency for International Development (AID) of the U.S. Government, visited Egypt in June 1977, and noted that the American commitment to sustaining this massive aid program will continue for "a long time to come." "Egypt is a powderkeg," Gilligan added. "If we don't help to stabilize the economic situation here we're going to have a highly volatile situation over a long period of time."[7]

Egypt's prospects for establishing a favorable trade balance remain uncertain. Before the 1967 war, revenues from tourism and the Suez Canal offset most of the trade deficit. But tourism declined for awhile, then showed signs of picking up again in the mid-1970s. Cairo, with help from outside private investors, was launching a decade-long program of major hotel construction.

The canal was reopened in June 1975 after eight years of no business. Ship passage was brisk the first month, but estimates of toll collections the first year of operation varied widely, from $125-million to $500-million. Whatever the sum, the money probably would go back into the waterway, as Cairo has embarked on a six-year canal improvement program that would cost an estimated $1-billion.

Egypt had been receiving an annual subsidy of $250-million from Saudi Arabia, Kuwait and Libya—sort of compensation money to Egypt for bearing the brunt of the wars with Israel. But with needs ranging from $15- to $25-billion over the next 5 to 10 years, Egypt has been able to argue for more aid with some success. In addition to the larger amounts of American aid, assistance from the Arab oil-producing countries by 1977 was substantially in excess of $1-billion yearly. In July, Sadat announced that Saudi Arabia had agreed to assist Egypt in building up its army over a five-year period.

Early History

Egypt's recorded history, the longest continuous account in the world, dates from 3200 B.C. From then until 333 B.C., Egypt was a united kingdom under various dynasties of pharaohs. During that period, the great pyramids were built to propel the pharaohs, believed to be divine, back to heaven. After 333 B.C., the country's territory alternately increased and diminished according to the conquests of foreign occupiers.

Egypt's choice location made it a prime target for foreign invaders. It has been successively occupied by Asian barbarians, Assyrians, Persians, Alexander the Great, Greco-Egyptian Ptolemies, Romans, Moslems, Turks, Tunisians, Ottoman warriors and Napoleon Bonaparte.

Although the invasion of Egypt was a phase of Napoleon's war against the English, France's domination of the country from 1798 to 1801 ushered in the modern period of Egypt, which emerged from a long, dark age. Bonaparte's army included scientists, medical doctors and teachers. A French officer unearthed the famed Rosetta Stone, and Jean Francois Champollion, the first Egyptologist, deciphered its hieroglyphics, providing the key to Egypt's ancient glories—about which Egyptians for many centuries had known very little. Thus many Egyptians look upon the French emperor as the father of modern Egypt.

Political disorganization after French withdrawal in 1801 gave rise in 1805 to the reign of Mohammed Ali, founder of the last Egyptian dynasty, who continued the modernization movement.

Suez Canal

The first survey for the Suez Canal was made by French army engineers directed by Napoleon himself, who spent several days on the Isthmus of Suez setting up the study. After that, little was done for half a century until another Frenchman, Ferdinand de Lesseps, who had served as a diplomatic officer in Cairo, undertook to make the long-discussed canal project a reality.

He had to overcome major financial and political obstacles. Initially de Lesseps raised half the capital for construction in France. Ironically, de Lesseps' principal opponent was the British government, which, suspicious of French intentions, fought the canal project with every means at its disposal, short of war. Finally, after 10 years of construction work, the canal was opened on Nov. 17, 1869, creating the short route between Europe and the Indian Ocean and western Pacific. The canal soon became one of the world's busiest waterways. Great Britain turned into the chief guardian of the canal, a vital defense and trade lifeline between the mother country and India and other Asian colonies.

British Occupation

The ruling dynasty was interrupted in 1882 by British occupation of Egypt. In 1914, occupation eased and Egypt became a British protectorate. In 1922, Egypt was returned to a kingdom under King Fuad and later King Farouk. Ties to Britain remained, however. Under a 1936 mutual defense treaty, Britain maintained a military base in the Suez Canal Zone and, together with Egypt, continued to administer the Sudan.

World War II

One of the most important campaigns of World War II was fought on Egyptian soil. Germany's elite Afrika Corps, under Field Marshal Rommel, drove across North Africa into Egypt, reaching El Alamein, only 60 miles from the huge British naval base at Alexandria and, beyond it, the Suez Canal. After vicious fighting in the summer and fall of 1942, the British stopped Rommel's advance and subsequently chased him out of Egypt. El Alamein was the high mark of German expansion. After that campaign, Hitler's armies knew only retreat.

Egypt itself did not declare war on Germany and Japan until 1945. Egypt was among the founding members of the United Nations and became a member of the Arab League in 1945.

7 The Washington Post, 28 June 1977, p. 1.

Creation of Israel

Egypt was among the Arab states that bitterly rejected the U.N. decision in 1947 that partitioned Palestine, creating Israel and leading to the first Arab-Israeli war in May 1948. The armistice agreement of 1949 gave administration of the Gaza Strip to Egypt.

Republic Proclaimed

Arab failure to eradicate the new Jewish state, coupled with ongoing minor disputes with Britain, fed a wave of opposition to King Farouk. The underlying tension erupted Jan. 26, 1952, when dissidents burned and looted foreign establishments in Cairo and killed many British residents. On July 23, 1952, a military junta seized the government. The "free officers" led by Lt. Col. Gamal Abdel Nasser forced the abdication of Farouk. On June 18, 1953, the monarchy was abolished and Egypt was proclaimed a republic.

The new government was headed by Gen. Mohammed Naguib, a war hero from the Palestinian invasion. Disagreement among the ruling group of military officers eventually brought about the ouster of Naguib. Nasser assumed leadership of the government April 18, 1954, and became president Nov. 14, 1954.

The new regime's reformist, socialist outlook and the personal magnetism of Nasser quickly gained esteem from the Arab world. To set the stage for Arab unity along a moderate, socialist line, the government undertook measures to divide aristocratic landholdings, raise the cultural and economic level of the farmer and urban worker, increase industrialization and reduce the degree of foreign participation in the country's commercial enterprises and other affairs.

The last move led to Egypt's rejection of Western efforts to formulate a Middle East defense strategy. The West then turned to Egypt's chief rival, Iraq, and the Baghdad Pact resulted in 1955. The United States never joined the alliance, although it promoted it, because of U.S. hesitance to alienate Egypt. Egypt countered the Baghdad Pact, signing an arms agreement with Czechoslovakia and mutual defense treaties with Syria, Saudi Arabia and Yemen, each opposed to Iraqi policies, and recognizing Communist China.

End of Occupation

Strong Egyptian nationalist feelings also led to Egyptian pressure on Britain to evacuate the Suez Canal. On July 27, 1954, Britain agreed to withdraw all its troops within 20 months. The 74-year occupation ended June 13, 1956. The diplomatic antagonism and disagreement over plans for the construction of the Aswan Dam led the United States July 19, 1956, and Britain July 20 to withdraw their promises of aid for the dam. On July 26, 1956, in retaliation, Nasser nationalized the Suez Canal, controlled by the British and French.

Egyptian relations with Britain and France rapidly deteriorated while, on another front, tension mounted with Israel. Israel invaded the Sinai Oct. 29, 1956, and soon after, Britain and France attacked and bombed Egypt, with the announced intention of ending Egyptian-Israeli fighting. The Soviet Union threatened to intervene when the canal was blocked; and combined Soviet and United States pressure forced the British, French and Israelis to withdraw.

Egypt interpreted President Eisenhower's policy of assistance to Middle East nations to maintain their in-dependence as an attempt to block Arab unity under Egyptian leadership. Accordingly, U.S.-Egyptian relations declined. Meanwhile, civil conflict in Syria between nationalists and Communists increasingly diverted Egypt's attention to its Arab neighbor.

Syrian unrest was resolved when Syria, attracted by Nasser's leadership and Egypt's support of Arab nationalism, agreed to unite with Egypt in the United Arab Republic. The union was established Feb. 1, 1958. Economic and political incompatibility undermined the U.A.R., leading to Syria's secession Sept. 30, 1961, after a Syrian military coup. Egypt alone continued to carry the name U.A.R. until 1971.

Formation of the U.A.R. raised fears in Jordan and Lebanon of Egyptian expansion. That fear, coupled with Nasser's hostility toward Lebanese President Camille Chamoun's pro-West policies and Lebanese civil unrest, led to Lebanon's request for U.S. Marines July 14, 1958. Marines arrived July 15. Meanwhile, British forces were sent to Jordan in response to a similar request. The Lebanese unrest eventually subsided, and American and British forces were able to withdraw by November.

1967 War

After 1957, there was relative quiet along the Egyptian-Israeli border. In 1966, tensions once again began to mount as retaliation raids increased. In May 1967, Nasser ordered U.N. forces out of the cease-fire zone set up in 1956, closed the Strait of Tiran to Israeli cargoes and moved troops into the Sinai. *(Maps, pp. 74, 75)*

War broke out June 5 between Egypt and Israel. Jordan, Syria and Iraq soon joined the battle against Israel. Within six days, Israel made sweeping conquests, defeating Egypt and occupying the Sinai Peninsula, the Gaza Strip

---

"...Full, Complete Peace With Israel"

In a May 1977 interview with the editor of *The Middle East*, Egypt's ambassador to the United States, Ashraf Ghorbal, repeatedly emphasized that his country was now striving for a permanent peace and normalization of relations with Israel.

"If we talk about peace we are not doubletalking," Ambassador Ghorbal stated. "We know what peace entails. And we know what peace leads to. And it is only normal to expect what normal conditions lead to." Reflecting on the years immediately after the 1973 October war, Ghorbal noted that "in building the two disengagement agreements that Henry Kissinger helped us to build with the Israelis, we were determined to make of these a beginning of the establishment of real peace." Ghorbal went on to stress that Egypt was now "ready for a full, complete peace with Israel. And Israel is a Jewish state, we're not quarreling about that."

What Egypt is demanding is Israeli return to the 1967 borders and Israeli recognition of Palestinian national rights, probably in the form of a West Bank and Gaza Strip Palestinian State. "I am telling them the fruits of peace for both of us in this generation," Ambassador Ghorbal concluded.

The entire interview with Ambassador Ghorbal—along with a similar interview with Israeli Ambassador Simcha Dinitz—appeared in *Worldview* Magazine, July-August 1977.

and areas in Syria and Jordan. The Suez Canal was closed. By June 11 all parties had accepted a U.N. cease-fire.

Sadat's Succession

The Arab world was shaken by Nasser's death following a heart attack Sept. 28, 1970. Anwar Sadat, vice president under Nasser, was elected to succeed him in a nationwide plebiscite Oct. 15, 1970. Sadat pledged to pursue Nasser's policies.

A merger between Egypt and Libya, announced during the summer of 1972 to take effect Sept. 1, 1973, ran aground because of Egypt's hesitancy about association with Libya's increasingly extremist policies. In July 1973, Sadat firmed up an alliance with Saudi Arabia and chose close relations with his eastern neighbor instead of ties to Libya.

Sadat surprised the world when he emerged as a strong leader. He embarked on new directions—reversing Nasser's politics of state socialism, moving Egypt away from dependence on the Soviet Union and toward friendship with the United States and dropping the holy war call for annihilation of Israel as the only basis for a lasting Middle East peace.

1973 War

In a surprise attack Oct. 6, 1973, Egyptian forces crossed the Suez Canal into the Israeli-occupied Sinai, and Syrian troops invaded the Golan Heights. Israelis, caught unaware while observing the holy day of Yom Kippur, suffered substantial initial losses. Two weeks later, Israel, however, had turned the tide of the battle and driven back its Arab enemies. U.N., American and Soviet efforts brought about a cease-fire Oct. 25, but not until the situation had posed the danger of a U.S.-Soviet confrontation. Egypt and Israel signed a six-point armistice Nov. 11.

Although neither the Arab states nor Israel emerged clearly victorious from the war, the successful Arab oil boycott weakened Western support for Israel and appeared to increase Egypt's chances of regaining the Israeli-occupied lands lost in the 1967 war.

Egypt re-established its presence in the Sinai for the first time since 1967 and looked upon the crossing of the canal and the initial Israeli reverses as a vindication of Egyptian arms. All this gave Sadat new stature. He was emboldened enough to have his deputies sit down with Israeli officials, for the first time since 1949, to sign the armistice agreement and then a disengagement of forces accord in January 1974. The Egyptian foothold on the East Bank enabled Sadat to clear the Suez Canal of mines, sunken ships and other debris and to reopen it on June 5, 1975.

Government

Egypt's government is headed by a strong president who in turn appoints vice presidents, a prime minister, his cabinet and the governors of Egypt's 24 governorates, or provinces. When Nasser held the office between 1954 and his death in 1970, he dominated nearly all aspects of Egyptian life, but Sadat has not equaled that wide-ranging dominance.

Under the constitution, approved by the people Sept. 11, 1971, executive authority rests with the president, who is also chief of state and supreme commander of the armed forces. He is popularly elected to a six-year term after nomination by the People's Assembly. Sadat has been elected twice, most recently in September 1976, but in unopposed elections where the people were only asked to validate the choice of Egypt's single party, the Arab Socialist Union. After his latest victory, though, Sadat began experimenting with a more liberal, democratic system. Three "forums" were developed within the ASU representing left, right and center. Cautiously these forums have taken on some characteristics of parties, but there is still a long way to go. Recent signs of instability in Egypt have caused Sadat to slow down the process, and it remains uncertain whether the country will develope a genuine multiparty system of government.

Besides the power of veto, the president holds emergency powers to issue binding decrees when the People's Assembly delegates such authority by a two-thirds majority.

The People's Assembly is a unicameral legislative body elected for an indefinite term by universal adult suffrage. It serves mainly as a forum for discussion and for automatic approval of government proposals.

Foreign Policy

As a leading Arab country, Egypt's policies affect inter-Arab affairs from the Atlantic Ocean to the Indian Ocean. Internationally, Egypt has been active as a Third-World and non-aligned nation, and wishes to remain so.

Egypt's ties to the West were weakened by the 1956 Suez crisis and the 1967 war, and Communist ties were correspondingly increased. Major goals of Egyptian foreign policy during these Nasserist years included Arab nationalism, Arab socialism and the victory of Arab states over Israel.

Although Sadat signed a 15-year treaty of friendship with the Soviet Union, he astonished the world on July 18, 1972, when he summarily expelled 20,000 Russian military advisers and technicians from Egypt and placed all Soviet bases and equipment under Egyptian control. In a four-hour speech, Sadat said Moscow's "excessive caution" as an ally led him to the decision. A few years later the friendship treaty was totally abrogated by Egypt and Egyptian-Soviet relations have turned very sour.

In response, the Soviets worked to build up Iraq and Syria as rivals to Cairo's influence and established cordial relations with Libya President Muammar Qaddafi, who turned vociferously against Sadat after Egypt backed off from union with Libya. Sadat lashed out at Russian-Libyan arms deals, complaining that he did not understand Soviet policy, which cut off arms supplies to Egypt and refused Sadat a period of grace in amortizing a $5-billion debt for previous military purchases.

As Sadat cooled relations with Moscow, he turned more and more to Washington. He publicly embraced Secretary of State Kissinger after the latter's "shuttle diplomacy" resulted in the Israeli-Egyptian disengagement of forces on the Suez Canal front in January 1974. On Feb. 28, 1974, Egypt and the United States resumed diplomatic relations, broken off during the 1967 war. Sadat began to rely on the good offices of the United States to bring about a general Middle East settlement, rather than on the Geneva Conference, of which the Soviet Union was a cosponsor.

The Arab use of the oil weapon in the 1973 war stimulated American interest in a Middle East settlement. When President Nixon toured the Middle East in June 1974, he made Cairo his first call. Sadat and Nixon signed a friendship pledge. In June 1975, Sadat and President Ford met in Austria to explore possibilities for a peace settlement, with Sadat always hopeful of prying the United States loose from unqualified support of Israel.

But presidential parleys were only window dressing for the key role the United States was to play in mediating the Sinai accord of 1975. After Kissinger's shuttle diplomacy failed to produce a new interim agreement in March, the dejected secretary returned to Washington, where he and Ford launched a "total reassessment" of U.S. Middle East policy. They wondered publicly about what would have happened had Israel been "a little more flexible."

The United States went to work to persuade Israel to accept a new pact as a stepping stone to eventual peace, in the process relieving some of the tensions that again had built up after the spring diplomatic setback. Pleased with the new approach, Sadat, for one, was willing to give the United States another chance. Accordingly, Kissinger went back to the Cairo and Jerusalem negotiating tables in August and ended that round triumphantly with a new Sinai agreement, signed Sept. 4 at Geneva.

Sadat visited the United States in both 1975 (when he addressed a joint session of Congress) and 1977, the first Arab head of state to visit Jimmy Carter's Washington. Before his latest journey he spoke out in a *Parade* magaine interview. "I want the American people to know," he declared, "that never before have the prospects for peace been better. Not in the last 28 years—since Israel was created—have we had a better chance for a peace settlement in the Middle East."[8] For the first time Sadat has begun speaking of a complete normalization of relations with Israel, including eventual diplomatic relations.

But during his visit not only the Arab-Israeli predicament was on Sadat's mind. He courted President Carter on both economic aid and arms. He further discussed how American and Egyptian interests in Africa were similar and how Egypt could play a role in holding back Communist advances on the continent. Indeed, within weeks of his return to Cairo, President Sadat dispatched Egyptian pilots to Zaire to help repulse Soviet and Cuban-supported forces invading Zaire's Shaba province from neighboring Angola.

Forecasts for Egypt are mixed. There are those who think Sadat has chosen a tough but ultimately rewarding road. With Saudi financing and American political support, coupled with improvements in the Egyptian economy, some see signs of better times ahead for the people of Egypt. Others are more cautious. *The New Republic's* Morton Kondracke, for instance, gives this assessment: "Perhaps Sadat will survive, no matter what. Yet, he has raised great expectations in his people by aligning himself so absolutely with the United States, and it is hard to see how these expectations will be fulfilled. He comes to us saying that he has thrown away his weapons for us, given up his friends and put his head down on a block—and that it is now up to us to save him from the axe. It is a posture of self-imposed weakness that he has assumed, and in the cruel world of international maneuver, it does not usually lead to victory."[9]

Still, Sadat may have more than traditional forms of "victory" in mind. He appears to truly want to be rid of the Israeli issue. He needs a settlement; for without it his economic and political plans are sure to eventually collapse and his friendship with the U.S. will be in doubt. Furthermore, Sadat knows how unstable and unpredictable the Arab world remains. After the second Sinai disengagement, Egypt's relations with Syria plummeted as the Lebanese civil war mushroomed and engulfed the entire region. Not until the Riyadh and Cairo summits in late 1976 was some semblance of unity restored among the Arab confrontation states. And everyone is aware that, unless progress can be made toward an overall Middle East settlement, disunity and instability will resume.

In July 1977 serious fighting broke out between Libya and Egypt. Sadat vowed to teach Libya's leader, Muammar Qaddafi "a lesson he will never forget." For the past few years Sadat had repeatedly accused Qaddafi of attempting to overthrow Sadat's regime. Sadat had also expressed considerable anxiety about the quantities of Soviet arms being bought and stored in Libya; but according to Qaddafi he was only serving as the arsenal for eventual war against Israel. Both countries remain, to some degree, dependent on each other—200,000 Egyptians are employed in Libya and Libya maintains large deposits in Egyptian banks.

---

8 "An Interview With Egyptian President Sadat," *Parade* magazine, Feb. 6, 1977, p. 8.

9 Kondracke, "Sadat's Sorry State," *The New Republic*, 19 March, 1977, p. 16.

EVANGELICAL BLEND OF ORTHODOX AND RADICAL

Col. Muammar al-Qaddafi has pursued Arab unity with singular fervor, turning Libya into a suitor whose persistence only repels less ardent Arab states. The colonel has turned east to Egypt, then west to Tunisia, but has failed to forge a union of his nation with either one. Although both of his neighbors would welcome a share of Libya's oil wealth, they are scared off by Qaddafi himself, at once aggressive, radical and erratic, and never shy about heaping scorn on Arab leaders who spurn his overtures.

During the past few years, relations between Egypt and Libya have become tense. In August 1976 President Sadat of Egypt moved troops to guard his border with Libya. In April 1977, Egypt convicted and hanged Libyan terrorists said to be involved in assassination plots and sabotage. That same month a Libyan mob sacked the Egyptian consulate in Benghazi. Actual fighting broke out between Egypt and Libya in July.

Qaddafi has come to fear the possibility of an Egyptian military attempt to overthrow him. He has failed to command leadership in Arab affairs, because, among other things, he breaches no softening of his blend of strictly orthodox Islam and radical socialism. He will consider only the annihilation of Israel as a state (a policy that helped kill the proposed 1974 merger with moderate Tunisia). Qaddafi has made the concept of Arab unity nearly a religion, but he has failed miserably in bringing it about. Sadat has said he is a "madman," a "lunatic." President Nimeiry of the Sudan says he has "a split personality, both of them evil." The Shah of Iran has said Qaddafi is "crazy."

On Feb. 10, 1974, Qaddafi was quoted as saying that he would arm and train revolutionaries to overthrow the governments of Tunisia, Egypt and Algeria if Arab unity could not be achieved by "normal means"—a statement his aides said was misinterpreted, but one accepted at face value by leaders in Tunis, Cairo and Algiers. In July 1974, Qaddafi was accused of aiding an unsuccessful military plot in neighboring Sudan, also an Arab state. On May 23, 1975, Palestinian newspapers reported that "all of Libya's arms" were at the disposal of the commando movement in Lebanon. In June 1975, Libyan Premier Abdel Salam Jalloud was in Baghdad, where he said he was forming with Iraq an Arab Struggle Front aimed at rejecting a negotiated Middle East settlement and the recognition of Israel. It has even been alleged that Qaddafi has meddled in faraway, strife-torn Northern Ireland. All this has continued in the mid-1970s, with the encouragement and support of the Soviet Union, according to Egypt's Sadat.

After the collapse of the Egyptian and Tunisian unions, it was reported April 6, 1975, that Qaddafi was being shorn of political, administrative and traditional duties, but was continuing as armed forces chief. However, later reports had him relinquishing only some ceremonial duties. At any rate, Qaddafi has remained Libya's strongman.

Terrain

Located on the north central coast of Africa, Libya has a coastline of about 1,100 miles on the Mediterranean Sea.

Libyan Leader, Col. Muammar al-Qaddafi

It is bounded by Egypt on the east, Sudan on the southeast, Tunisia and Algeria on the west and Niger and Chad on the South.

Libya's area of 679,360 square miles—about 2½ times the size of Texas—is approximately 95 per cent desert or semi-desert. There are two small areas of hills and mountains in the northeast and northwest and another zone of hills and mountains rising more than 10,000 feet, in the Saharan south and southwest.

Only 2 per cent of the total land area, largely in the narrow coastal strip, is arable; another 4 per cent is semi-arid grazing land. Libya has no permanent rivers, although large subterranean water reserves are fairly widespread. Rainfall, always scanty, falls about once every three years. A hot, dry, dust-laden wind called the *ghibli* is a recurrent threat to crops.

People

Libya had an estimated population of 2.6 million in 1977, with an average annual growth rate of 3.7 per cent. Ninety per cent of the people live in less than 10 per cent of the area, chiefly in the coastal regions. About 20 per cent of the people live in the co-capitals of Tripoli and Benghazi.

Libya's population is 97 per cent Arab and Berber, with some Negro stock. The population includes some Greeks, Maltese, Jews, Italians and Egyptians. About 97 per cent of the people are Moslems, nearly all of whom belong to the orthodox Sunni sect. Arabic is the official language.

Economy

Oil is Libya's principal product and main source of revenue, accounting for 99 per cent of its exports. Since the first major discovery of oil in 1959, Libya's economy has expanded rapidly. But by 1974, oil revenues alone had jumped to $7-billion from $1.6-billion in 1972, making Libya one of the richest countries in the world on a per capita basis, even though petroleum production fell from an average of 2.2 million barrels per day to 1.5 million.

Proximity to Europe and the closing of the Suez Canal in 1967 were major factors in Libya's rapid rise as an oil producer. Unlike other Middle East oil states that gave one huge concession to a major foreign oil company, Libya granted concessions to many companies of different nationalities—mainly U.S., British and West German. On Sept. 1, 1973, Libya nationalized 51 per cent of the assets of all foreign oil companies operating in the country. On Feb. 11, 1974, Libya totally nationalized three U.S. oil firms—Texaco, California-Asiatic Oil Company, a subsidiary of Standard Oil of California, and Libyan-American, a subsidiary of Atlantic-Richfield Company.

Although overshadowed by oil, agriculture is the second-largest sector in the Libyan economy. Libya is not self-sufficient in many kinds of food. Major crops include barley, wheat, olives, dates, citrus fruits and peanuts. Sheep and goats are the chief livestock. An estimated 36 per cent of the labor force was engaged in agriculture in 1971.

History

Libya was conquered by the Arabs in the eighth century, ruled by Turkey from 1553 to 1911 and administered as an Italian colony from 1911 until World War II. After Italian forces were expelled from the area in 1943, Libya was administered by the British and French.

Libya was the first nation to achieve independence under the auspices of the United Nations. In November 1949, the U.N. General Assembly adopted a resolution calling for the independence of Libya not later than Jan. 1, 1952. Libya declared its independence Dec. 24, 1951, as a constitutional and hereditary monarchy under King Idris I, a local ruler who had led Libyan resistance to Italian occupation.

King Idris was deposed Sept. 1, 1969, by a military junta which abolished the monarchy and proclaimed a Socialist Libyan Arab Republic. Col. Qaddafi, the leader of the coup, became chairman of the Revolutionary Command Council, the nation's chief governing body. A provisional constitution promulgated in December 1969 proclaimed Libya "an Arab, democratic and free Republic which constitutes a part of the Arab nation and whose objective is comprehensive Arab unity."

One of the early objectives of the new government was the withdrawal of all foreign military installations from Libya. In 1970, the British withdrew from their military installations at Tobruk and El Adem, and the United States withdrew from Wheelus Air Force Base near Tripoli.

Qaddafi in April 1973 proclaimed a new "popular revolution" involving a five-point reform program. The program called for suspension of existing laws and promulgation of new ones; implementation of Islamic thought, distribution of arms to loyal citizens; purging of political deviationists, including those who preached communism and capitalism, and a campaign against bureaucratic inefficiency. "Popular committees" were set up throughout the country to serve as the main instrument of the revolution.

Statistics on Libya

Area: 679,360 square miles.
Capital: Tripoli, Benghazi co-capitals.
Population: 2,607,000 (1977 est.).
Religion: 97 per cent Moslem.
Official Language: Arabic.
GNP: $11.9-billion, $4,920 per capita (1975).

Foreign Relations

Under Qaddafi, two themes have dominated Libyan foreign policy: Arab unity and anti-imperialism.

In September 1971, Libya joined Egypt and Syria in a loose political confederation. On Aug. 2, 1972, Qaddafi and Egyptian President Anwar Sadat announced that "unified political leadership" had been established between their two countries and pledged to achieve a full merger by Sept. 1, 1973. But Qaddafi and Sadat differed on major issues, including their approach to Israel, and on Aug. 29, 1973, the two leaders signed a compromise agreement under which they pledged only to establish a union by gradual stages.

But as the merger plan collected dust and eventually died, relations between Egypt and Libya grew increasingly hostile. In February 1974, Qaddafi said he would turn Libya into "a school where we will teach how a people can take up arms to stage a revolution." His threat was aimed at Sadat, among others. In the spring of 1975, Tripoli began threatening to sever relations with Cairo, and the controlled Libyan press described Sadat and his wife as a "20th century Anthony and Cleopatra." Sadat countered the insult by saying Qaddafi was "100 per cent sick and possessed by the devil." In June, Cairo newspapers reported that Egypt had imposed new restrictions on travel to Libya, requiring special permits and banning tourist travel. Since then the relationship has progressively deteriorated.

Plans to implement a merger between Libya and Tunisia also foundered. Tunisian President Habib Bourguiba declined a merger offer from Qaddafi in December 1972, but on Jan. 12, 1974, Libya and Tunisia announced an agreement to unite their two countries in a single nation to be known as the Islamic Arab Republic.

Algeria and Morocco strongly opposed the decision, however, and Tunisia subsequently announced that the proposed union required further consultations with Libya. It also invited Algeria, Morocco and Mauritania to join the union. A referendum on the merger, planned for Jan. 18, 1974, was postponed. As in the case of Egypt, Qaddafi's hard-line policy toward Israel appeared to be a major stumbling block to union. Finally, on March 29, 1975, Bourguiba explained away the union, saying that Tunisia could enter a merger only if a popular referendum were held and that the Tunisian constitution did not provide any referendum procedures.

In its relations with the superpowers, Libya has been critical of both the United States and the Soviet Union, but has benefited by various relationships with both. It has denounced American support for Israel, and it refused to go along with the March 18, 1974, decision of seven other Arab oil-producing states to end the oil embargo against the United States. (Libya lifted the embargo at the end of 1974.) Libya has criticized the Soviet Union, as well as the

United States, for practicing imperialism. It also has charged the Soviets with duplicity in their arms deals with Arab states, while Marxism has been attacked as an atheistic doctrine that is incompatible with Arab thought. Still, Libya's relationship with the Soviet Union has caused Sadat in Egypt to see Libya as one of Moscow's agents in attempts to penetrate Africa with Soviet influence.

Politically, Libya has pursued a policy of neutralism between East and West, but its major trading partners lie in the Western bloc: West Germany, Italy, the United States, the United Kingdom and France. France, the Soviet Union and Czechoslovakia are its chief arms suppliers; U.S. arms sales to Libya were suspended in September 1969.

During a May 1975, visit to Libya by Soviet Premier Alexei N. Kosygin, the two countries signed a major arms deal, variously estimated at $800-million by the Russians, $1-billion by U.S. intelligence sources and $4-billion by the Egyptians. Libya and the Soviet Union have expanded this arms relationship in recent years. Egypt has begun openly expressing fear that Soviet influence in Libya and Ethiopia might seriously threaten the Sudan, Egypt's ally.

But Qaddafi insists he is only stockpiling arms to serve as an arsenal for the Arab states in case of a new Arab-Israeli war.

In mid-1977 Qaddafi began attempts to mend his fences with Arab leaders. Libya and Egypt, for some months, halted propaganda attacks against each other and Libya ended the expulsion of Egyptians living there. Qaddafi also appealed to the United States to send a full ambassador to Tripoli and release several C-130 military transport planes purchased by Libya but later blocked from delivery by the State Department. The United States, however, replied that before better relations could be established Libya would have to stop aiding international terrorism and begin to seek a basis of cooperation with the United States in efforts to achieve an Arab-Israeli settlement. In July, Egyptian patience with Qaddafi abruptly ended and large-scale fighting broke out on the border escalating into air and ground battles.

Reports in mid-1977 indicated that Qaddafi had been receiving help from a small number of Cuban security experts to build a special anti-coup force under his personal command—a step reportedly taken after the execution of a number of Libyan officers on charges of sedition during the early part of 1977.

Libyan money has been finding its way to a variety of causes. It is generally believed that Libya contributes substantially to Palestinian groups most opposed to any settlement with Israel. Libya has also endowed a chair at Georgetown University in Washington, D.C. Other funds were used to support the film "Mohammed, Messenger of God." In short, Qaddafi has mixed oil, petrodollars and Islamic fanaticism into a blend of political and cultural chauvinism.

CENTURIES OF CONFLICT IN THE GARDEN OF EDEN

A 14-year revolt by the Kurdish minority against Iraq's central government collapsed abruptly in March 1975, ending the rebels' hope for an autonomous Kurdistan. The fate of the Kurds was sealed by a public kiss of peace between Vice President Saddam Hussein of Iraq and the Shah of Iran on March 6 at Algiers—an embrace that choked off Iranian support of the Kurdish cause.

The fiercely independent Kurds, within two weeks, fell victim to international politics far beyond their reach, largely the Shah's ambitions to end border disputes with Iraq and to bring about regional cooperation in the Persian Gulf that would exclude outside powers. Iraq's long dependence on Soviet arms was expected to be reduced by the termination of the Kurdish civil war.

The Kurds are a non-Arab Moslem people who inhabit the mountains of northern Iraq, northwestern Iran and southeastern Turkey and who spread out into tiny pockets of Syria and the Soviet Union. They make up about 20 per cent of Iraq's population. Ever protective of their own identity and traditions, the Kurds, beginning in 1961, launched intermittent guerrilla warfare that fought Iraqi forces to a standstill. In 1970, the rebels won from the Baghdad government a promise of Kurdish autonomy within four years. In March 1974, the autonomy proposals were presented by the Kurds and rejected by the Iraqis.

The Kurds demanded virtual veto power over legislation in Baghdad affecting Kurdistan and the inclusion of rich oil fields in their region. The central government called the demands tantamount to secession.

Fighting erupted on an even bloodier scale, reportedly involving thousands of casualties. Baghdad employed tanks and jet aircraft, and the Kurdish "Pesh Merga" forces led by Gen. Mustapha Mulla Barzani were aided by Iran, which supplied the insurgents with wire-guided antitank missiles and sent two artillery battalions into Iraq. By his action, the Shah hoped to undermine the left-wing Baathist regime in Baghdad, with which Iran feuded almost constantly.

But a year later, the Iranian-Iraqi rapprochement isolated Barzani, who, on March 22, 1975, with his army crumbling around him, said, "We are alone with no friends." On March 30, the general fled into Iran. By early April, Iraqi forces had completed their takeover of rebel mountain strongholds, meeting no further resistance. In a report published May 3 in Teheran, Barzani was quoted as saying, "The battle for Kurdish autonomy is futile, and the struggle will never be resumed."

Even with the Kurdish revolt under control, Iraq remains a country which is not yet politically fully unified. A 1976 study on the Middle East concluded: "Although the government seems more firmly in control of society than any previous one, political life still exists within a space narrower than that of Iraq as a whole. There is still more than one sense in which Iraq is not yet a unified political society. Political activity is mainly concentrated in the towns of the Tigris valley.... Even within the cities, the consensus which supports the government is limited and weak. Power tends to lie, as in Syria, in the hands of new men from

Iraqi President Ahmed Hassan al-Bakr

small provincial towns who have come up through the main channel of social mobility, the army; in the present regime, in those of a group from the northern Tigris town of Tikrit. They rule in uneasy partnership with technocrats, civilian ministers, and officials of high education and special qualifications, but with no independent basis of political power."[1]

Terrain

Iraq occupies a strategically significant position in the Middle East. In addition to having vast oil resources which made it in 1974 the fourth-largest producer in that region, Iraq is located on the Persian Gulf and shares borders with two powerful non-Arab countries—Turkey and Iran. Iraq's other neighbors are Kuwait, Saudi Arabia, Jordan and Syria.

Historically known as Mesopotamia—"the land between the rivers"—Iraq includes the twin river system of the Tigris and Euphrates. Eighteen per cent of Iraq's approximately 170,000 square miles is agricultural; 10 per cent is seasonal and other grazing land, and 4 per cent is forested; the remainder is primarily desert.

People

Arabs, the dominant group in the Iraqi population of more than 11 million, occupy most of the central, western and southern regions of the country. Other ethnic groups besides the Kurds include Assyrians, Turkomans, Iranians, Lurs and Armenians. The bulk of a once-significant Jewish community which had totaled about 150,000 emigrated from Iraq to Israel in the late 1940s.

About 90 per cent of Iraqis are Moslems. It has been estimated that approximately 75 per cent of the Arabs belong

1 A. L. Udovitch, ed., *The Middle East: Oil, Conflict and Hope* (Lexington, 1976), p. 285.

to the Shia sect, along with 50-60 per cent of the Kurdish population. The remainder belong to the Sunni sect. Other religious communities include Christians, Jews, Bahais, Mandaeans and Yezidis.

Economy

Oil dominates the Iraqi economy. Though no official statistics have been published, it is estimated that Iraq's oil reserves are second only to those of Saudi Arabia among members of OPEC. In 1974, Iraq ranked fourth among Middle East oil producers, behind Saudi Arabia, Iran and Kuwait. In 1977, oil revenues were estimated at about $8-billion, approximately the GNP just three years before.

In June 1972, Iraq nationalized the Western-controlled Iraq Petroleum Company, which had dominated almost the entire oil production of the country. In February 1973, the Iraqi government announced an agreement with the company on terms of the nationalization and compensation. Later, in 1973, Iraq nationalized two American oil firms, Mobil and Exxon, to chastise the United States for its support of Israel in the 1973 war.

Services, agriculture, industry and trade—in that order—contribute to Iraq's GNP in addition to oil. Though oil dominates the country's revenues and provides nearly all the foreign exchange, agriculture occupies most of the population. About 55 per cent of Iraq's population has been estimated as living by agriculture or stock-rearing. Besides being the largest supplier of dates in the world, Iraq also produces wheat, barley, rice, millet, cotton and tobacco.

Early History

Mesopotamia is one of the famous place names of ancient history. The Sumerians founded city-states there about 3000 B.C. Other civilizations which flourished between the Tigris and the Euphrates Rivers were the Babylonian and the Assyrian.

After 500 B.C., Persia, and later Macedonia, dominated the area. In the seventh century A.D., Mesopotamia was overrun by the Arabs, who established a capital at Baghdad, storied setting for *The Tales of the Arabian Nights.* Mongol invasions followed in the 12th and 15th centuries. In the 16th century, the area fell to the Ottoman Turks, under whose rule it remained until World War I.

It is held that the Garden of Eden once stood between the two rivers, and it was there that Nebuchadnezzar created the splendors of Babylon. King Hummurabi established the first recorded code of justice.

Iraqi Independence

Iraq began its move toward modern statehood during World War I, when the Turkish Empire, a belligerent on the Austro-German side, began to fall apart. Turkey's entry into the war in late 1914 prompted a British expedition to the Turk-controlled territory which was to become the state of Iraq. The move eventually led to British occupation of the area and Britain's receipt of a League of Nations mandate for the territory in April 1920.

Hashemite King Amir Faisal, driven by the French from his throne in Syria in July 1920, was elected to the Iraqi throne by a referendum and crowned in Baghdad in August 1921. Iraq became independent Oct. 3, 1932, but the Hashemite family continued to rule Iraq as a constitutional monarchy, following a generally pro-British, pro-Western policy.

In 1955, Iraq signed a mutual defense treaty—called the Baghdad Pact—with Britain, Iran, Pakistan and Turkey. Although the United States promoted the pact and provided members with assistance, it never became a formal partner. The main reason for this ambivalence was that Iraq and Egypt were rivals for leadership of the Arab world; and the United States, in the hope that Egypt could eventually be persuaded to join the pact, did not want to alienate Egypt by formally allying itself with Iraq. An alliance with Iraq also would have created problems for U.S. relations with Israel.

Military Coups

The Hashemite monarchy was ousted July 14, 1958, by a leftist military coup in which King Faisal II was killed. The leader of the coup, Brig. Gen. Abdul Karim Kassim, seized control and began reversing the country's policies, including renunciation of the Baghdad Pact, establishment of relations with the Communist bloc countries and purchasing of Soviet military equipment.

Kassim was assassinated in another military coup in February 1963, when a group of "free officers," primarily members of the Arab Socialist Resurrection Party (Baath Party), took over. The Baath regime was unable to consolidate its power and was ousted in another coup in November 1963. The new regime pursued a neutral East-West policy and turned its efforts to strengthening relations with other Arab states, especially Egypt.

The Baath Party came back to power when yet another military coup took place in July 1968. A Revolutionary Command Council took control and chose Maj. Gen. Ahmed Hassan al-Bakr as president and prime minister.

Ultimate power was granted to the command council. A Council of Ministers was set up with administrative and legislative responsibilities.

The provisional constitution adopted in July 1970 called for the establishment of a national parliament, but set no date for its creation. Bakr crushed a coup on July 30, 1973, which aimed at overthrowing the Baath regime. After a swift trial, the government executed 36 coup leaders. In the trial's aftermath, the president announced plans for the formation of a 100-member National Council, to be implemented later and to which the Revolutionary Command Council was to appoint the membership and turn over its legislative powers.

Iraq's vice president, Saddam Hussein, emerged as the new strongman of Baghdad politics. It was he who helped bring about the detente with Iran in the spring of 1975.

Foreign Relations

Iraq broke off diplomatic and consular relations with the United States on June 7, 1967, after the outbreak of the Arab-Israeli war. Iraq retained several diplomats in Washington in an "interest section" established under the aegis of the Indian embassy. The United States had the same option since the 1967 break but did not exercise it until July 1972, when two foreign service officers were sent to Baghdad to open a similar "interest section." During May 1977, a step was taken which could lead to resumption of full diplomatic relations between the two countries. Philip Habib, top political affairs officer of the U.S. State Department, met Iraqi Foreign Minister Saadoun Hammadi in Baghdad to pursue this possibility.

Iraq has maintained close ties with the Soviet Union, its principal arms supplier. In April 1972, Iraq signed a 15-

Statistics on Iraq

Area: 167,924 square miles.
Capital: Baghdad.
Population: 11,759,000 (1977 est.)
Religion: 90 per cent Moslem, 10 per cent other.
Official Language: Arabic; Kurdish minority speaks Kurdish.
GNP: $16-billion, about $1,400 per capita (1977 est.).

year treaty of freindship and cooperation with the Soviets, similar to the 1971 Moscow-Cairo treaty (abrogated in 1976). Although details of the treaty have been vague, it was generally considered to be a significant development in the Persian Gulf region, with the Soviets acquiring a strong foothold for naval operations in the Indian Ocean. On Aug. 31, 1974, President Ford named Umm Qasr in Iraq, at the head of the gulf, as one of the ports the Soviets were using for their Indian Ocean fleet. Moscow has repeatedly denied it was building up bases in the region. In June 1977 Iraq agreed in principle to buy from France more than 70 French Mirage F-1 fighter-bombers in a move to diversify its Soviet-equipped air force.

Iraq's leftist Baath regime and its alliance with Soviet Russia have been eyed suspiciously by its conservative neighbors—Iran, Saudi Arabia and the gulf emirates—which have been steadfast in their determination not to be swept up by missionary socialism. Both Kuwait and Iran have fought limited border wars with Iraq.

During 1975 and 1976 Syria moved infantry units and tanks to its border with Iraq because of Iraq's unhappiness over Syrian intervention in Lebanon and the long-standing dispute between the two countries over the waters of the Euphrates.

Iran and Iraq each won major concessions as the two patched up their differences in the March 1975 agreements. The Shah summarily dropped his support of the Iraqi Kurds, enabling the Baghdad government to crush their war for independence. In turn, Iraq yielded to Iran on the borders issue, notably rights to the Shatt al Arab waterway, where the Tigris and Euphrates converge before entering the Persian Gulf. Iraq acceded to Iran's demand that the boundary between the two countries follow the midcourse of the Shatt.

In the spring of 1973, Iraq occupied some Kuwaiti border posts in clashes with that sheikdom, causing Saudi Arabia to send 20,000 troops into Kuwait for a short period in order to bolster the state's defense. At the time, Iraqi Foreign Minister Murtada Abdul Baki said there was no legal basis for the existing frontier between Iraq and Kuwait, adding that Iraq "wants to become a Persian Gulf state—that is the crux of the border dispute with Kuwait." Later, Iraq claimed it had withdrawn from the outposts, an assertion denied by Kuwait.

Iraq has had a 13-year argument over Euphrates water with Syria, which also has a radical Baath regime. Although the matter should be technical, the approach of the two countries has often been spiteful. In April 1975, the Arab League Council referred the water dispute to a seven-nation technical committee. The gulf's conservative states looked with favor on Iraq's new willingness to arbitrate differences, hoping that Baghdad, growing richer from its increasing oil production, was temporizing its missionary zeal and that the gulf region was entering into a tranquil phase.

Iraq, like Libya, has remained ideologically opposed to any settlement with Israel. In November of 1976 Iraq hosted an "International Symposium on Zionism" attended by persons from 46 countries including U.S. Senator James Abourezk (D S.D.)[2] There, according to one reporter, "Zionism was baked, boiled, fried and roasted in an unrelenting ideological tirade,"[3] After the right-wing victory of Menahem Begin's party in Israel in May, Iraq took the lead among Arab countries in urging abandonment of all negotiations with Israel and a stepped-up military effort. Though still sharply at odds with Syria, Iraq indicated its willingness to set aside all intra-Arab differences in the interest of confronting Israel with unity. Information Minister Tariq Azziz stated in May 1977 that "The Arab nation can afford to struggle for many years against Zionists and still live well." And Iraq's powerful Saddam Husayn bluntly stated in an interview the same month, "We will never recognize the right of Israel to live as a separate Zionist state."[4]

2 For the 19-point declaration issued at the end of the symposium see *The Washington Post*, 22 November 1976, p. A13.

3 For the story on the symposium see Judith Miller, "Letter from Iraq," *The Washington Post*, 15 November 1976, p. A8.

4 "Iraq: Determined to Upset U.S. Plan for Mid-east Peace," *U.S. News & World Report*, 16 May 1977.

CAUTIOUSLY PURSUING A "GREATER SYRIAN" DREAM

By mid-1977, once-turbulent Syria was approaching seven years of stability under President Hafez al-Assad—no mean trick in a land that had counted a dozen coups between Syrian independence in 1945 and Assad's own armed seizure of power a quarter century later in November 1970. Despite its tumultuous record—or perhaps because of it—Syria claimed a place in the front ranks of pan-Arabism and of implacable hostility to the Jewish state.

During 1976 Syria became deeply involved in the Lebanese civil war. First, it appeared, Syria was aiding the side of the Moslems, reformers, leftists and Palestinians. But when the left's position became dominant, Syria appeared to switch and began aiding the Maronite Christians. Syrian relations with other Arab countries (except with Jordan) became strained as her participation in the Lebanese conflict escalated. President Sadat of Egypt was especially critical as he was still chafing from Assad's assaults on Egypt for agreeing to the September 1975 second-stage Sinai agreement with Israel. As late as September 1976, Sadat was describing Syrian intervention in Lebanon as "a black chapter in Arab history."[1] But by late 1976 Syrian relations with most Arab countries, especially with Saudi Arabia and Egypt, had dramatically improved.

Paradoxically, the Syrians have been jealous of their national identity; yet they succumbed to the enchantments of Gamal Abdel Nasser and joined with Egypt to create the United Arab Republic in 1958. Under Nasser's leadership, Syria became the junior partner, and Damascus was relegated to the role of provincial capital with power emanating from Cairo. In 1961, the Syrian army, in another coup, took the country out of the union.

In its pursuit of militant Arab nationalism, Syria often has reached out with an abrasive hand to sister Arab states. Then a self-appointed protector of the Palestinians and their cause, Syria sent armored units into Jordan in 1970 when King Hussein sought a showdown with the Palestinians, but, fearing U.S. intervention, Syria withdrew quickly. Syria's leftist military regimes and its dominant Baath Party long vowed to "liberate" Saudi Arabia by destroying its "reactionary" Saudi family.

But much has changed in the Arab Middle East—former enemies, such as Saudi Arabia, are now friends; and former friends, such as the PLO, are now uneasy about Syrian intentions.

Embittered by the triumph of Israeli arms in the 1948 war, Syria became the first Arab nation to use oil as a weapon eight years later. When Anglo-French-Israeli forces attacked the Suez, Syria responded by blowing up its own oil pipelines. Syria suffered its most devastating blow from Israel in the 1967 war, when Israel destroyed two-thirds of the Syrian Air Force on the ground and seized the Golan Heights.

1 The Washington Post, 29 September 1976.

Syrian President Hafez al-Assad

Benefiting from a degree of domestic calm and relative political stability, Assad moved Syria from a posture of belligerence to one of pragmatism in dealing with other Arab countries. He prayed beside King Faisal in Damascus' Omayad Mosque in January 1975 and accepted a $250-million gift from the Saudis. Assad has received some airplanes from Persian Gulf princes. Since 1975 Saudi aid has increased.

After the 1973 war, Syria was six months slower than Egypt in coming to any terms with the Israelis; but, after keeping Secretary of State Kissinger hopping from capital to capital for a full month, Syria and Israel signed a disengagement of forces agreement on May 31, 1974—the first time the two enemies had put pen to the same document since 1948. However, Syria still adamantly calls for return of all occupied territories, especially Syrian territory known as the Golan Heights. And though Syria has battled the PLO in Lebanon, she continues to champion her cause internationally insisting that Israel must deal with the PLO at any resumed Geneva conference. Along with Egypt and Saudi Arabia, Syria has for the past few years been engaged in what has been described as a peace phase—symbolized by regular extensions of the U.N. peacekeeping force mandate, which separates Syrian and Israeli troops on the Golan Heights; and by talk of a limited peace with Israel. In the case of Syria, though, President Assad has emphasized this peace would not include actual recognition of Israel or any form of economic or cultural relations.

Syria has long avowed its neutralism between East and West, yet has inched into the Soviet orbit by its acceptance of massive amounts of Russian weaponry, only to make openings to the West again in the aftermath of the 1973 war. Assad did some serious soul-searching before agreeing to welcome President Nixon to Damascus in June 1974. The visit resulted in the two countries' establishing full diplomatic relations, broken since the 1967 war. Since then, small amounts of U.S. aid have been granted Syria.

Terrain

Syria contains some of the most arable land in the Middle East, a growing industrial sector and a refining facility that, although damaged during the 1973 Arab-Israeli war, is important to Iraqi as well as Syrian oil. However, successive government upheavals have upset the cohesiveness of Syrian development programs.

A land of Middle East contrasts, Syria extends southward and eastward from the Mediterranian Sea over alternate stretches of fertile valley, plain land, desert and mountains. Lebanon and Israel are to the west, Jordan to the south. Iraq lies to the east and Turkey to the north. In earlier times, greater Syria included Jordan, Lebanon and Israel. About the size of North Dakota, modern Syria covers 72,000 square miles, including 500 square miles occupied by Israel. Approximately 48 per cent of the land is arable, and 29 per cent is used for grazing. Forests cover 2 per cent of the country and deserts the remaining 21 per cent.

The country is divided into seven distinct regions. A fertile belt, site of Syria's major port, Latakia, borders the Mediterranean. Further east, mountains extend southward to the Anti-Lebanon range, peaking with Mount Hermon at 9,232 feet. The central region of Syria includes the fertile valley of the Orontes River and the arable plains of Aleppo, Hama and Homs. The Syrian desert occupies the southeast portion of the country, site of historic Palmyra. The Euphrates River valley extends northward from the desert; the rich Ghutah valley, surrounding Damascus, extends southward. At the southernmost tip of the country lies the black, hilly region of Jebel Druze, where mountains reach 5,900 feet. The Trans-Arabian pipeline cuts through Jebel Druze.

Syria's standing as one of the most cultivable lands in the Middle East results from its access to major sources of water—the Euphrates and Orontes Rivers and their tributaries.

People

In 1977, Syria's population was estimated at 7,733,000. Nearly all the people, 90 per cent, are Arabs, descendants of the Arab branch of the Semitic family. Another 9 per cent are Kurds, a group racially akin to the Iranians. Other ethnic groups include Armenians and nomadic Bedouins. The great majority of Syrians are concentrated in the fertile areas around Aleppo, Hama and Homs, around Damascus and on the coastal plains of Latakia.

Most Syrians are Moslems, 70.5 per cent of the people following Sunni Islam. Another 16.3 per cent belong to other Moslem sects, primarily Shia and Ismaili. President Assad is a member of the Alawite sect of the Shias. Christians, mostly Greek, Armenian and Syrian Orthodox, comprise 13.2 per cent of the population. Jews have dwindled in recent years to a negligible group.

Economy

Syria's economy is considerably more balanced than that of most Arab states. In 1971, services, including oil refining, accounted for 57.5 per cent of total gross national product, agriculture 23.3 per cent and industry 19.2 per cent. In Syria's two million labor force, 67 per cent were involved in agriculture, 12 per cent in industry and 21 per cent in services. A majority of Syrian laborers are unskilled, and there is a shortage of skilled labor.

A warm climate, much like that of Arizona, and a rainy season from November until April enhance Syria's agricultural capability. Main crops are cotton, wheat, barley and tobacco. Cotton is the single largest source of foreign exchange, comprising about 20 per cent of Syria's export earnings. Fertile plains provide pasture land for sheep and goat raising, another major occupation.

Syria has stepped up efforts to improve its agricultural output. An agreement was signed with the Soviet Union in April 1966 to provide assistance for a giant dam on the Euphrates River to be used for irrigation and power. Programs are under way to increase agriculture through other irrigation plans and through various agrarian reform measures.

Industrial growth occurred after World War II, following the declaration of Syrian independence in 1945. Today, major industries include textiles, petroleum, food processing, beverages and tobacco. Among the nation's fastest-growing industries are flour milling, oil refining, textile production and cement. Syria's current five-year plan places major emphasis on industrialization.

Syria issued nationalization decrees Jan. 3, 1965, in an effort to halt the flow of capital from the country. Another decree, issued March 4, 1965, nationalized six Syrian oil companies, the American-owned Socony-Mobil Oil Company and the Standard Oil Company of New Jersey and a Royal Dutch Shell affiliate. That effort, which was successful in controlling 95 per cent of industrial production, 50 per cent of commercial transactions and 15 per cent of the agricultural land by July 1965, soon lost momentum. When President Hafez al-Assad came to power in 1970, the trend was reversed. Small entrepreneur firms have sprung up with government encouragement.

A petroleum industry has grown up around Syria's modest oil reserves, located in the northeast parts of the country. When negotiations between the government and American, French and German oil companies failed to develop a plan for exploitation, the government stepped in to perform the task. With Soviet assistance, the government has developed the fields at Karatchouk, and an Italian-built pipeline carries the crude oil to Tartous for refining.

The country's major refinery, At Homs, was originally built by the Czechs. Syria's production of 120,000 barrels a day provides most of the country's own oil needs. Some petroleum products still must be imported, however. Drilling and exploration expeditions are seeking new oil resources. Syria also refines oil brought in from Iraq, and considerable revenues are derived from that service. In addition, the Trans-Arabian pipeline cuts through Syria en route to Mediterranean ports. But Iraq cut oil supplies to Syria in April 1976.

In 1975, Syria's GNP was $2.4-billion, or about $330 per capita. Military expenditures have been a continuing drain on the economy, accounting for nearly half of the ordinary yearly budget. To meet budgetary expenses, the government relies mainly on taxes and pipeline transit fees. The government is also encouraging tourism with an eye to increasing its revenues. The economy has suffered a lack of cohesiveness from the successful upheavals in government, but the Syrian pound is strong and becoming more valuable

Statistics on Syria

Area: 71,586 square miles, including about 500 square miles occupied by Israel.
Capital: Damascus.
Population: 7,733,000 (1977 est.).
Religion: 87 per cent Moslem, 13 per cent Christian.
Official Language: Arabic, also Kurdish, Armenian, French and English.
GNP: $2.4-billion, $330 per capita (1975 est.).

with respect to the dollar as the Syrian economy improves and foreign investment increases.

In a full-page advertisement in the January 30, 1977, issue of *The New York Times*, Assad noted that since he assumed the presidency in 1970 "the Syrian economy has moved forward with vigor and vitality unprecedented in the modern history of our country." Assad has opened the door to private and foreign investment, and the fourth five-year plan (1976-1980) is focused on expansion of basic industries; growth of food processing, textile, chemical and engineering industries; along with continuing infrastructural improvements. French journalist Eric Rouleau reported from Damascus in mid-1976 that "The country is experiencing an extraordinary boom as the result of a tenfold expansion of investment since Assad took office" and "impressive headway is being made in setting up the country's infrastructure."[2] But he also noted the growth of an exploitative, bourgeois middle-class while much poverty remains unchecked.

Trade

Syria exports cotton, fruits, vegetables, grain, wool and livestock, mostly to the Soviet Union, Italy and Lebanon. Syria's exports to the United States have been confined largely to small quantities of carpet wool. Its imports, machinery and metal products, textiles, fuels and foodstuffs, come mainly from Lebanon, West Germany, Italy, the Soviet Union, Japan and France. Imports from the United States have been primarily pharmaceuticals, agricultural machinery and tires, although in drought years wheat shipments are significant.

Syria has suffered severe balance-of-payments deficits, fueled by unfavorable balances of trade. The difficulty dates from the drought years of 1958 to 1961 and was worsened by the recall of capital in fear of Socialist measures, nationalization of foreign and domestic firms and ambitious economic programs.

Early History

The history of Syria, one of the longest-inhabited areas of the world, reflects the country's strategic position between Africa, Asia and Europe. In ancient times, its location made it an important battlefield as well as a prime trade route. Its present capital, Damascus, first settled about 2500 B.C., was successively dominated by Aramaean, Assyrian, Babylonian, Persian, Greek, Roman, Nabatean and Byzantine conquerors.

In 636 A.D., Damascus came under Moslem occupation. Shortly thereafter, it rose to its historic peak of power and prestige as the capital of the Omayyad Empire,

which stretched from India to Spain from 661 to 750. From 1260 to 1516, Damascus was a provincial capital of the Mameluke Empire, before coming under a 400-year rule by the Ottoman Turks in 1516.

World Wars

The modern state of Syria began to emerge during World War I, when Turkey's entry into the war gave opportunity for the expression of Arab nationalist aspirations. Syria, still under Ottoman domain, became a military base and fell in 1918 to British forces. At the conclusion of the war, Syria temporarily was ruled by an Arab military administration. At this time, the Sykes-Picot Agreement of 1916, dividing the Middle East between Britain and France, and the Balfour Declaration of 1917, favoring the establishment of a Jewish homeland, were signed, documents whose intentions and pledges were to prove irreconcilable.

Following the San Remo Conference in April 1920, which gave France a mandate over Syria and Lebanon, French troops occupied Syria, dethroning King Faisal, later to become king of Iraq, who had assumed power in March 1920. The grant by Britain of independence for Egypt in 1922 and Iraq in 1932 put pressure on France to recognize Syria's independence. Negotiations with Syrian nationalists in 1936 resulted in such an agreement, but by 1938, France had still not ratified the treaty.

Free French and British forces seized Syria in June 1941. Arab nationalists, with British assistance, eventually procured a pledge of independence from the Free French forces. In April 1945, when the French troops withdrew, a republican government, under President Shukri al-Kuwatly, elected in 1943, assumed full control of the country. Meanwhile, Syria had already become a founding member of the United Nations and of the Arab League.

Military Coups

Over the next 25 years, Syrian government was marked by successive changes of leadership. President Kuwatly's government was overthrown by a bloodless military coup led by Col. Husni al-Zayim, chief of the Syrian army, on March 30, 1949. Zayim reasoned that the change of government had been necessitated by Syria's poor military showing in the war with Israel in 1948. Five months later, reaction to the repressive military rule resulted in a second coup, which yielded power to Syria's elder statesman and former president, Hashim al-Atasi. That government was ousted in December 1949 by a third military coup. The new, harsh regime, headed by Lt. Col. Adib al-Shishakli, failed to consolidate support, and on Feb. 25, 1954, the government was returned to a democratic regime under Hashim al-Atasi. Kuwatly was re-established as president on Sept. 6, 1955.

Kuwatly's ascension to power gave voice to groups favoring union with Egypt as a first step toward a united Arab state that would merge the Arabic-speaking Middle East states. On Feb. 1, 1958, President Kuwatly and Egyptian President Gamal Abdel Nasser signed an agreement in Cairo proclaiming the United Arab Republic. The two countries were united under one president, Nasser, and defended by one army. A Syrian plebiscite Feb. 21, 1958, gave overwhelming approval to the new merger. But geographic separation, Nasser's imposition of higher taxes and import limitations, nationalization of commercial firms, and Egyptian domination of the union eventually undermined the merger. A revolt by the Syrian army Sept. 28, 1961, established a new civilian government and led to Syria's secession from the U.A.R. and the establishment of

2 Eric Rouleau, "Assad's Calculated Risk," *The Guardian*, 27 June 1976.

the independent Syrian Arab Republic. The government faltered, troubled by an ongoing conservative-Socialist rivalry.

Baath Party

In March 1963, a pro-Nasser military faction, led by Lt. Gen. Louai al-Atasi, seized the Syrian government. The Arab Socialist Resurrection Party, the Baath Party, proclaiming "unity, freedom and socialism," assumed predominant control of the government, thereby ending its 20-year clandestine existence in Syria and other Arab countries. Lt. Gen. Amin al-Hafez became premier and de facto president. On April 17, 1963, an agreement was signed in Cairo for the union of Egypt, Iraq and Syria, but disagreements between the parties developed, and the tripartite federation failed to materialize. The fall of the Baath Party in Iraq in November 1963 doomed tentative plans for an alternative Syrian-Iraqi union.

President Hafez was ousted Feb. 23, 1966, by a dissident army faction within his own party that claimed Hafez had betrayed Baathist principles and assumed dictatorial powers. Following the "rectification," executive and legislative powers were assumed in full by the Baath Party.

Arab-Israeli Wars

Syria was an active participant in the Arab-Israeli war in 1967, although it was quickly defeated. Israeli forces overran Syrian positions and captured the Golan Heights. Syria joined other Arab states in refusing to negotiate a peace settlement with Israel. The nation remained a leader in Arab efforts to resolve the status of Palestinian refugees, 170,000 of whom are located in Syria. During the war, Syria severed diplomatic relations with the United States because of American support for Israel. It boycotted the conference of chiefs of Arab states, held Aug. 29 to Sept.1, 1967, believing more forceful measures should be taken to neutralize Israeli gains.

In November 1970, the ruling civilian Baath Party was supplanted by the ruling military faction of the party. Hafez al-Assad was brought to power by the coup and subsequently was elected president. This followed Assad's refusal, a month earlier when head of the Syrian air force, to order his men into action against Jordan in a skirmish that almost toppled King Hussein and resulted in a regional war. Israel (with U.S. approval) was poised to attack the Syrian tanks had they entered the Jordanian capital, Amman.

Since 1970 the new Baathist regime has embarked on a more pragmatic course than its civilian predecessor, pursuing a more liberal economic policy and playing down ideological tenets.

Syrian troops joined Egyptian forces in a two-pronged surprise attack on Israel Oct. 6, 1973. Initial Syrian gains along the Golan Heights were eventually offset by Israel. A week after the fighting began, Israeli forces had penetrated six miles into Syria, attacking Latakia and Tartous and heavily damaging the oil refinery at Homs. On Oct. 23, Syria accepted the U.N. cease-fire resolution of Oct. 22, conditional upon complete Israeli withdrawal from occupied lands. On Oct. 29, President Assad said Syria accepted the Oct. 25 U.S. cease-fire after the Soviet Union gave guarantees that Israel would withdraw from all occupied territory and recognize the rights of the Palestinians. Assad added that Syria would resume the war if these conditions were not met.

Syria was more reluctant than Egypt to take steps beyond the cease-fire, and confrontation between Israeli and Syrian forces on the Golan Heights continued for six months. The two belligerents engaged in sporadic air, artillery and tank duels. Finally on May 31, 1974, at Geneva, the two signed an agreement which set up a U.N.-policed buffer zone between the two armies and called for the gradual thinning out of forces. The two countries also exchanged prisoners of war. The fate of POWs had been an emotionally charged issue in Israel.

Government

Syria's government has been under a left-wing military regime since March 1963, when the Baath Party seized power. Executive power rests in the president and Council of Ministers, legislative power in the People's Assembly and judicial authority in special religious as well as civil courts, relying on Islamic and civil law. A new constitution was promulgated in 1973.

The Arab Socialist Resurrectionist Party, the Baath Party, dominates the political scene. The Baath Party Regional Command is particularly powerful. A "national front" cabinet was formed in March 1972, comprising mostly Baathists, with some independents and members of the Syrian Arab Socialist Party, the Arab Socialist Union and the Syrian Communist Party. Communists are mostly sympathizers and number only 10,000 to 13,000. Outside of the "national front" cabinet, little influence is exerted by groups other than the Baath Party. The greatest threat to the Baathist regime lies within the factionalism of the party itself.

Foreign Relations

Syria has long sought to maintain its independence from foreign influence. In line with this goal, successive governments have articulated a policy of nonalignment with Western alliances or the Communist world. But Western-Israeli ties, which Syria views as an obstacle to Arab unity, and Soviet arms supplies and assistance, which have greatly aided Syria in its ongoing conflict with Israel, have moved Syria away from neutrality and toward Soviet influence.

Syria is a major proponent of Arab unity, subject to the preservation of Syrian interests. It fully supports the immediate withdrawal of Israeli troops from lands occupied during the 1967 war and the restoration of Palestinian rights. Syria has frequently been on the side of more stringent Arab measures against Israel.

During the Six-day War in 1967, Syrian-American diplomatic ties were severed. In keeping with Assad's more pragmatic approach to external affairs, the two countries renewed full relations during President Nixon's visit to Damascus June 16, 1974.

Assad's more moderate approach is paying off economically. By mid-1975, there was a steady flow of investment from the conservative Arab oil powers, making ally Egypt envious and enemy Israel all the more apprehensive. With Saudi Arabian aid, Syria has nearly completed recovery from 1973 war damage—a bill that has run somewhere between $350-million and $1.8-billion. Saudi Arabian aid has continued to flow regularly to Damascus, and other Arab countries have also begun to contribute with aid and investment. A Kuwait corporation, for instance, has committed $200-million for Syrian tourism construction.

But by early 1977 Syria had become frustrated by unfulfilled promises. Finance Minister Mohammed Imadi indicated that Arab nation contributions were "negligible to the burden we carry, and less than we need, less than promised."[3] During 1976, Syria received about $400-million in foreign aid from both Arab and non-Arab sources and $750-million was anticipated for 1977.

Outside the Arab world, the United States, France, West Germany and Italy, as well as Syria's Soviet-bloc friends, are engaged in investment or aid projects in the country. The World Bank, among international organizations, is lending money to the Syrians.

Intervention with Syrian forces in the Lebanese conflict began in April 1976, and became massive in June. Within a few months, nearly 30,000 Syrian troops had been deployed in Lebanon in an effort to cripple the Palestinians and leftists under the leadership of Yassir Arafat and Kamal Jumblatt. Egypt was outraged and the Soviet Union turned on its friend. Soviet Leader Leonid Brezhnev, in a July 11 letter to Assad, stated, "We insist that the Syrian leadership should take all possible measures to end its military operations against the resistance and Lebanon's national movement."[4] Arafat, in turn, accused Damascus of "bloody butchery."[5]

Explanations of the Syrian "invasion" have been varied. Eric Rouleau has written that the Syrian gamble was undertaken to preserve Syrian influence in Lebanon in the face of a possible move by the battling forces to push Syria out.[6] This is quite understandable since the major unifying theme of Syrian policies under Assad was to attempt to expand Syria's influence throughout the area. Syrian aspirations have frequently been termed the "Greater Syrian" policy in recognition of Syria's historical territorial claims, especially to Lebanon, and her desire for regional leadership. These aspirations are especially worrisome to Israel, where it is felt the Syrians seek a kind of regional superpower role. Israeli scholar Shimon Shamir, head of the School of History at Tel Aviv University, writes that "the concept of Greater Syria, which would include the territories of Syria, Lebanon, Jordan and Israel (or Transjordan and Palestine), has been clearly asserted at times and left implicit at others, but it has always remained at the basis of the Syrian strategic outlook."[7] The Israeli embassy in Washington also widely circulated a 1976 *Chicago Tribune* report that it is "Syria's desire to use the Palestine Liberation Organization as the instrument for controlling all the land now ruled by the government of Israel."[8]

Assad's efforts toward regional influence have, in fact, met with considerable success. As of mid-1977, Syria's control over the Palestinians, especially of the important *Saiqa* faction of the PLO, as well as Syrian influence throughout Lebanon, were substantial. "In many ways, Damascus is the capital of Lebanon," *The Washington Post* reported in May 1977.[9] Furthermore, Syrian integrative efforts with Jordan, begun in August 1975, had moved forward to the point where economically, at least, the two countries are fast becoming one unit.

Relations with the United States have continued to improve since the Yom Kippur War even though Assad repeatedly has charged that Kissinger and the United States were attempting to divide the Arabs from each other to Israel's benefit. In 1977 Assad also charged the United States with fomenting the Lebanese civil war "in order to get all Arab countries drowned in it."[10] But Assad was the one Arab leader who preferred to meet President Carter on neutral ground, rather than in Washington. The meeting took place in Geneva during Carter's first international trip soon after taking office. Washington is well aware that Syria plays an important, possibly a crucial, role in influencing the outcome of the Arab-Israeli dispute.

---

3 Stuart Auerbach, "Syrians Say Oil States Stingy with Aid," *The Washington Post,* 26 February 1977.

4 See Mary Costello, "Arab Disunity," Editorial Research Reports, 29 October 1976, p. 792.

5 *U.S. News and World Report,* 21 June 1976.

6 Eric Rouleau, op. cit.

7 In A. L. Udovitch (ed.), *The Middle East: Oil, Conflict & Hope* (Lexington: 1976), p. 199. Also see Daniel Dishon, "Syria's Dilemma," *Jerusalem Post Weekly,* 8 June 1976.

8 Donald Kirk, "Syria's Political Ambitions," *Chicago Tribune,* 2 March 1976.

9 Auerbach, op. cit.

10 *The Washington Post,* 26 June 1977.

A PRO-WESTERN KINGDOM WITH A DURABLE LEADER

The mercurial fortunes of King Hussein, to whom as a boy the Hashemite crown was passed nearly a quarter century earlier, appeared uncertain in mid-1977. Though Jordan's ties with Syria and Egypt have been improved from what they were a few years ago, the fate of the occupied West Bank and the role of the Palestinians, who comprise a majority of the population even on the East Bank which Hussein still controls, present the country with much potential instability. Hussein's desert kingdom remains a primary, though at times subdued, participant in the still unresolved Arab-Israeli quagmire.

The hub of Hussein's turmoil has been the West Bank region—occupied and later annexed by the current monarch's grandfather, King Abdullah, after the 1948 war, and then occupied by Israel since the 1967 war. Since then, major elements within the Palestine Liberation Organization (PLO) have demanded that this area, along with the Gaza strip, become a separate Palestinian state in any prospective peace settlement. Even if this were to occur, some form of affiliation with Jordan would seem likely. The longer-range problem is thus whether the Palestinians might come to overpower the Hashemites and Bedoins who are the ruling elite of Jordan. Few in Jordan forget the 1970 civil conflict in which it took behind-the-scenes U.S. intervention and the threat of Israeli aid for Hussein to assure Hussein of his throne.

Jordanian-Palestinian Links

Long-standing Jordanian policy linking the Jordanian and Palestinian causes reached a nadir in October 1974, when 20 Arab heads of state, meeting in summit at Rabat, Morocco, relieved Hussein of all authority to negotiate for the return of the Israeli-held West Bank. Faced with such Arab unanimity, Hussein could do little but capitulate to the decision to elevate the PLO to quasi-government status. Two weeks after the Rabat decision, PLO chief Yassir Arafat went to New York in triumph to address the United Nations General Assembly. Hussein's 1972 proposal for a United Arab Kingdom which would allow for two separate regional governments with Hussein retaining overall federal authority appeared to have been decisively rejected.

But the 1975-76 Lebanese civil war saw a partial eclipse of the Palestinians who have since come under increasing pressures for more conciliatory policies from Egypt, Saudi Arabia and Syria. In May 1975, President Sadat became the first Egyptian leader to visit Jordan. A month later, Hussein was host to Syrian President Assad. Since then Syrian-Jordanian relations have blossomed and there has even been talk of a regional federation involving both countries. During 1976, President Sadat, in a clear step back from the Rabat decisions, declared that the contemplated Palestinian state would have to be linked in some way to Jordan. President Carter, in his discussions of a "Palestinian homeland," has also foreseen such a linkage. The partial reconciliation between Jordan and the PLO, which has included a symbolic meeting in Egypt between Hussein and Arafat, came in recognition of this shifting situation.

Jordan's King Hussein

During April 1977, King Hussein participated in the pilgrimage of Arab leaders to visit President Carter. This followed on the heels of the headlines which charged that the CIA had personally paid Hussein millions to allow American intelligence-gathering in his kingdom. Though President Carter immediately indicated that the payments would cease, the U.S.-Jordanian friendship seems firm and the U.S. appears committed to the continuation of the Jordanian state in the framework of an overall Middle East settlement which would include some provisions for recognition of Palestinian self-determination.

Jordan, like Israel, is one of the newest Middle East nations. Before it came under British administration following World War I, the region that now comprises Jordan had been part of various empires and countries, but never a separate political entity. Formal independence was achieved only in 1946. Under King Abdullah ibn-Hussein and his descendants, Jordan has struggled to strengthen its sovereignty, making considerable economic strides during the 1960s and staving off repeated civil unrest in the 1970s. The United States has periodically provided the country with economic and military assistance and is apparently prepared to continue to do so.

Terrain

The arid land of Jordan lies between Syria to the north, Iraq to the east, Saudi Arabia to the south and east and Israel to the west. Nearly landlocked, it has a 16-mile coastline on the Gulf of Aqaba, where its only major port, Aqaba, is located. It also shares the Dead Sea with Israel.

About the size of Indiana, Jordan covers 37,737 square miles of territory, 2,181 square miles of which were occupied by Israel during the 1967 war. Only 11 per cent of the land is arable, and another 1 per cent is forested. The remaining 88 per cent is primarily desert.

A great gorge divides the country, forming the Jordan River Valley in the north, the basin of the Dead Sea and Lake Tiberias in the center and the dry bed of the Wadie el Araba in the south. On the East Bank of the Jordan River, plateaus rise toward the vast Syrian desert, which covers most of the area. On the small West Bank, the land is hilly and receives sufficient rainfall for cultivation.

People

In 1977, Jordan's population was estimated to be 2,898,000, including the people living on the Israeli-controlled West Bank of the Jordan River and in East Jerusalem. Nearly all Jordanians, 98 per cent, are of Arab stock. Circassians and Armenians account for the other 2 per cent, although these ethnic groups long ago adapted to the preponderant Arab culture of the country.

Following the greatest concentration of rainfall, most of the people reside along the Jordan River, where sufficient rains allow for normal cultivation. Approximately one-third of the population lives in the Israeli-occupied West Bank, which was a portion of Palestine until Jordan seized the area during the Arab-Israeli war of 1948. More than 500,000 registered refugees live on the East Bank. The population is nearly evenly divided between rural and urban living, with 50 per cent of the people in the countryside, 44 per cent in the cities and 6 per cent nomadic or semi-nomadic.

Islam is the overwhelming majority religion in Jordan. Sunni Moslems comprise 95 per cent of the population, and Christians constitute the remaining 5 per cent.

Economy

Jordan's rapid economic expansion during the 1960s has been cited as an outstanding example of a well-planned and well-administered assistance program. With few natural resources and limited arable lands, Jordan as late as 1958 was a country with no industry, with an underdeveloped tourist trade, with a large unemployed or under-employed refugee population and with a need for adequately paved roads.

Using assistance from the United States, the United Nations, the International Bank for Reconstruction and Development, Great Britain and West Germany, Jordan expanded all major sectors of its economy, more than quadrupling its gross national product (GNP) between 1954 and 1967. Irrigation projects increased the amount of arable land. Light industries were established. Shipping and port facilities were set up at Aqaba. Phosphate deposits in the Dead Sea were exploited. And tourism was encouraged with the restoration of historic sites and the construction of better roads and more hotels.

The Arab-Israeli war of 1967 and the ensuing years of civil unrest were major setbacks to the economic strides of the 1960s. During the 1967 war, Jordan lost the fertile West Bank and hence agricultural income. As a result of the war, it also was burdened with additional Palestinian refugees. However, by maintaining employment levels and consumer demand and by stepping up government spending, Jordan was able to limit the war's impact. Economic revival and military expenditures fed a newly increased demand for imports. Agriculture improved also, except in the Jordan River Valley, where sporadic clashes between Jordanian and Israeli forces kept production down.

Fighting between the Jordanian army and Palestinian guerrillas in 1970 and 1971 also hampered economic growth. The government's steadfast policy against the guerrillas led to Libya's and Kuwait's withdrawal of respective $26-million and $39-million annual subsidy payments to Jordan. The United States moved to offset the loss.

Agriculture is the life source of the Jordanian economy, rivaled only by phosphate production. Although the country is not totally self-sufficient in foodstuffs, it produces considerable amounts of wheat, fruits, vegetables, olive oil and dates. Joran's exports, mainly fruits, vegetables and phosphate rock, fall far below its imports. In 1972, exports totaled $48-million, while imports soared to $274-million. Imports include petroleum products, textiles, capital goods, motor vehicles and foodstuffs. Trading partners are primarily the neighboring Middle East countries of Lebanon, Syria and Kuwait. Other major partners include Great Britain, West Germany, Japan and Yugoslavia. The Jordanian government has continually offset its unfavorable balance of trade and its deficit budget spending with subsidies from Arab neighbors and with assistance from the United States and the United Nations. Between 1949 and 1973, Jordan received $753-million in economic assistance and $235-million in military assistance from the United States, most of which was given in grants.

Early History

In ancient times, Jordan was part of various empires and civilizations, but not until after World War I did it exist as a separate nation. The Old Testament recounts its settling by Gilead, Ammon, Moab and Edom and by the Hebrews under Joshua in the 13th century B.C. Arabian Nabataeans, Greeks, Romans, Moslem Arabs and Crusaders successively held the area until the Ottoman Empire extended its domination over much of the Arabian Peninsula, including what is Jordan, in the 1500s.

British forces and their Arab allies ousted the Turks from Palestine and Transjordan in 1918, and the area came under the brief rule of King Faisal I as part of the Kingdom of Syria. Faisal's ejection by the French in 1920 placed Transjordan under loose administration by Britain. The regions presently known as Israel and Jordan had been awarded to Britain as the mandate for Palestine and Transjordan under the League of Nations mandate system. The French got similar mandates for Syria and Lebanon.

When Faisal's brother, Abdullah ibn-Hussein, moved into the area, threatening to vindicate Arab rights and attack the French in Syria, Britain created a partially autonomous kingdom for him, the Emirate of Transjordan. In May 1923, Britain recognized the independence of Transjordan under British tutelage within the mandate system of the League of Nations.

Under King Abdullah, Transjordan moved forward economically and politically and created the Arab Legion under British command. By arrangement, Transjordan accepted financial assistance from Britain and agreed to accept advice on financial and foreign affairs and to allow Britain to station forces in the country.

The British mandate over Transjordan ended May 22, 1946. On May 25, 1946, the kingdom became the independent Hashemite Kingdom of Transjordan, taking its name in part from the Hashemite family whose original patriarch was a disciple of Mohammed and the "guardian of Mecca."

When the British mandate over Palestine ended May 14, 1948, and the state of Israel was proclaimed, Trans-

Statistics on Jordan

Area: 37,737 square miles, including 2,181 square miles occupied by Israel.
Capital: Amman.
Population: 2,898,000 including approximately 760,000 under Israeli occupation on the West Bank and in East Jerusalem (1977 est.).
Religion: 95 per cent Sunni Moslem; 5 per cent Christian.
Official Language: Arabic.
GNP: $1.2-billion, $440 per capita (1975 est.).

jordan forces joined the Arab attack on the new nation. Its forces took control of areas of central Palestine, including the West Bank and the Old City Sector of Jerusalem. In addition, nearly 500,000 Palestinian Arabs fled to Transjordan and Transjordan-occupied territories.

The armistice agreement of April 3, 1949, between Transjordan and Israel left the Arab nation in control of about 2,165 square miles of the original territory of the Jewish state. The country was renamed the Hashemite Kingdom of Jordan in 1950 to include the occupied lands annexed by King Abdullah.

On July 20, 1951, King Abdullah was assassinated by a Palestinian extremist in Jerusalem. His eldest son, Talal, was proclaimed his successor on Sept. 5, 1951, but mental disorders led to his removal by Parliament in August 1952. Talal's eldest son, Hussein, assumed power under a regency until his 18th birthday, May 2, 1953, when he was crowned king.

During 1955 and 1956, anti-Western sentiment in Jordan led to the country's severance of several ties to the West. The British commander of the Arab Legion was dismissed, and a mutual defense treaty with Britain was ended by mutual consent.

Scattered rioting erupted against Jordan's joining the Baghdad Pact, and Jordan later decided to remain outside the defense alliance. British troops left the country by 1957.

Domestic turmoil struck the country in the spring of 1957. King Hussein survived a military plot to overthrow his government, but only after public rioting, an extended cabinet crisis and rumors of foreign intervention had disrupted the country. Political parties were dissolved at that time by royal decree.

Federation with Iraq

On Feb. 14, 1958, the Arab Union of Iraq and Jordan was proclaimed as a countermove to Egypt's merger with Syria. The Iraqi revolution of July 1958 spelled the end of the federation, however, and the Arab Union was formally dissolved Aug. 2, 1958. The Iraqi coup led King Hussein to accuse Communists and the United Arab Republic of conspiring to overthrow his regime and to undermine the independence of Jordan. Hussein requested British and American assistance to meet the threat. Accordingly, British troops were stationed in Jordan from July 17 to Nov. 2, 1958, and American economic assistance was greatly increased.

Toward the end of 1966, Jordan faced another crisis, due to Palestinian guerrillas, notably from the Syria-based al-Fatah or al-Asefa organizations, who streamed across the Syrian-Jordanian border en route to raiding Israel. Israeli reprisal raids on Jordan put pressure on the country to cut off the flow of commandos. Clashes on the border with Syria resulted from Jordan's attempt to stem the infiltration. Relations with Syria were eventually broken off over the issue in May 1967.

Notwithstanding the guerrilla problem, Jordan maintained its support of the Arab cause against Israel. It affirmed its solidarity with Syria when Israeli forces massed on the Syrian border later in 1967. Jordan also committed itself to a defense pact with Egypt.

1967 War

Jordan's defense treaty with Egypt committed the country to join the fight against Israel when the 1967 war erupted June 5. Within two days, Jordan lost all of the West Bank to Israeli forces, including the whole city of Jerusalem. Among the Arab states, Jordan suffered the heaviest losses of the war: more than 6,000 dead or missing and an influx of 200,000 additional Palestinian refugees. Although Saudi Arabia, Kuwait and Libya stepped in with economic assistance to offset the losses, Jordan's economy suffered a major setback from the war.

Following the war, guerrilla commandos of the Palestinian resistance movement, the *fedayeen,* expanded their activities within Jordan, Syria and Lebanon. In Jordan, particularly strong activity was aimed at the overthrow of King Hussein in order that the war against Israel could be continued. By the beginning of 1970, tension between the government and the guerrillas had led to sporadic fighting.

An agreement, whereby Jordan agreed to fully support the continued war effort against Israel, and the guerrilla groups agreed to restrain their members and honor Jordan's internal security, soon proved unsuccessful. Two weeks of fighting in June 1970 came to an end only through the efforts of an Arab mediation committee. Occasional clashes continued.

In September 1970, tensions erupted again when the *fedayeen* members held captive in Switzerland, West Gerdan, Egypt and Israel. As part of the effort, one *fedayeen* group, the Popular Front for the Liberation of Palestine, hijacked three commercial planes belonging to the United States, Britain and Switzerland and flew the planes and 400 hostage passengers to a desert airstrip outside Amman. Several days later, the *fedayeen* blew up the planes and released the hostages in exchange for the release of fellow *fedayeen* members held captive in Switzerland, West Germany and Britain.

Heavy fighting broke out in mid-September 1970 between the Jordanian army and the *fedayeen* in Amman, the new headquarters of the guerrilla movement. A Syrian tank force crossed into Jordan Sept. 18 to aid the *fedayeen,* but fears of an enlarged conflict subsided several days later when the Syrian forces, repelled, withdrew.

Foreign ministers of surrounding Arab states met in Cairo Sept. 22 to try to resolve the crisis. King Hussein and guerrilla leader Yasir Arafat signed an agreement in October calling for substantial concessions by the *fedayeen,* but tensions continued into 1971. By April 1971, King Hussein had strengthened his position and ordered the *fedayeen* out of Amman. Despite commando protests, other Arab governments refused to intervene, and in July 1971, the Jordanian army crushed the last guerrilla strongholds.

The Palestinian guerrillas threatened reprisals for Jordan's action. Members of the Black September organization claimed responsibility for the assassination of Jordanian Prime Minister Wasfi al-Tal in November 1971 and for an unsuccessful attempt on the life of Jordan's ambassador to London.

Later, Hussein appeared to have mitigated his strong feelings against the Palestinian guerrillas. In a royal decree issued Nov. 13, 1974, he freed 100 political prisoners held in Jordanian prisons—an apparent gesture to Arafat, who addressed the United Nations General Assembly the same day. Most of the prisoners were members of Arafat's Al Fatah.

On May 6, 1975, during a U.S. tour, Hussein said at the Citadel in Charleston, S.C., that Israel, as a condition for a permanent Middle East peace, must accept "the legitimate rights of the Palestinians in their homeland." The king repeated the assertion during other U.S. stops.

During the past few years Hussein has continued to play an ambiguous role on the Palestinian issue. While constantly supporting the rights of the Palestinians, he has expressed the hope that whatever Palestinian entity emerges will be linked to Jordan and not become an irredentist force in the area. As Hussein told journalist Stanley Karnow in an interview in early 1975 not long after the Rabat decision, "My feeling is that the ties between Palestinians and Jordanians are very, very strong, and they will be involved with each other if there is peace."[1]

Proposed Federation

King Hussein March 15, 1972, set forth plans for a "United Arab Kingdom." Under the federation, the Gaza Strip, the Israeli-occupied West Bank and the East Bank would have been united under Hussein. Each bank would have had a parliament under a federal parliament. While Israel's reaction was mixed, the Arab world thoroughly denounced the proposal at that time. The federation never materialized, but the continuing discussion of a Palestinian entity linked in some way to Jordan could possibly result in a rethinking of some aspects of this plan.[2]

1973 War

At the outbreak of the Yom Kippur war on Oct. 6, 1973, Jordanian troops were put on the alert. But when the forces did not engage in the fighting, King Hussein came under dual pressure to modify his position. While the United States asked for Jordan's assistance in quickly ending the war, Arab leaders urged Hussein to send troops to fight Israel. Jordan finally entered the war Oct. 13 after an appeal from Saudi King Faisal, who had heavily contributed to Jordan's budget.

On Oct. 17, 1973, King Hussein called for peace in the Middle East, although he affirmed his intention of not yielding an inch of territory to Israel and of continuing to work for the recovery of occupied Arab lands. Jordan Oct. 22, 1973, accepted the U.N. truce and began to disengage its forces.

Government

Jordan is a constitutional monarchy under King Hussein I. The constitution was promulgated in January 1952. Executive authority rests with the king and the Council of Ministers. The king signs and executes laws and retains veto power unless overridden by a two-thirds vote by both houses of the National Assembly. His other powers include appointment and dismissal of all judges, approval or rejection of all constitutional amendments, declaring war and commanding the armed forces. Cabinet decisions, court judgments and the national currency are issued in his name. The king is also immune from all liability for his actions.

Foreign Policy

Jordan follows a consistent pro-West foreign policy, although it has never recognized Israel and supports the Arab cause against the Jewish state. Since the 1967 war when Israel occupied the West Bank, a primary objective of Jordan's foreign policy has been the recovery of that land.

Since the Rabat decision in 1974 elevating the PLO to spokesman for the occupied West Bank, King Hussein has followed a dual policy. He has retained ties to the West Bank leadership while endorsing the need for restoration of Palestinian rights. Whatever comes of the West Bank, Hussein is striving to protect his country from any regional forces which could threaten his rule. There are many who argue that an independent West Bank, PLO-run Palestinian state would represent more of a threat to Jordan than to Israel. After the Rabat decision, Hussein reorganized the Jordanian government to exclude most Palestinian representatives. When the National Assembly temporarily reconvened in February 1976 with West Bank representatives, the PLO charged Hussein was attempting to undermine it.

Jordan receives considerable American military and economic aid. A controversial Hawk missile sale was finally approved in Sept. 1976 after Hussein, at one point, visited Moscow to discuss alternatives. During his Washington visit in April 1977 the king spoke of "generous assistance programs" from the U.S. which had helped transform his desert kingdom "without oil, without peace, and, at times, without the support of *The Washington Post.*" This last reference was to the *Post's* front-page handling of the story about CIA funds to Hussein.

More than other Arab leaders, Hussein has expressed a cautious outlook about hopes for settling the Arab-Israeli conflict. "Those who lead Arab opinion to believe that peace can be restored in 1977 are playing with fire," he told the French newspaper *Le Monde* just before the right-wing victory in Israel's election on May 17. Before leaving the United States, Hussein further warned that Israel was continuing a "dead-end policy" but that "the Arab parties, while seeking the end of occupation and redress for the Palestinians, are ready for the obligations of peace."

King Hussein's two major foreign policy goals determine most of his other policies. First, he has emphasized the need to maintain a strong linkage with the United States. Second, he must ensure himself against those forces who wish to topple his kingdom. This means keeping a working alliance with Syria, Egypt and Saudi Arabia while working for a relationship with the Palestinians that will produce stability rather than turmoil in the area. In an interview in *Newsweek* in March 1977, Hussein looked to the uncertain future: "My duty is to develop my country, its human resources, armed forces and security and intelligence services, to face the tremendous threats that lie ahead," he said. "The road is strewn with mines to sabotage the peace process.... Reality shows that Geneva is not a panacea. And over-optimism will be our greatest problem."[3]

---

1 Stanley Karnow, "An Interview with King Hussein," *The New Republic*, 22 February 1975, p. 20.

2 See Chapter 1 in John K. Cooley, *Green March Black September* (Frank Case: London, 1973) for a detailed discussion of this plan and of Hussein's motives.

3 "Hussein on his CIA Money," *Newsweek*, 7 March 1977, p. 18.

REBUILDING A WAR-RAVAGED COUNTRY

Lebanon, where orderly government a few years ago rested on a delicate balance between its Christian and Moslem populations, was plunged into a ruinous civil war in April 1975. By the time the war was brought under control in November 1976, it had resulted in approximately 50,000 deaths (including the American ambassador, Francis E. Meloy in July 1976), the economic destruction of a country frequently described as the "Switzerland of the Middle East" and the domination of the country by neighboring Syria.

By mid-1977, the political divisions which had led to the brutal conflict had not yet been resolved. "The only major change," *The Washington Post* reported in April 1977, "has been in the attitudes of the most influential Arab nations, which have made peace with each other so that Lebanese factions are no longer used as proxies for feuding Arab powers."[1]

What began as the coming apart of a fragile and, many thought, outdated form of political accommodation between Christians and Moslems, became part of the larger inter-Arab struggle when "Syrian military intervention and the participation of other Arab states transformed Lebanon's factional strife into a kind of Arab civil war that had far-reaching implications for regional peace."[2] Until late 1976 the Lebanese conflict brought already strained Egyptian-Syrian relations (due to Egypt's acceptance of a separate agreement with Israel in September 1975) to a new low and resulted in a respite for Israel from the mounting diplomatic pressures to make major territorial and political concessions to the Arabs.

Complex Lebanese Politics

Though comprised of three major groups—Christians, Moslems, and Palestinians—the Lebanese political landscape is complex, with each group itself subdivided into numerous factions which at various times during the civil war fought with each other. A Beirut newspaper editor has commented, "...here we are with three armies, two police forces, 22 militias, 42 parties, nine Palestinian organizations, four radio stations, and two television stations."[3]

For decades the Phalange Party led by Pierre Gemayel has been the strongest, most cohesive political group. These Christian rightist forces have sought the preservation of their privileges stemming from the 1926 Constitution and the unwritten 1943 National Covenant. They also have sought to rid the country of the Palestinian guerrillas. The Maronite Catholics, in particular, have been resentful of the Palestinians whom they had blamed for Israel's massive retaliation strikes against Lebanon, including a number of commando assaults on Beirut, in the early 1970s. The

Lebanese President Elias Sarkis

hatred was expressed in a sign posted outside the Palestinian Tal Zaatar refugee camp near Beirut which was to be destroyed in an historic 52-day siege during the summer of 1976: "It is the duty of every [Christian] Lebanese to kill a Palestinian."[4] Gemayel and other Christian leaders have called for the redistribution of the approximately 400,000 Palestinians in Lebanon among the less populated Arab countries. Former President Suleiman Franjieh, in his September 19, 1976, farewell address, placed total blame for the war and the country's destruction on the Palestinians.

Another important Christian leader is former President Camille Chamoun, whom President Franjieh appointed as foreign minister in June 1976, in a move characterized by Moslem Premier Karami as "unconstitutional and groundless" and designed to thwart a smooth transfer of power to the new President, Elias Sarkis.[5]

The leftist alliance of Moslems and Palestinians is even less cohesive than the rightist coalition. As late as summer 1977, competing Palestinian forces were still battling each other. At least 24 Palestinian guerrillas were killed on July 17, when Syrian-controlled Al Saiqa forces clashed with Iraqi-backed PFLP (Popular Front for the Liberation of Palestine) forces who oppose any kind of compromise settlement with Israel and are consequently termed the "rejection front."

Until his assassination on March 16, 1977, Druze sect leader Kamal Jumblatt, a wealthy feudal boss over much of Lebanon's mountain country, led the nation's socialist-

1. *The Washington Post*, 13 April 1977, p. A8.
2. Norman F. Howard, "Tragedy in Lebanon," *Current History*, June 1977, p. 1. For background information on the Lebanese situation through mid-1976, see "Middle East Peace Prospects," Hearings before the Subcommittee on Near Eastern and South Asian Affairs of the Senate Foreign Relations Committee, May, June, July 1976, pp. 121-183.
3. Quoted in Norman F. Howard, *Ibid.*, p. 1.
4 *Ibid.*, p. 2.
5 *Ibid.*, p. 2.

leftist-Palestinian coalition. Jumblatt's forces sought various political changes which would have resulted in a more equitable distribution of power between Christians and Moslems and a foreign policy more in line with Arab national goals, the main concern of the Palestinians. The Palestinians, comprising about four-fifths of the military force available to the leftists, were somewhat reluctantly drawn into the war late in 1975 in an effort to protect their privileges in a country in which they did not really belong.

One analyst gave the following description of the Palestinian entrance into the war; an action that precipitated, within a few months, the Syrian intervention:

"Despite PLO efforts to stay out of the conflict, the PLO leadership under Yasir Arafat reasoned, not illogically, that if the Christians defeated the indigenous left, the Palestinians would be their next target. Not surprisingly, therefore, by late 1975 the radical Popular Front for the Liberation of Palestine (PLFP) had become deeply implicated in the war on the leftist side. Arafat's Fatah joined the struggle openly in 1976, with disastrous results both for the Palestinians and for Lebanon. In March, 1976, for instance, Arafat reported that 16,000 Palestinians had been killed and 40,000 wounded. Rivalries among various Palestinian groups also intensified.... Many Palestinians apparently lost sight of their central goal: Salah Khalaf, Fatah's second-in-command, for instance, declared that the road to Palestine (Israel) ran through Jounieh, the Christian center north of Beirut. By summer, 1976, the PLO had suffered severe blows to its political prestige, and its forces were on the verge of total defeat at the hands of Syria."[6]

Syrian Intervention

During the early part of 1976, Syria became more and more politically involved in the Lebanese conflict. On January 22, the battered Christian forces accepted a cease-fire along with a Syrian-sponsored peace plan and reform package. President Franjieh at the time stressed that the reforms—which provided some additional power for the Moslems—could only be accepted after the Palestinians had in effect disarmed and started to adhere to the 1969 Cairo agreement which was designed to regulate their activities in Lebanon.

But within six weeks the peace plan failed. Franjieh came under heavy pressure from factions of the disintegrating army and the leftists to resign, but he refused. It was at this point that Syria became involved on the Christian side by adamantly refusing to support Jumblatt's mounting demands that Franjieh step down. A Saiqa communique revealed Syria's feelings that "Jumblatt wants to rekindle the fire of sedition and sabotage of Syrian [peace] initiative."[7]

On April 10, the parliament voted 90-0 to amend the Constitution to permit the early election of Franjieh's successor. After some delaying tactics by Franjieh; the election took place on May 8, some two months before the scheduled balloting.

The Syrians pressed hard for their candidate, Elias Sarkis, governor of Lebanon's central bank. Jumblatt supported Raymond Edde, who was outspoken in his opposition to Syrian involvement in Lebanon. A general strike was called by the leftists in the hopes of preventing Sarkis' election. The election eventually took place, with mortar rounds falling near the temporary parliament building.

Franjieh, however, refused to leave office and Jumblatt urged "popular resistance" to Sarkis' election which he termed "a flagrant distortion of the wishes of the Lebanese people."[8]

As the year went on, little was resolved politically and the war intensified. A de facto partition of the country gradually came into effect as the Christian militias controlled a small enclave extending from eastern Beirut north to Tripoli. Sarkis took over the presidency on September 23, and by early October Syrian and Lebanese Christian forces had succeeded in overcoming leftist resistance in most areas of Lebanon.

Syria had gradually shifted her support to the Christians during 1976 when the left showed signs of victory. In April Syrian army regulars entered Lebanon, and on June 1 the Syrians intervened in force with a build-up to 30,000 men and 500 tanks. In June, a senior Syrian official called Arafat "a fool" for opposing the Syrian intervention, and Syrian President Assad placed the entire blame for the war on the PLO in July.[9] In September the Syrians opened a general offensive to crush the Palestinians and end the fighting. It was at this point that the Saudis and Egyptians decided a major Arab peace effort had to be made.

Saudi Influence

Arab leaders went to Riyadh, Saudi Arabia, on October 17 and 18 for a meeting, sponsored by the Saudi king, which was to mark the beginning of the end of the civil war in Lebanon. One major impetus behind their action was the desire of the Arab leaders to settle their own problems and prepare for resumption of negotiations with Israel following the U.S. presidential elections in November.

At the Riyadh meeting, Assad and Egyptian President Sadat agreed on the establishment of a definitive cease-fire commencing October 21 and to be maintained by a 30,000-man Arab "deterrent" force under control of President Sarkis but which in fact would consist mainly of the Syrian forces then in Lebanon. Libya and Iraq wanted only limited Syrian participation in the force, but the plan was approved during an Arab League summit meeting in Cairo late in October. By November the war had largely ceased, though periodic fighting continued into 1977. The assassination of Jumblatt in March 1977 was a particular shock to the left and Palestinians.

Some estimates of the death toll during the 19 months of heavy fighting go as high as 60,000. Proportionally, this would be 75 times as many people killed during the civil war as were Americans killed during the eight years of fighting in Vietnam. The hatreds that existed before the conflict have thus been exacerbated, since nearly every family has experienced personal tragedy. For the left, the situation is especially bitter, for the country is now dominated by Syria, and the Maronite majority remains in control of the social, economic and political systems.

But the Christians, too, face a very anxious future. They remember their near defeat in early 1976 before the Syrian intervention. To preserve their way of life, they are prepared, if necessary, to partition the country—a step Syria, the leftists in Lebanon and the entire Arab world opposes. Nevertheless, the Christians in 1977 were busy constructing an international airport (the Pierre Gemayel International Airport) in territories they controlled, and they had already set up the skeleton of a telephone and telex

---

6 Ibid., p. 3.
7 Ibid., p. 4.

8 Ibid.
9 Ibid., p. 30.

system for use should the Moslems ever cut off the regular facilities.

"Regionalization," a plan for de facto partition of Lebanon, is being pushed by the more extreme Camille Chamoun followers, and less so by the Phalangists. Chamoun's son, Dory, second in command of the Liberal Party, explained in April that "You can't force these people [the Lebanese] to live together now; there is too much tension between them. The tension must be reduced, and the only way it can be done is by creating regions that govern themselves under a federal system. It's the only way to preserve national unity," he added.[10]

Southern Lebanon

Nov 76

While attempts were being made in 1977 to rebuild Lebanon politically and economically, fighting continued in the southern part of the country near the Israeli border. The Israelis had let it be known they would not tolerate the presence of the Syrian army near their border with Lebanon, nor would they accept the return of Palestinian guerrilla forces to what has been called "Fatah Land."

To prevent the return of the Palestinian forces, Israel entered an alliance with Christian forces in the area, mainly Shiite Moslems who had attempted to avoid involvement in the war. The alliance became symbolized by what the Israelis called "The Good Fence." Israel not only supplied Christian forces with medical aid, food, and in some cases even jobs in Israel, but also with arms to be used against the Palestinians in the later stages of the war. In March the Palestinians went on the offensive in southern Lebanon; apparently getting a green light from the Syrians after Chamoun's forces rejected Syrian demands that the Christians break off their cooperation with Israel.

By mid-1977 there was talk of rebuilding the Lebanese army to assert authority in the south. In June a 3,000-man nucleus of a new army had been organized. On July 18 Damascus radio confirmed reports that the PLO and Syria had reached an agreement to bring Lebanese army forces into the area to halt the fighting between Christians and Palestinians. One reason the Palestinians seemed eager for some kind of end to the southern Lebanon fighting was fear of a possible Israeli attack against the Palestinian guerrillas, especially after the election of the Likud Party in Israel.

Rebuilding the Country

Direct economic losses from the war—that is, physical damage alone—were estimated at nearly $3-billion, with indirect losses mounting to as much as $15-billion. Though Saudi Arabia and other oil-rich Arab countries had promised considerable aid, little had been received by mid-1977, and it was thought there would have to be a real political accommodation between the country's torn factions before the aid would begin flowing. Whether Beirut would ever again become a financial center for international businessmen and a playground for the jet set of the Arab world remained a major question, even were the country to be structurally rebuilt.

While efforts to stabilize the situation continued, the Syrians imposed a strict press censorship law ending the tradition of press freedom for which Lebanon had been known in the West. "One era has ended and another begun for the Lebanese press," the leading daily, *An Nahar,*

────────────
10 *The New York Times,* 10 April 1977.

Statistics on Lebanon

Area: 4,000 square miles.
Capital: Beirut.
Populatiion: 2,454,000 (Jan. 1977 est.)
Religion: 55 per cent Christian, 44 per cent Moslem and Druze (based on 1932 official census); by the mid-1970s, Moslems believed to constitute a slight majority.
Official Language: Arabic; French is widely spoken.
GNP: $3-billion or $730 per capita (1973 est.).

---

stated. In a note to the government from the newspaper owners' association, it was charged that "the law in effect imposes permanent censorship."[11]

In mid-1977, it appeared that the Syrian presence in Lebanon might last a long time and that Syrian influence, even after Syrian troops are withdrawn, will be much greater than before the war. *The Washington Post* reported on April 13, "There is just one power in Lebanon today—the might of Syria and its army.... Generally, what Syria wants in Lebanon, Syria gets."

Terrain

Lebanon is situated on the eastern shore of the Mediterranean Sea and is bounded on the north and east by Syria and on the south by Israel. With an area of about 4,000 square miles, it is smaller than the state of Connecticut. It has a maximum length of 135 miles and is 20 to 35 miles wide.

Lebanon's terrain is predominantly mountainous. Behind a narrow coastal plain rise the high Lebanese Mountains, which are separated by the fertile Beqaa Valley from the Anti-Lebanon Mountains along the Syrian-Lebanese border. Lebanon's only important river, the Litani, flows into the sea north of Tyre. About 64 per cent of the land is desert, waste or urban; 27 per cent is agricultural and 9 per cent forested.

People

In January 1977 Lebanon had an estimated population of 2.45 million[12]—down about one million as a result of the war. The population is 93 per cent Arab and 6 per cent Armenian. Before the war, about one-third of the total population lived in the capital city of Beirut. Population figures do not include registered Palestinian refugees, who officially numbered 182,000 in December 1971. It is estimated that there are nearly half a million Palestinians now in Lebanon.

Religious divisions are of great political importance. By official estimates, the population is 55 per cent Christian and 44 per cent Moslem and Druze. But this is based on a census conducted during the French mandate in 1932, which permitted the French to give a controlling position to Lebanese Christians, who traditionally supply the president of the nation while Sunni Moslems provide the prime minister.

No official census has been conducted since then. Because of the higher Moslem birth rate, Moslems are now

────────────
11 *The New York Times,* 5 July 1977.
12 Central Intelligence Agency estimate in *National Basic Intelligence Handbook,* January 1977.

believed to constitute a slight majority of the population. This led to Moslem demands for a stronger voice in government, a demand which helped spark the civil war.

Christian sects include the Maronites, who are affiliated with Rome and make up the largest Christian group; Greek Orthodox, and Armenians. Moslems include members of both Sunni and Shia sects. The Druze, an offshoot of Shia Islam, constitute a significant minority. Arabic is the official language, although French is widely spoken.

Economy

In 1973, Lebanon had an estimated gross national product (GNP) of more than $3-billion, over $700 per capita. Two-thirds of the GNP was drawn from services, primarily banking, commerce and tourism. (What follows about the Lebanese economy deals with the pre-civil war situation.)

Beirut is a free port, and transit trade is a major element in Lebanon's free-enterprise economy. There are 75 banks, and the country counts 120 legitimate labor unions, organizations generally banned in other Arab countries. Beirut also is an important gold and foreign exchange market. Although no oil has been found in Lebanon, Tripoli and Saida are terminals of oil pipelines from Iraq and Saudi Arabia. Beirut, the country's chief seaport, also has a modern jet airport. International communications facilities include a satellite ground station and a cable linking Lebanon and France.

Lebanon has the highest literacy rate in the Arab world, 86 per cent, and a high percentage of skilled labor, but about 50 per cent of the labor force still derives its living from agriculture. Principal products are cereals, vegetables, fruit and livestock. Lebanon is not self-sufficient in food and depends on imports for a substantial part of its food supply.

Except for a few large oil refineries and cement plants, industry remains small in scale. Major industrial activities include food processing, textile manufacturing and production of building materials. Beirut is a major Arab publishing center.

Lebanon's imports far exceed exports, but its trade deficit is covered by large net receipts from such sources as transportation and tourism and private capital inflow. Exports—primarily fruit, other foodstuffs and textiles—totaled $360-million in 1971; exports went to Iraq, Syria, Saudi Arabia and Jordan. Imports totaling $700-million in 1971—including foodstuffs, machinery and metals—came primarily from Western Europe and the United States.

History

In ancient geography, much of present-day Lebanon was part of Phoenicia. In later centuries Lebanon served as a Christian refuge, and during the Middle Ages it became part of the crusaders' states. It was absorbed into the Ottoman Empire in the 16th century. After World War I, Lebanon and Syria were divided into separate entities administered under French mandate. Lebanon gained its independence in 1943. French troops were withdrawn in 1946.

Since that time, Lebanon's history has been marked by recurrent political unrest and friction with Syria. President Bichara el-Khoury (1943-52) was deposed by a popular movement in 1952. Moslem opposition to the pro-Western policies of President Camille Chamoun (1952-58) led to an open revolt in May 1958, which was aggravated by Lebanon's Arab neighbors. At the request of the Lebanese government, the United States sent troops into Lebanon in July to help restore order. The revolt dwindled, and the U.S. forces were completely withdrawn by Oct. 25, 1958, after the inauguration of Fuad Chehab as president and the formation of a coalition government.

Both Chehab (1958-64) and his successor, Charles Helou (1964-70), pursued neutralist policies that were generally acceptable to both Arabs and the West. An attempt to overthrow the government by the Syrian Popular Party, which sought union with Syria, was put down in December 1961. President Franjieh was elected Aug. 17, 1970.

Although Lebanon participated in the 1948 Arab war against Israel, it generally has tried to keep its distance from the Arab-Israeli conflict. But the Palestine problem has contributed to internal instability.

In December 1968, an Israeli airliner in Athens was attacked by Arab terrorists who were said to have come from a Palestinian refugee camp in Lebanon; Israeli commandos retaliated with a raid on the Beirut airport. Lebanon's effort to restrict commando activities led to armed clashes in the country in 1969. Continued commando raids in 1970-71 brought further Israeli reprisal raids.

An Israeli raid on Palestinian commando groups in Beirut and Saida April 10, 1973, resulted in the resignation of Lebanese Premier Saeb Salam. He was succeeded by Amin Hafez, a former Palestinian who was reported to have strong ties with guerrilla organizations. Hafez, who resigned briefly during fighting between the Lebanese army and Palestinian commando groups in May 1973, finally gave up his post June 14, 1973, under pressure from Sunni Moslems who charged that his government did not adequately represent their sect. Franjieh June 21, 1973, called on Rashid al-Solh to form a new government. He survived until May 1975, when major riots forced his resignation and Franjieh called on Rashid Karami to form a cabinet that would appease warring factions.

Government and Politics

Lebanon is a parliamentary republic whose governmental system is based on the Constitution of May 26, 1926, and the National Covenant of 1943, an unwritten agreement providing for the distribution of public offices among the country's religious groups.

The Chamber of Deputies, Lebanon's unicameral parliament, has 99 members who represent the country's nine religious communities in proportion to their national numerical strength. Members are elected by universal adult suffrage. Elections are held every four years or within three months after parliament is dissolved by the president. The last parliamentary elections were held in April 1972.

Executive power is vested in a president, who is elected by a simple majority of the Chamber of Deputies for a six-year term. He appoints members of the cabinet, who are subject to a parliamentary vote of confidence.

Political activity in Lebanon is organized along sectarian lines, and political parties generally are vehicles for powerful political leaders whose followers are often of the same religious sect. Political stability is dependent on the maintenance of balance among religious communities.

Lebanon had one of the largest Communist parties in the Middle East. The party, which was legalized in 1970, had an estimated 6,000 members and sympathizers before the civil war. Palestinian refugees and Arab guerrilla organizations continue to exert substantial political pressure. But, Lebanese politics since 1976 have been dominated by Syria.

Foreign Relations

Lebanon's dependence on trade demands good relations with other countries. Lebanese Christians are especially interested in friendly ties with Western countries, particularly the United States, and Lebanese foreign policy has been generally pro-Western. In its relations with the Communist bloc, Lebanon has pursued a more or less neutral line.

Although Lebanon is an Arab state and a member of the Arab League, it has sought to avoid major involvement in the Arab-Israeli conflict. Many Lebanese Christians would welcome peace with Israel—partly for the trade which peace would permit, but also as a contribution to Lebanon's internal stability. How the de facto alliance between Lebanese Christians and Israel will affect Lebanon's future was uncertain in mid-1977.

Under the National Covenant of 1943, Christians are prohibited from seeking foreign protection or attempting to bring Lebanon under foreign control or influence, while Moslems are prohibited from trying to bring Lebanon into any form of Arab union.

On Jan. 17, 1975, the U.S. State Department revealed that the United States was supplying TOW anti-tank missiles to Lebanon. The agreement, which was made in May 1974, specified 18 launchers and a slightly larger number of missiles. The United States said the deal was aimed at giving the small Lebanese army "psychological support in dealing with pressures from Syria."

But much has changed in a few short years as a result of the civil war. In most ways, Damascus, Syria, has become the capital of Lebanon. The United States was planning to provide some aid for rebuilding the country, but most of the economic as well as political support for reconstruction will have to come from the oil-rich Arab states and from Syria.

A CONSERVATIVE ISLAMIC BASTION, BUILT ON OIL

On the surface, Saudi Arabia is an Alaska-sized wasteland. But that vast, sun-scorched lid covers at least 165 billion barrels of oil, nearly one-fourth of the world's proven reserves. The guardian of the land and the mind-staggering treasure beneath it is the Saudi king, supported by a royal family that counts some 4,000 princes.

The fact that governing Saudi Arabia is a family affair was tested when an assassin's bullet killed 69-year-old King Faisal on March 25, 1975, the birthday of the prophet Mohammed and a holy day for the devout monarch, a champion of Islam and the Arab cause. Continuity was assured when the austere, shrewd Faisal was succeeded within minutes of his death by Crown Prince Khalid, his half-brother. The new monarch, in turn, elevated another half-brother, Fahd, to the crown princedom. The new order was rapidly agreed to by the five senior royal princes, then by the entire royal family.

Faisal was shot down by a nephew, 27-year-old Prince Faisal Ibn Musaed, while the king was holding court in his palace. The family announced quickly that the murderer was "deranged" and that he acted alone, thus assuring the country that there was no plot to incite a general insurrection. Later, a religious court found Prince Faisal to be sane, and in keeping with the precepts of the Koran, the Moslem holy book, he was beheaded on June 18, 1975, in a public square in Riyadh, the Saudi capital.

Not along ago it was written that Saudi Arabia was a remote feudal kingdom, a 20th century anachronism. Faisal ruled by the Koran alone, and he held a weekly court during which his subjects could approach him to present their problems. But by the time of his death, ending a 10-year reign, Faisal had turned his country from a poor, nomad-populated desert into a financial giant bursting into the modern world while clinging to the most conservative Islamic principles.

Since the oil-price rises, engineered by the Organization of Petroleum Exporting Countries (OPEC) beginning in 1973, Saudi Arabia has become a politically active and powerful country. The country's growing income—far outstripping its ability to spend on modernization—has given Saudi Arabia a financial power which it exercises through foreign aid (more than 10 per cent of the country's GNP, mainly channeled to other Arab states) and through investments (mainly in Western, industrial states).

Saudi income from oil in 1976 approached $30-billion. Even with the $142-billion five-year modernization plan approved in May 1975, the country is blessed with billions in "excess" income as spending projects are delayed by manpower and technical limitations. Journalist Edward R. F. Sheehan has termed the Saudi predicament "the epidemic of money."[1]

In 1977 the Saudis continued to look primarily to the West to invest what some experts expect might be as much as a $17-billion surplus. In 1976, for instance, the government appropriated $31-billion, of which $21-billion was for capital projects under the five-year plan. But only $22-

Saudi Arabian King Khalid

billion was spent, of which nearly $13-billion went for capital projects. The rest went into what the Saudi Arabian Monetary Agency terms "foreign assets," primarily short-term investments in the West. As of mid-1977, the Saudis had few direct investments and were just beginning long-term ventures. As is the case with their faith, the Saudis approach financial matters most conservatively and are careful in making any changes from past practices. Total Saudi monetary reserves at the end of 1976 were nearly $53-billion—second in magnitude only to West Germnay—and it was estimated that half was invested in the U.S. When this figure is compared to the 1972 GNP, only $4.2-billion, the staggering fortunes of the Saudi society stand out all the more.

With accumulating petrodollars replacing depleting oil (though each year more and more reserves are proven) Saudi Arabia has developed the financial clout to substantially affect regional and even world politics. Previously in the background of the Arab-Israeli conflict, Saudi Arabia had by 1977 assumed an activist role, influencing Arab governments and the Palestine Liberation Organization (PLO) towards moderation and encouraging the United States to put pressure on Israel to agree to the kind of settlements the Arabs were advocating. The Saudi government has strenuously denied any oil link or petrodollar link to politics—but there is increasing recognition that oil and the resultant wealth have brought stupendous influence.

Crown Prince Fahd—heir apparent and leader of the kingdom in political affairs—visited the United States in late May 1977. He was the last of the major Arab leaders who came to Washington early in the Carter administration to consult with the new President. But though he came last, the relationship between Washington and Riyadh is ex-

---

[1] *New York Times Magazine,* 14 November 1976, p. 31.

tremely important. For the Saudis control the oil the United States needs and the oil American allies in Western Europe and Japan depend on for survival. President Carter stated during Fahd's visit that "The future of Saudi Arabia and the future of the United States are tied together very closely in an irrevocable way." The following month, in a June 6 *U.S. News & World Report* interview he added, "My experience as President so far with the Saudi Arabians has been one of the most pleasant things that I've witnessed since I've been in this office. They have gone out of their way to be supportive and friendly to us, and they've been highly responsible in every action they've taken."

Terrain

Saudi Arabia extends over four-fifths of the Arabian Peninsula in a strategic location stretching from the Persian Gulf to the Red Sea and the Gulf of Aqaba. It shares borders with Jordan, Iraq, Kuwait, Bahrain, Qatar, United Arab Emirates, Oman, Yemen Arab Republic and the People's Democratic Republic of Yemen. Several of these borders are undefined. It faces non-Arab Iran across the Persian Gulf and Egypt, the Sudan and Ethiopia across the Red Sea. And it faces the Israeli-occupied Sinai Peninsula across the Gulf of Aqaba.

Saudi Arabia's 1,560-mile coastline provides the only access to water for the otherwise arid wasteland. Recently dams have trapped floodwaters for irrigation and city needs. Extinct volcanoes and lava beds scar much of the terrain. Only 1 per cent of the territory is arable, cultivated or pasture land. Less than 1 per cent is forested.

Saudi Arabia has five major regions. The holy cities of Islam, Mecca and Medina and the diplomatic capital of Jidda are located in Hijaz, adjoining the Red Sea. The mountainous Asir, where peaks rise above 9,000 feet, lies to the south along the Red Sea. Najd occupies the central part of the country where the capital, Riyadh, is located. The Eastern Province, known as al-Hasa, is the area containing the country's rich oil fields. The main feature of the rugged Northern Provinces is the trans-Arabian pipeline which crosses that area, Jordan, Syria and Lebanon en route to the Mediterranean Sea.

In the north, the Syrian desert stretches downward toward the 22,000 square miles of the Al-Nufud desert. In the south is the vast Empty Quarter, over 250,000 square miles of sand, a nearly uninhabited desert.

People

Although no official census figures are available Saudi Arabia's population has been estimated to be as high as 9.4 million.[2]

Previously, nearly all Saudis were nomadic or semi-nomadic, but rapid economic growth following the development of the country's oil resources has lowered that figure to an estimated 20 per cent.

Saudis are ethnically Arabs, although there are some small groups of non-Arab Moslems. These minorities are Turks, Iranians. Indonesians, Indians and Africans who came on pilgrimages and remained in the Hijaz Province. Saudis recognize no color line, and mixed racial strains are common.

Nearly all Saudis are Moslems practicing the puritanical Wahhabi interpretation of Sunni Islam. A small Moslem minority, belonging to the less rigid Shia sect (which dominates Iran), resides in the Eastern Province.

Economy

When it comes to oil, Saudi Arabia is the land of superlatives. In 1977, it ranked as the world's largest oil exporter, holding by far the largest proven oil reserves, mostly located in its barren Eastern Province. Employing 10 per cent of the labor force, the oil industry accounts for nearly all of the country's exports and foreign exchange. More than 90 per cent of government revenues are derived from oil royalties and taxes. As Dr. Mansoor Alturki, Deputy Minister of Finance and National Economy, noted in an interview, "Oil is our only resource."

Oil was discovered in Saudi Arabia during the 1930s, but exploitation of the rich fields was delayed until after World War II. Standard Oil of California was granted a 66-year concession in 1933 for the Arabian-American Oil Company (Aramco). Later, the company was jointly owned with Texaco, Standard Oil of New Jersey and Mobil Oil on a 30-30-30-10 basis. In 1949, another concession in the Saudi Arabian-Kuwait neutral zone was granted to Pacific Western Oil Company, later Getty Oil.

During 1950, two major developments in the oil industry strengthened the significance of the product for the country. First, a 753-mile pipeline was opened to carry crude oil out of the country across Jordan and Syria to the Mediterranean, where it could be easily shipped abroad. Second, Saudi Arabia and Aramco signed an agreement to share Aramco's profits on a 50-50 basis, thereby greatly increasing Saudi Arabia's return on its natural resource. This agreement was the first instance of what later became a common arrangement between oil companies and Middle East governments.

Oil production expanded rapidly in the early 1970s to meet rising world demand. In 1969, oil flowed at 3 million barrels a day. By 1974, that figure had nearly tripled to 8.48 million barrels. The country's production capability rose to about 10 million barrels a day during 1975. Moreover, with Saudi Arabia's proven oil reserves at 165 billion barrels, production could continue to increase even after the output of other states peaks sometime in the 1980s. But since the Saudis have begun to indicate a strong desire to conserve this precious resources, especially since they have no need for further current income, it is becoming less and less likely that Saudi Arabia will expand its production capabilities to meet the demands of the 1980s. A 1977 CIA study publicly cited by President Carter to formulate his energy proposals reached this conclusion and accordingly forecast a serious worldwide shortage of oil within a few years.[3]

This remains the cases even though the Saudi Ambassador to the U.S., Shaikh Ali A. Alireza, told a New York business audience that "The kingdom is still discovering more new oil reserves each year than the amount being taken out. And there are strong grounds for anticipating major new petroleum and other mineral reserves will be found with the more deliberate approach now being taken to exploration."[4]

In 1973, Saudi Arabia signed an agreement with Aramco, under which Saudi Arabia acquired a 25 per cent share of the ownership of the oil company. By 1975 the Saudi government assumed 60 per cent ownership, and the

---

2 Estimates of Saudi Arabia's population vary from a low of 5 million to this high figure of over 9 million. Discrepancy is partly due to uncertain number of foreign workers.

3 *The International Energy Situation: Outlook to 1985*, Central Intelligence Agency, April 1977.
4 Speech before the National Foreign Trade Council, Waldorf Astoria, New York City, 16 November 1976.

country by 1977 was in the process of taking over complete ownership of the company.

Regardless of ownership, Aramco continues to dominate the Saudi Arabian oil industry and is crucial to the economies of the Western countries and Japan. Aramco represented the largest single American investment in any foreign country; as of 1977, tens of thousands of Americans remained to manage the company.

The presence of Aramco has radically transformed the Saudi Arabian economy. Before the exploitation of oil, the economy revolved around simple exchanges between the nomads who raised camels, sheep and goats and the sedentaries who were mostly farmers and merchants. The revenues derived from tourists on pilgrimages to the holy cities yielded the greatest single source of income. Farming was poor due to extremely hot and dry conditions.

Aramco has stimulated the kingdom's economy and given rise to an industrial class. Government revenues from the oil companies have been funneled into internal improvements of transportation, communication and health and educational facilities. Agriculture has improved somewhat with the control of locust invasions. Dams have been built to retain seasonal valley floodwaters. Some irrigation systems have been built, and fishing practices are being modernized. Several industrial plants have also been built. The five-year plan offers hope that Saudi Arabia will develop a diversified economic base which will allow the country to remain modernized even when the oil runs out or the world shifts reliance to another energy source.

A serious shortage of Saudi manpower has been the main obstacle to rapid growth. As a result, more than a million and a half foreigners have flocked to Saudi Arabia or have been recruited and lured with tempting financial offers. In a 1976 report on U.S. interests in the Middle East it was pointed out that "All the hard manual work involved in the frenzy of construction which now grips the country is done by imported laborers, mostly Yeminis, while the architectural, technical and other required skills are supplied by Europeans, Americans, 'northern Arabs' or Pakistanis. The schools are staffed by Egyptians and the oil industry is run by Americans."[5]

In 1970, nearly half the population was still engaged in farming and stock-raising. Farming remains the principal occupation. For internal use, Saudi Arabia produces grains, vegetables, livestock and dates, which are a significant part of the diet.

Saudi Arabia must still import essential foodstuffs such as rice, flour, sugar and tea. Its other imports are machinery and building equipment, acquired mainly from the United States, Japan, West Germany and Great Britain. Besides oil and natural gas, also found in the oil fields, its exports include dates, skins, wool, horses, camels and pearls. Saudi Arabia exports mostly to the Common Market, but significantly also to Japan and the United States.

With the massive amounts of capital being expended in the five-year plan, Saudi contracts are eagerly sought by numerous companies and governments. The Saudi ambassador to the United States, Alireza, noted late in 1976 that "American companies have been awarded contracts within just the last year for over 27 billion dollars worth of work and goods to be completed in the years ahead.... And that is only a beginning."[6] The five-year plan includes such undertakings as a $24-billion desalinization program and an increase by 40 per cent in the country's irrigable acreage.

A United States-Saudi Arabian Joint Commission on Economic Cooperation was created in 1974. Jointly managed by the U.S. Treasury Department and the Saudi Ministry of Finance and National Economy, technical projects involving hundreds of millions of dollars have been agreed upon. They involve nine departments of the U.S. government and open the door for increased American exports of goods and services to the desert kingdom.

The American-Saudi relationship is based not only on oil, but also on the accumulating petrodollars, Saudi geography and the Saudi family's political moderation. Indeed, the Washington-Riyadh axis has become a central one in world politics and finance. As the Saudi ambassador in Washington described the relationship that has evolved, "The strategic significance of the partnership between our two countries has grown with each decade.... Indeed, it has now become one of the key links in the prosperity and stability of the entire international economy and the strength of the Free World's security system."[7] And Richard Nolte, a former American ambassador to Egypt, serving in 1977 as executive director of the Institute of Current World Affairs, concluded that "The amicable 44-year relationship with Saudi Arabia has become a matter of prime importance to the United States. This is true for economic and political reasons, and most crucially for reasons of United States security. Our national interest requires that this relationship...be supported and strengthened."[8]

Both the Saudi ambassador and Nolte went on to express fears that congressional efforts to pass anti-boycott legislation could seriously injure American relations with Saudi Arabia and cost the United States billions of dollars in contracts and increasing unemployment. But in the spring of 1977 a compromise was reached between Jewish organizations, business organizations and the Carter administration on legislation which appeared to be reluctantly accepted by most Arab states, including Saudi Arabia.

History

Saudi Arabia takes its name from the Saudi family, which has ruled it patriarchically, with only a few lapses, since the mid-1700s. To check the kingdom's rapid expansion in the early 1800s, Egypt invaded Saudi lands and occupied areas of the inner Arabian Peninsula from 1818 to 1840, until internal problems drew the Egyptian forces home. Ottoman-Egyptian occupation gave rise to the rival Rashid family, which ruled the area in the late 1800s, sending the Saudi king and his son into exile.

The son, who came to be known as Ibn Saud, returned in 1902 to reconquer his patriarchal lands. He waged a successful 30-year campaign and eventually took all of the inner Arabian Peninsula, proclaiming himself king. He drew his support from fanatic Wahhabi Moslems, whom he urged to fight in the name of their faith. But their ties to the Wahhabi traditions were to create obstacles for Ibn Saud.

On the eve of World War I, Ibn Saud conquered what is now the Eastern Province, a move that led to close contact with Great Britain. In 1915, the situation compelled him to sign an agreement which placed Saudi lands under Britain as a protectorate. But the country took no part in the war; and another treaty, signed in 1927, canceled the protec-

5 A.L. Udovitch (ed.), *The Middle East: Oil, Conflict and Hope* (Lexington Books, 1976), p. 433.
6 Speech before the National Foreign Trade Council, New York City, 16 November 1976.
7 *Ibid.*
8 "The Saudi Connection and the Arab boycott," *New York Times*, 18 February 1977.

torate and declared Saudi Arabia's absolute independence.

When Ibn Saud was unexpectedly even-handed with the non-Moslems his armies had conquered, and even took a moderately favorable view of modernization, introduced by contact with the West, he was opposed by some of the conservative Wahhabi Moslems who rejected Western ways and technology. In 1929, leaders of the armies he had led accused him of betraying their faith and traditions and launched attacks against Saudi tribes. The civil war was brought to an end in 1930, when British forces captured the rebel leaders in Kuwait and delivered them to Ibn Saud. The king then set out to proclaim the complete independence of Saudi Arabia and the supremacy of Islam within the kingdom.

Role in Conflicts

The modern kingdom of Saudi Arabia dates from 1932. On Sept. 24, 1932, Ibn Saud issued a royal decree, unifying the dual kingdoms of Hijaz and Najd and their dependencies. Since 1934, the kingdom has known only peace and security.

During the second world war, Saudi Arabia retained a neutral stance, but its pro-Allies inclination was readily apparent. In 1945, Ibn Saud nominally declared war on Germany, an action which ensured Saudi Arabia's charter membership in the United Nations. Also in 1945 the importance of Saudi Arabia's oil and location were recognized when President Roosevelt met with Ibn Saud on board an American warship in the Mediterranean. The same year, Saudi Arabia joined the Arab League, but the country's religious conservatism limited its interaction with other Arab states in the pursuit of Western modernization.

Ibn Saud's death in 1953 placed his eldest son, Saud, on the Saudi throne, and his second son, Faisal, was named heir apparent. The dissimilarity between the two sons, Saud's traditionalism and Faisal's penchant toward moderate modernization, fed an increasing rivalry between them. Faisal was granted executive powers in March 1958. On Nov. 2, 1964, Saud was deposed by a family decision, and Faisal became king.

Saudi Arabian relations with Great Britain were strained between 1952 and 1955 over a possibly oil-rich oasis in Muscat and Oman, a sultanate with alliances to Britain. When Saudi tribal forces moved against the oasis at Al Buraymi, British-led forces from Oman countered. An attempt to settle the dispute broke down, and British-led forces from Muscat and Oman and Abu Dhabi reoccupied the oasis. During the 1956 Suez crisis, Saudi Arabia broke off ties with Britain, and relations were not resumed until 1963.

During 1961, Saudi Arabia sent troops into Kuwait in response to a request from its ruler after Iraq claimed sovereignty over the country. The troops remained there until 1972.

Continued hostility between Saudi Arabia and Egyptian President Gamal Abdel Nasser dominated Saudi-Egyptian relations during Nasser's tenure. In 1958, Nasser accused King Saud of plotting his assassination. Thereafter, Egyptian propaganda vehemently attacked the Saudi royal family and the country's form of government. Tension again resulted in 1962, when Egypt and Saudi Arabia backed opposing sides in the Yemeni civil war.

Since World War II, the United States has been the dominant foreign influence in Saudi Arabia, although the United States has refrained from involvement in the

---

Statistics on Saudi Arabia

Area: 829,995 square miles; some borders poorly defined.

Capital: Riyadh; diplomatic capital located at Jiddah.

Population: 7,517,000 (1977 est.—highly uncertain).

Religion: 100 per cent Sunni Moslem.

Official Language: Arabic.

GNP: $41-billion (1975 est.), $5,530 per capita.

---

country's internal affairs. Between 1952 and 1962, the United States maintained an air base at Dhahran on the Persian Gulf. That arrangement was not renewed in 1961, partly due to Saudi Arabia's opposition to American aid to Israel. Both King Saud and King Faisal took strong stands against communism, warning against its influence in Arab and Moslem countries.

Egypt's defeat by Israel during the Six-day War in 1967 apparently did not disappoint King Faisal, due to strained Egyptian-Saudi relations. But aware of Arab nationalist sentiments, Faisal called for the annihilation of Israel and sent troops to Jordan, although they did not engage in extensive fighting.

On Sept. 4, 1973, Saudi Arabia foreshadowed its major role in the Yom Kippur War, when King Faisal said via U.S. television that American support of Zionism "makes it extremely difficult for us to continue to supply the United States' petroleum needs and even to maintain our friendly relations." On Oct. 8, after the outbreak of fighting, Saudi forces were placed on alert. A small contingent of forces crossed into Syria, although they were never reported as fighting.

Saudi Arabia's major role in the war was its lead in employing its rich oil resources as a political weapon. During the 1967 war, Saudi Arabia had reluctantly joined the Arab effort to withhold oil only after extensive pressure from Egyptian President Nasser, and lifted the embargo as soon as possible. But it joined with 10 other Arab oil-producing nations Oct. 17, 1973, in reducing by 5 per cent each month the amount of oil sent to "unfriendly" countries. The next day, Saudi Arabia independently cut oil production by 10 per cent to bring pressure on the United States. On Oct. 20, it announced a total halt of oil exports to the United States, action which only Abu Dhabi already had taken. This step—which caused some discussion in the United States of the need for possible military intervention to keep oil flowing—followed President Nixon's announcement that the administration was requesting an unprecedented $2.2-billion in emergency aid for Israel. This new Saudi willingness to use the oil weapon has been attributed, first, to the country's additional latitude gained from increasing wealth and, second, to its increased ability to guide the oil weapon following the death in 1970 of Nasser and the subsequent improvement in Egyptian-Saudi relations.

However, after the United States acted as a mediator in bringing about troop disengagement accords between Egypt and Israel in the winter of 1974, and while Secretary of State Henry A. Kissinger was spending much of that spring in quest of a similar agreement between Syria and Israel, the Saudi Arabians assumed a moderate position in the Organization of Arab Petroleum Exporting Countries (OAPEC). At a meeting in Vienna on March 18, 1975, Saudi

Arabia was joined by Algeria, Egypt, Kuwait, Abu Dhabi, Bahrain and Qatar in lifting the five-month oil embargo against the United States. Only Libya and Syria dissented from the decision. Simultaneously, Sheikh Yamani announced Saudi Arabia would hike his country's oil output by one million barrels a day and ship the increase to the United States.

Saudi Arabia has continued to argue against large OPEC price rises insisting that the stability of the Western world's financial system in the mid-1970s could be severely undermined. In December 1976, Saudi refusal to go along with the full amount of an OPEC price rise resulted in a two-tier pricing system and Saudi increases in the amount of oil produced. But oil minister Yamani accompanied this small break with OPEC with a warning: "Don't be too happy in the West" he declared. "We expect the West to appreciate what we did, especially the United States." This close linking of oil and politics, of oil with the Arab-Israeli conflict, raised considerable anxiety in the U.S. and renewed concern about Arab blackmail. "The oil business," the liberal *New Republic* noted, "no longer is a business at all. It has become a political enterprise ruled by a council of clever extortionists."[9]

Clement Martin, chief of the Federal Energy Administration's international division, summed up Yamani's message to the West in this way: "They are definitely sending a signal to the United States that the political use of oil is still a major arrow in their quiver.... They have said, 'Look, over the past couple of years we've been very supportive of your efforts to bring about Middle East peace and stability. We have bank-rolled the moderates in Egypt; we have bankrolled the Jordanians; we have encouraged the Syrians in the Lebanon situation.... Now it's your turn, United States, to take the next step.... Unless we have some real progress in a year or so the moderates will be in real trouble.' "[10]

Since Yamani's warning, the Saudis have grown more cautious and have often denied the linkage between oil, petrodollars and politics. Nevertheless, it is a linkage that poses many problems for the United States and one which is sure to remain central in the coming years.

Government

Saudi Arabia's government is patriarchal, though major decisions are usually made by consensus. There is no parliament or formal constitution, although the Koran, the basis of Islamic law, restrains Saudi kings, as does unwritten tribal law and custom. The king reigns as chief of state and as head of government. The judiciary branch consists of the Islamic Court of First Instance and Appeals. King Faisal acted as his own prime minister and foreign minister.

The new King Khalid, at age 61 ascending the throne upon the assassination of Faisal on March 25, 1975, has long been regarded as a weaker figure than his half brother, Faisal. Khalid underwent open heart surgery in Cleveland, Ohio, in 1974 and, until his ascension had not been active in state affairs in recent years. For these reasons Crown Prince Fahd, another half-brother, has become power behind the throne in the new reign. Fahd was serving as interior minister at the time of Faisal's murder. Khalid's continuing poor health could result in Fahd assuming the throne at any time.

9 "Not So Friendly Saudis," *The New Republic*, 1 & 8 January 1977, p. 3.
10 *Ibid.*, p. 4.

Saudi Arabia on Peace Prospects

During a May, 1977 visit to Washington Saudi Arabia's Deputy Minister of Finance and National Economy, Mansoor Alturki, discussed the possibilities of peace in an interview. "It's an unfortunate thing that happened between the Jews and the Arabs," he commented. "Having a Jewish state is not at issue. It's a common belief. You see, if they just want a Jewish state I wouldn't see any problem. But the Zionists, the way I understand it, they always want to expand. That is what we are afraid of. That is why there is a lack of confidence in Israel's desire for a settlement."

Dr. Alturki, who was in the U.S. in his role as co-ordinator of the U.S.-Saudi Arabian Joint Economic Commission, went on to emphasize the Saudi desire to reach a full normalization of relations with Israel. "Permanent peace" is the goal. When this is achieved it "means we will not consider Israel as an enemy" any longer. He further suggested that Saudi Arabia would welcome invitations from American Jewish groups to discuss the Middle East situation and U.S.-Saudi relations.

See Mark Bruzonsky, "The Saudi-American Partnership," *Christian Science Monitor*, 25 May 1977.

Not all 4,000 Saudi princes on the civil list play an active role in governing the nation, but at least a few hundred of them do. Saudi monarchs have been clever in splitting military power between the 47,000-man regular army and the 33,000-man national guard. Prince Sultan heads the army and Prince Abdullah runs the militia. Both are half-brothers of King Khalid. The Saudi family always has been careful to cultivate its bonds with the two other great families in the country, the Sudairis and the as-Sheikhs, who helped the Saudis overrun most of the Arabian Peninsula between 1750 and 1926. Members of the two lesser families usually hold important posts in the government.

The Saudi state in 1977 appeared to rest on three major pillars. The first was the alliance between the house of Saud and the important families of the major tribes. The second was Aramco, the company whose ownership was being completely taken over by the Saudi government but which continued to symbolize the U.S.-Saudi partnership. And the third was religion. Every Saudi leader has assumed not only the kingly throne but also the office of *imam* of the Wahhabiya—thus becoming the spiritual as well as the temporal leader of the Saudi people.

Foreign Policy

Saudi Arabia maintains close relations with the United States, and its oil is of great importance, especially as American energy needs grow. Aside from disagreements over American support of Israel, Saudi-U.S. relations have been friendly since they were established in 1940.

Maintaining no diplomatic relations with any Communist states and opposing Communist influence in the Moslem world, Saudi Arabia has historically turned to the United States for assistance in modernizing its army and developing its resources.

Only on the question of Israel have relations been occasionally and markedly strained. Saudi Arabia has sought to encourage the United States to spur Israel's withdrawal from occupied lands. Saudi Arabia has always valued its bonds with the United States, holding them to be important for voicing Arab interests vis-a-vis Israel. Thus, unlike other Arab states, Saudi Arabia did not break off diplomaic relations with the United States as various Arab-Israeli crises erupted. Saudi Arabia was both a leader in putting teeth into the 1973 war's oil embargo and in ending it after the belligerents quieted their guns.

The extremist policies of Libyan President Muammar al-Qaddafi led to expanded Saudi cooperation with Egypt in an effort to keep Arab policies in a moderate line and to enhance the late King Faisal's personal leadership role in the Arab world. During the summer of 1973, when Egypt still faced a possible merger with Libya, scheduled to take effect Sept. 1, 1973, Saudi Arabia and Egypt firmed up their alliance. The merger did not materialize. Saudi Arabia has subsidized Egypt for revenues lost with the 1967 closing of the Suez Canal, which was reopened to shipping June 5, 1975. In 1977 the Saudis provided aid to many Arab countries, including Egypt and Syria, and made a five-year commitment to subsidize the Egyptian army's development.

Since 1975 Saudi foreign policy has become more activist, under the leadership of King Khalid and Crown Prince Fahd. It was Saudi influence in late 1976 which helped resolve the Lebanese conflict. And it has been Saudi influence which has helped create a climate of potential peace with Israel. Like the Egyptians, the Saudis have started to speak of "complete, permanent peace" and a "normalization" of relations with Israel. "Palestinian leaders are ready to accept a peaceful solution" if it involves "establishment of a Palestinian state on the West Bank and in Gaza," Fahd stated in May 1977.

Still, potential strains in the U.S.-Saudi relationship do exist. Continuing Middle East stalemate might discredit the moderate leadership. Another oil embargo is already being talked about as a possibility and oil pricing and supplies themselves can take on political dimensions. Further, Saudi Arabia's immense financial reserves might be manipulated in ways designed to pressure for political goals.

Saudi Arabia's "new and special relationship" with the United States was first pointed out by Theodore A. Rosen of the Treasury Department's Office of Saudi Arabian Affairs as far back as 1974.[11]

As of mid-1977, the Saudi-U.S. friendship, which has become one of the most important in world politics and economics, appears secure. As long as the Saudi leadership remains stable and there is progress toward an Arab-Israeli settlement, it is a friendship that is likely to continue to grow and become even more central to the politics of the Middle East.

_____

11 Remarks to the 28th Annual Conference of the Middle East Institute in Washington, 12 October 1974.

THE EMIRATES: DESERT POVERTY TO OIL RICHES

Not long ago, the desert sheikhdoms that dot the western and southern shores of the Persian Gulf[1] were remote and backward lands peopled by camel herders and pearl fishermen. The place names of Kuwait, Qatar, Bahrain, Abu Dhabi, Dubai and Sharjah had minuscule impact on world consciousness.

The British knew the names and places, for they became links in the British Empire when it was growing to its zenith under Queen Victoria. About the time Britain quit the region in 1971, philatelists came to recognize Sharjah and Fujairah, for several of the tiny emirates began to flood postage stamp markets with gaudy issues—bearing the likenesses of such non-Arab heroes as George Washington and Joe DiMaggio. The stamps were a revenue-raising scheme.

Whether or not the stamps brought in any cash is a minor point. For, where once there were only camel trails, overnight there were roads—to corrupt an old cliché—paved in black gold. In less than a decade, the pumping of oil in vast commercial quantities cast Kuwait, Qatar and the United Arab Emirates in significant roles on the world stage. For 75 years, Britain had bossed the Kuwaitis. But in 1973, Britain was filling 23 per cent of its oil needs from Kuwait alone. When Kuwait joined the other Arab states in shutting off oil supplies to the West that autumn, Britons found themselves knitting sweaters to keep warm and queueing up in long lines at gasoline stations.

Westerners who took a high school or college course in world history were sure to run across such ruling families as the Bourbons, the Bonapartes and the Hapsburgs. They did not learn about the al-Khalifahs, the al-Sabahs and the al-Thanis. But those desert tribesmen in the 1970s became entrepreneurs, loaning money to the city of Paris for its airports and to the Austrian Central Bank in Vienna.

In 1964, the gulf sheikdoms were counting sheep. By 1974, they were counting 11 billion barrels of oil. Kuwaitis and Abu Dhabians argue over which of the two has the highest per capita income in the world. It is a Middle East truism that native-born Abu Dhabians are too rich to work—that labor is done by the 75 per cent of its population that is foreign.

Texas oil money created arch rivalries, Dallas and Houston vying to erect the tallest skyscraper in the state. Now Abu Dhabi, Dubai and Sharjah compete to build the most imposing clock tower rising from the sands of the United Arab Emirates or the most impressive international airport. All this generates a flood of jokes about the comic-opera side of the gulf states. Yet gulf Arabs, like many *nouveaux riches* before them, enjoy the last laugh—on the way to the bank.

Bahrain

Area: 240 square miles.
Capital: Manama.
Population: 269,000 (1977 est.).
Religion: Moslem, predominantly the Sunni sect in urban areas, predominantly the Shia sect in rural areas.
Official language: Arabic.
GNP: $400-million or $1,690 per capita (1974).

Terrain

The emirate consists of about 35 desert islands in the Persian Gulf—with a total territory one-fifth the size of Rhode Island—which lie 15 miles east of Saudi Arabia and the same distance west of the emirate of Qatar. Bahrain is the main island. Other important islands are Muharraq, Sitra, Umm Na'san and the Howar group. The beaches are mud flats.

People

The majority of the 269,000 persons are Moslem Arabs. There are 35,000 non-Bahrainian Arabs, Indians, Iranians and Pakistanis, and a colony of 3,000 Europeans and Americans. Most of the population inhabits Bahrain, the largest island and the one from which the emirate derives its name. The largest cities are Manama, the capital (150,000), and Muharraq (38,000). The predominant Moslem religion is evenly divided between the Sunni and the Shia sects. The ruling family is Sunni.[2]

Economy

Bahrain discovered oil in 1932, one of the first Persian Gulf states to do so, but it is one of the poorest in that precious resource in the gulf region, pumping only 68,000 barrels per day in 1974. The production was hardly enough to cover expenses until the price of oil was quadrupled in the wake of the 1973 Arab-Israeli war.

The oil field is at Jabal Dukhan, in the middle of Bahrain Island. The Bahrain Petroleum Company, an

---

1 Many of the Arab states term the sea between Iran and Saudi Arabia the "Arab Gulf". Popular usage in the U.S. continues to refer to the area by the title "Persian Gulf."

2 For a revealing description of life in Bahrain see Part III in Linda Blandford's *Superwealth, The Secret Lives of the Oil Sheikhs* (Morrow, N.Y.: 1977).

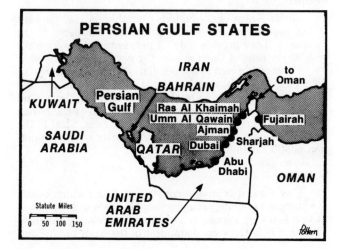

PERSIAN GULF STATES

IRAN
BAHRAIN
to Oman
KUWAIT
Persian Gulf
Ras Al Khaimah
Umm Al Qawain
Ajman
Fujairah
SAUDI ARABIA
QATAR
Dubai
Sharjah
Abu Dhabi
OMAN
UNITED ARAB EMIRATES

Statute Miles
0 50 100 150

American syndicate of Standard Oil of California and Texaco, holds the oil concession.

A large oil refinery on Sitra Island is an important industry, having an annual capacity of 10 million tons. It refines what's left of Bahrainian oil, but mostly oil moved through underwater pipeline from the Aramco fields in Saudi Arabia. About two-thirds of the 9,000 persons employed by the oil industry are Bahrainians.

Manama was declared a free transit port in 1958, and trade is important to the economy. The international airport at Muharraq is the most important in the Persian Gulf. British Airways and Gulf Air have headquarters there.

About 100 European and American companies have offices and installations in Bahrain. An American engineering and construction firm, Brown and Root Inc. of Houston, Texas, maintains the largest headquarters, supplying offshore oil-drilling rigs to oil companies. Other important foreign companies include Gray Mackenzie shipping lines and Grindlay's, a British bank.

The only agriculture is on the northern end of Bahrain Island, where dates, rice and vegetables are grown. The fields are irrigated from underground springs. Once-important pearl fishing is now in decline. The building of dhows and the making of sailcloth and reed mats are minor but old industries.

History

Archaeologists and historians relate Bahrain Island to the ancient Sumerian civilization center of Dilmun, known as the "earthly paradise," dating back to 2600 B.C. The archipelago is mentioned for its strategic and commercial importance by Assyrian, Persian, Greek and Roman geographers and chroniclers.

Portuguese sailors captured the islands from local Arab rulers in 1521. After Portuguese dominance ended in 1602, Bahrain was conquered by a series of Arab tribes until a paramount family, the al-Khalifahs, established themselves as sheikhs of Bahrain in 1783. The family still rules.

Isa bin Sulman al-Khalifah

In 1820, the British established hegemony over the islands, taking over responsibility for defense and foreign policy. In return for this protection from large Arab tribes, the sheikhs promised to refrain from "prosecution of war, piracy or slavery."

In August 1971, Bahrain declared independence from Britain and signed a treaty of friendship with the former mother country. Bahrain joined the United Nations and the Arab League. Choosing to remain totally independent, Bahrain rejected a proposed federation with neighboring Qatar and the United Arab Emirates.

Government

The ruler is Emir Isa bin Sulman al-Khalifah, born in 1933. He enjoys almost absolute power. In 1970, an 11-member Council of State was created, responsible to the emir and serving at his pleasure. Elections were held to a small National Assembly in December 1973, but the assembly was dissolved within two years.

The government, getting 70 per cent of its national income from oil royalties and the remainder from custom duties, provides extensive social services including free medical treatment and free primary, secondary and technical education. Bahrain has a reputation for having the only good schools in the region, one of the reasons the islands attract Western companies.

Foreign Relations

Bahrain still has strong ties to Britain, which, with Saudi Arabia, is its main trading partner.

The U.S. Navy had docking facilities at Jufair. In October 1973, Bahrain suspended the naval agreement, charging Washington with ignoring its warning concerning the United States' "hostile stand against the Arab nations." After quiet negotiations, the Bahrainians allowed the Navy to return in 1974. But the arrangement ended in 1977.

Quickly running out of oil, Bahrain by 1977 aspired to be a banking and service center for the region. With a working infrastructure, good communications, and an educated work force, Bahrain has a head start. A controversial causeway, with funding offered by Saudi Arabia, is being considered to link the country with the Saudi mainland.

Kuwait

Area: 6,886 square miles.
Capital: Kuwait.
Population: 1,101,000 (1977 est.).
Religion: About 99 per cent Moslem, 1 per cent Christian.
Official language: Arabic, with English the second language.
GNP: $13.9-billion or $13,500 per capita (1975 est.).

Terrain

Kuwait's nearly 7,000 square miles make it about the size of the combined areas of Connecticut and Delaware, the third- and second-smallest American states. Kuwait is located on the northeastern corner of the Arabian Peninsula, bordered on the north and west by Iraq, on the east by the Persian Gulf and on the south by Saudi Arabia. The terrain is mainly flat desert with a few oases. There are no rivers. The heat is torrid in summer; winter brings severe dust storms.

People

Native Kuwaities, who today number only 40 per cent of the population, are Arabs, principally Moslems of the Suni sect. They provide only 20 per cent of the work force and 70 per cent of them are on the government payroll.

About 60 per cent of the population are immigrants who do not enjoy Kuwaiti nationality. Included in this number are about 270,000 Palestinians, many Indians and Pakistanis, some Egyptians, Arabs from neighboring countries, and a few Britishers and Americans. Most of the population lives in Kuwait town. A quarter of the population is still nomadic.

Economy

Tiny Kuwait is an opulent strip of desert which boasts of a standard of living second to none in the world, thanks to its one and only export, crude oil and petroleum products. The country is the second smallest in the Arab League and the richest—with an estimated per capita income of more than $13,000 in 1975.

The Kuwait Oil Company, the ownership of which is equally divided between British Petroleum Company and Gulf Oil Corporation of the United States, obtained a concession covering all the country except the neutral zone of 2,200 square miles. (The zone was created by a treaty in 1922 with Najd, now a province of Saudi Arabia.)

Oil was first struck in 1938, but development was delayed until 1945, after World War II ended. Progress then turned spectacular. Oil was struck in commercial quantities in the neutral zone in 1953. Getty Oil and the American independent oil companies hold the concessions in the zone. More oil was discovered in northern Kuwait in 1955. Kuwait possesses one-fifth of the world's known oil reserves.

In 1974, oil production hit an estimated 9.12-billion barrels, and known reserves are expected to last another 50 years. Unspent earnings amounted that year to $5-billion. Long before the price of oil increased fourfold as a result of the 1973 war, Kuwait was quietly making investments abroad—in securities, real estate and banks.

Oil accounts for about 60 per cent of the gross national product, and large supplies of natural gas are derived from the oil fields. Still, the largely automated oil industry accounts for only 5 per cent of Kuwaits more than 200,000 jobs. Other GNP factors include government services, 20 per cent; non-oil industries (chemical fertilizers, building materials, fishing), 3 per cent, and trade, 9 per cent. Only 1.5 per cent of the land is cultivated, producing fruits and vegetables.

History

The name Kuwait is derived from "kut," meaning a small fort. In the early 18th century, the territory was settled by three Arab clans, and in 1756 Sheikh Sabah Abdul Rahim founded the al-Sabah dynasty, which rules to this day.

As one recent visitor amusingly tells the story, "After a shocking drought in the desert in 1710 some of the more aristocratic Bedouins went down to the Gulf for a sight of water. They wandered around until they settled on a convenient spot, Kuwait. Someone had the bright idea of sending one of their clever young men off to see the nearest Turkish governor to get permission to stay put, on condition they kept quiet. His mission was a success; his name was al-Sabah. His family has been top dog ever since; everyone else who was around at the time is 'heritage'.[3]

The ruling prince of the nominally Ottoman Turkish province made a treaty with Great Britain in 1899 because he feared the Turks wanted to make their authority effective. One British aim was to prevent the German kaiser from building the terminus of the Berlin-Baghdad railway on Kuwaiti territory on the Persian Gulf. When war broke out with Turkey in 1914, London recognized Kuwait as an independent state under British protection.

British troops twice have gone to Kuwait's rescue when it was troubled by larger neighbors, the first time repelling attacks from Najd (now Saudi Arabia) by Wahhabi fanatics. A 1922 treaty settled the matter and set up the neutral zone southeast of Kuwait—later a rich oil area (with 4 per cent of the world's oil reserves) over which Kuwait and Saudi Arabia share sovereignty.

Great Britain granted Kuwait full independence in 1961, which led to Iraq's claiming the tiny state as a

province. When Baghdad threatened to invade, the Kuwaitis asked for British military aid, which was promptly dispatched, and Iraqi action was deterred. Iraq recognized Kuwait's independence in 1963, the same year the sheikhdom was admitted to the United Nations.

Government

Kuwait is a constitutional monarchy with Sheikh Sabah al-Salem al-Sabah as emir. He is assisted by a cabinet of 14 ministers (10 appointed and four elected) headed by a prime minister. The constitution, dating from 1962, provides for a 50-member National Assembly, elected by all natural-born, literate Kuwaiti males over 21. In August, 1976, the National Assembly was suspended. The emir promulgates laws and decrees, and has taken over full legislative powers.

Life Styles

Because of Kuwait's vast oil riches, poverty hardly exists in the country. Kuwait is a curious mixture of capitalist intake of money and socialist output of services. (Actually, the government is planning for a complete takeover of the British-American-owned Kuwait Oil Company.) Citizens are provided free education, free medical services, free telephones and virtually free housing, and they pay no taxes.

Sabah al-Salem al-Sabah

However, Kuwait often is an economic paradise for the native-born only, even though foreigners comprise 60 per cent of the population and 80 per cent of the labor force. Each Kuwaiti citizen is guaranteed a job and a minimum annual income of about $8,500, a spectacularly high figure outside the Western world. Yet the immigrants—Palestinians, Iraqis, Egyptians, Indians, Iranians and Pakistanis—often earn no more than $2,500. Only Kuwaiti citizens receive subsidized housing and can own land on which a house can be built. It is exceedingly difficult for foreigners other than Arabs to obtain Kuwaiti citizenship. However, most aliens complain little, probably because they do better in Kuwait than they would have done in their home countries.

Kuwait has a higher proportion of the latest American automobiles than any other country outside the United States, and they are driven over wide, well-surfaced roads. The government built the modern city of Rikka in the desert 25 miles south of Kuwait town. The city has 5,000 duplex housing units valued at $13,000 each.

The sheikhdom boasts of twice as many air-conditioning units as people and one millionaire for every 230 persons. In short, modern Kuwait is a garish, *nouveau-riche* land.

Kuwaiti efforts to employ its oil riches to transform a hot, sleepy land into a modern state are largely the work of Sheikh Abdullah al-Salem al-Sabah, who died in 1965 and who was succeeded by his younger brother, Emir Sabah. Abdullah believed that the country could not develop unless it shed its tribal structure and its wealth was distributed among its people. The current emir is continuing that policy.

---

3 *Ibid.*, p. 212. See Chapter IV in Blandford's book for a description of Kuwait today.

Foreign Policy

During the 1967 Arab-Israeli war, Kuwait joined in the ineffectual oil embargo ordered by Arab states against Western nations. Kuwait cut off shipments to the United States and Britain on June 6, 1967. At the time, Britain was getting 23 per cent of its oil requirements from Kuwait. However, Kuwait did not break off diplomatic relations. Kuwait, a member of the Organization of Petroleum Exporting Countries (OPEC), also took part in the effective 1973 oil boycott.

Kuwait's conservative ruling family has expressed alarm over Soviet inroads in the Persian Gulf—the arming of Iraq by Moscow and the use of an Iraqi port at the head of the gulf by Soviet warships. Yet Kuwait has diplomatic relations with both the Soviet Union and China and has recently begun acquiring Russian arms and some military advisers.

Some years ago Kuwait considered forming a federation of nine neighboring sheikhdoms as a means of thwarting the possibility of the sheikhdoms toppling to revolutionary movements. But Kuwait had not by mid-1977 taken such steps. Rather the government uses is monty to support such causes as that of the Palestinians, hoping, in this way, to not become a target of revolution itself. Because there is some danger of instability in the country, the once-free press and the National Assembly have both been bridled.

Qatar

Area: 4,757 square miles.
Capital: Doha.
Population: 183,000 (1977 est.).
Religion: Most Qataris are Sunni Moslems of the Wahhabi sect.
Official language: Arabic.
GNP: $1.8-billion or $10,000 per capita (1975 est.).

Terrain

Qatar, pronounced "gutter" in Arabic, occupies a thumb-shaped desert peninsula stretching north into the Persian Gulf. On the south it borders the United Arab Emirates and Saudi Arabia. Qatar is about the size of Connecticut. The peninsula is a low, flat, hot, dry plain, consisting mostly of limestone with sand on top. On the west coast is a chain of hills, the Dukhan Anticline, and beneath it is oil, which gives Qatar its importance.

People

The population is composed of Qataris and other Arabs from neighboring states, especially Iranian. About 80 per cent of the people are located in and around the capital city of Doha on the east coast; Doha has an estimated population of 130,000. There is a sprinkling of Bedouin nomads.

Economy

The oil resource accounts for nearly all of Qatar's national income. Discovered in the Dukkan range in 1940, oil was brought into production in 1947. Development of offshore fields and improved techniques in the 1960s greatly increased output. The country produced an estimated 189 million barrels in 1974.

One-fourth of the oil revenue is reserved for the ruler. The rest is spent on public development and services. In 1975, oil royalties totaled more than $1.5-billion.

Using oil monies, the government has built a seawater distillation plant, doubling water supply, and two power stations. There have been some government-aided exploitation of the gulf fishing potential, experimentation in natural gas recovery and the introduction of domestic refining. The government is constructing a steel mill at Musay'id (Umm Sa'id) and expects to import iron ore from Australia to support that industry. Various shipping lines call at Musay'id, and the port is being expanded to nine berths. There is also some vegetable growing and herding of camels and goats.

History and Government

The Qataris long were ruled by the Persians, and they paid the governor of Bushire an annual bounty for the right to fish for pearls. Qatar became independent of Persia in the 19th century under Thani, found of the al-Thani dynasty.

Thani was succeeded by his son, Mohammed, who signed an agreement in 1868 with Great Britain which provided protection to the sheikdom. Mohammed's son, Qasim, died in 1913 at the age of 111, and Qasim's son, Abdullah, renewed the accord with the British in 1916.

From 1947 to 1960, Qatar was ruled by Emir Ali, who was forced from his throne by his son, Ahmed, with the help

Khalifa ben Hamad al-Thani

of a British gunboat in Doha harbor. In turn, in 1972, Ahmed, who spent most of the country's fortunes on himself and the large royal family, was deposed in a bloodless coup by his cousin, Emir Khalifa ben Hamad al-Thani.

Qatar declared its independence on Sept. 1, 1971, after the British announced they were pulling back their forces from the Persian Gulf by the end of that year. Along with Bahrain, Qatar considered entering into a large federation with the United Arab Emirates, but the plan eventually was dropped.

Emir Khalifa is assisted by a 23-member advisory council, including 20 elected members. The state is run under a strick religious law, but it is not as severe and puritanical as that in Saudi Arabia. Qatar, for example, permits movie houses and women drivers, two items banned by its larger neighbor. Qatar has introduced a free educational system and free medical services.

Foreign Relations

After the British abandoned their protectorate status over Qatar, the two countries signed a new treaty of friendship. In September 1971, Qatar joined the Arab League and the United Nations.

In 1975 a three-year, $250-million industrial plan was announced to lessen the country's total dependence on oil. Qatar continues to be interested in joining some kind of regional federation. In March 1976, the emir issued an appeal for a Gulf common market to prevent duplication of economic development in the area and to coordinate the trade of Gulf nations in world markets.

Development projects that are currently underway include additional desalination plants, an enlarged harbor at Doha, refrigerated storage facilities for the fishing industry, and a gas liquefaction plant.

United Arab Emirates

Membership: A federation of the sheikdoms of Abu Dhabi, Ajman, Dubai, Fujairah, Ras al-Khaimah, Sharjah and Umm al-Qaiwain.
Federal capital: Abu Dhabi.
Area: 32,278 square miles.
Population: 656,000 (1977 est.).
Religion: Moslem.
Official language: Arabic.
GNP: $4-billion or $16,600 per capita (1974 est.).

Terrain

Six of the emirates are strung along the southern rim of the Persian Gulf. Fujairah faces on the Gulf of Oman, a part of the Arabian Sea. The United Arab Emirates (UAE) is bordered by Qatar, Saudi Arabia and Oman. The approximately 32,000 square miles of largely undefined borders are the size of South Carolina. Most of the region is a low, flat desert where temperatures sometimes reach 140 degrees Fahrenheit. At the eastern end, along the Oman border, are the Western Hajar Mountains.

People

The population is made up of Arabs, Iranians, Baluchi and Indians. Only 25 per cent of Abu Dhabi's population of 235,000 is native. The UAE's total population, including foreigners, is nearing 700,000.

Characterization

Abu Dhabi, the largest of the seven states, is mostly desert, but it has the largest oil deposits. Settlements include the towns of Abu Dhabi, Dalma' and Das Island, and Al Jiwa and Al Burayami oases. There are airfields at Abu Dhabi and Das Island.

Dubai, including the town of Dubayy, has a population of about 200,000, and possesses one of the few deepwater ports in the area, Port Rashid. It has a police force, municipal council, elementary schools and a small hospital.

Sharjah includes Sharjah town (Ash Shariqah), and has a population of about 20,000. There are a medical mission, schools, steamer and air services. There are red oxide mines on Abu Musa Island.

Ajman, which has only 100 square miles, is the smallest of the emirates. The population is 22,000.

Ras al-Khaimah is comparatively large and fertile, and has a town and several villages. No oil has been discovered there.

Umm al-Qaiwain (22,000) includes Falaj al 'Ali oasis.

Fujuairah (26,000) consists of a few small villages.

Economy

The emirates are highly competitive among themselves. Abu Dhabi and Dubai, the two most important, are the chief rivals.

Until oil was discovered in Abu Dhabi in the early 1950s, Dubai ranked first, because it was a commercial center, home of the entrepot trade, which included the smuggling of illicit gold to India and Pakistan. Dubai billed itself as "the Hong Kong of the Middle East."

Dubai began pumping oil in 1969, joining Abu Dhabi in production of the resource. In July 1974, Sharjah became the third of the seven states to produce oil. The emirates pumped nearly 620 million barrels in 1974. That year Abu Dhabi made $3.5-billion from oil, and managed not to spend $1.5-billion of it. By 1976, the UAE's oil revenues reached $7-billion.

Before oil, Abu Dhabi lacked paved roads, telephones and apartments.

After oil, the emirate began building a city, including a wide boulevard atop a new sea wall along the coast. Abu Dhabi built an international airport. After oil, Dubai started construction of its own international airport. After oil, Sharjah began its own international airport, not far from the other two. Abu Dhabi, Dubai and Sharjah all are building their own cement factories. Each of the seven states has its own radio station, and there are two television stations in the federation.

Aside from oil, the economy is based on fishing and pearling, herding, date-growing and trading.

Several of the emirates—Ras al-Khaimah, Ajman, Sharjah, Fujairah and Umm al-Qaiwain—have produced large and varied issues of postage stamps aimed at catching the eye of the world's stamp collectors. Colorful sets highlight and honor events and personalities with no connection to the desert sheikdoms—American baseball greats Babe Ruth and Ty Cobb, historical and contemporary figures Benito Juarez and Charles De Gaulle, the Olympic games, antique automobiles.

Banking is an important business. In a 1974 survey on the Persian Gulf, *Newsday,* the Long Island newspaper, reported there are 165 branch banks in the UAE, about one for every 1,400 residents. Tiny Ajman has six banks alone.

Newsday also found that Abu Dhabi in 1974 put $675-million into bonds—guaranteed loans at 8, 8.5 or 9 per cent—to the Spanish Highway Authority, the Korean Development Bank, the Austrian Central Bank, the Industrial Fund of Finland, the Republic of Ireland, the Paris Airport Authority, the European Investment Bank and the World Bank. Morgan Guaranty Trust of New York handles a large portion of Abu Dhabi's $240-million in direct investments in telephone companies, textiles, gas and electric companies, food companies and merchandising companies. Of course, investments since 1974 have leaped.

History

The emirates were formerly known as the Trucial States or the Trucial Oman because of their ties to Great Britain by truce and treaty.

The dominant Arab tribe was the Qawasim, made up of competent sea traders until they were overrun by Wahhabis from the interior of the Arabian Peninsula in 1805. They then became pirates. The British sent in a naval fleet after two of their commercial ships were plundered, leading to the general treaty of peace in 1820. The sheikhs also signed with Britain the perpetual maritime truce in 1853 and the exclusive agreements in 1892, which gave Britain control over their foreign policy.

After Great Britain in 1968 announced its intention to withdraw the Union Jack from the gulf by the end of 1971, the Trucial States, Qatar and Bahrain initiated plans to form a confederation. But the latter two sheikhdoms finally decided in favor of independent sovereign status.

In December 1971 the United Arab Emirates was formed. Ras al-Khaimah did not join the federation until February 1972.

Government

The president of the UAE is Zayed ibn Sultan al-Nahayan, emir of Abu Dhabi, who became head of the

Zayed ibn Sultan al-Nahayan

federation at its inception. Emir Rashid ben Saeed of Dubai is the vice president.

The Supreme Council, made up of the seven ruling emirs of the member states, is the highest legislative authority. There is also a Federal Council, in which each state is represented and which holds executive powers. Matters that are not of federal concern are left to the individual rulers.

The UAE is a loose federation which shares common roads, telephone and postal systems. Concerned about their own identity and the possible future ambitions of larger Arab and non-Arab nations in the region, the seven states joined together chiefly to form a common foreign and defense policy. The nucleus of the federal defense force was created from the British-trained Trucial Oman Scouts; however, Abu Dhabi has a larger army. Before accepting a second term as President, Sheikh Zayed demanded greater authority to rule effectively over the fledgling country.

Foreign Relations

The UAE boundaries are vague, and there have been border disputes with Saudi Arabia. But relations between the two have remained good. After independence in December 1971, the UAE was admitted to the United Nations and the Arab League.

The United States, Britain and the European nations are the UAE's principal trading partners. The federation joined in the oil boycott against the industrial nations in October 1973. Since then the UAE has basically followed Saudi policy on oil matters.

Like Saudi Arabia, the UAE has a special relationship with the United States. In December 1976, celebrating its fifth birthday, a public relations campaign was carried out to acquaint the American people with the history and aspirations of the UAE. A major section appeared in *Time* magazine and full-page ads appeared in major newspapers.[4]

In May 1977 Sheikh Zayed restated that on OPEC matters he would always support the Saudi positions. On the Arab-Israeli conflict, he had this to say in light of Menahem Begin's victory in Israel earlier in the month: "The Arabs fought four wars against those who are now described as moderates. Will the hawks be more extremists than them?" Asked how he now viewed prospects for a Middle East settlement, he answered: "I am not pessimistic.... The more obstinate the enemy becomes the more unified the Arabs tend to be. Arab calculations can be more realistic and with less wishful thinking. I told the PLO representative who [recently] handed me a message from Yassir Arafat that maybe such a disease will cure the whole body."[5]

4 For instance *The New York Times,* 19 December 1976, p. E 5.
5 *Events,* 3 June 1977, p. 5.

ISSUES AND POLICIES

SEARCHING FOR A COMPREHENSIVE SETTLEMENT

For three decades—ever since Israel's creation in the aftermath of the Second World War—the emerging Arab world and the small Jewish state have been at war. Even in the beginning, the Great Powers were involved in a competition for influence and economic advantage while the United Nations was debating various plans and principles. Though the United States gave strong political support to the creation of Israel, America was not eager to become directly involved in the Middle East situation. Israel fought its war of independence alone—with much sympathy but little concrete assistance. The Arabs, on the other hand, felt betrayed. They had expected national independence after World War I only to find themselves victimized by British and French imperial aspirations. After World War II, the Arabs living in Palestine expected to rid themselves of foreign control and to halt the attempt by Jewish immigrants to transform the area into a Jewish state. Arab leaders thought President Franklin Roosevelt had agreed, when he met with Saudi Arabia's King Ibn Saud in 1945, not to support a separate Jewish state in the area if opposed by the Arabs.

Israel's Jews and Palestine's Arabs had come from different historical experiences and had differing, unfortunately competing aspirations. Neither side was really able to understand and appreciate the other. Zionism, the search for Jewish national existence in the historic homeland of Eretz Israel, and Arab nationalism, a natural political orientation in the 20th century, experienced a head-on collision, and the various damage claims which resulted have yet to be settled.

If the ongoing discussions of a comprehensive settlement are to succeed, they will have to incorporate these decades of claims and counterclaims which have emersed the parties in a web of historical arguments. But today, there are added dimensions to the conflict. The warring parties themselves have all become dependent as never before on outside actors, especially the United States, making the Middle East conflict one central to international politics and economics. And, because of oil and petrodollars, the United States has itself become dependent on the Arab Middle East.

Within months of taking office, the new American Secretary of State, Cyrus Vance, announced that "The search for a just and lasting peace in the Middle East is one of the highest priority items on the foreign policy agenda of our country." And the new American President, Jimmy Carter said, "To let this opportunity [for a settlement] pass could mean disaster not only for the Middle East, but perhaps for the international political and economic order as well." *(See U.S. Policy chapter, p. 80)*

Nevertheless, a complete settlement of the Middle East conflict seemed most unlikely to many of the experts in the participant countries who at best were hoping for a gradual process leading to lowered tensions. "The Arab-Israeli dispute is perhaps the most difficult conflict of this generation in terms of the polarization of the positions of the adversaries and the complexity and variety of factors intensifying

it," wrote Israeli scholar Shimon Shamir during 1976. "A complete solution to the conflict within the foreseeable future is inconceivable."[1] Even a "comprehensive settlement," Shamir pointed out, "is not necessarily identical with a historical termination of the...dispute and a complete normalization of relations between the two societies—for those can be achieved only through a protracted process which a diplomatic act is more apt to initiate than to conclude."

Over the years, America's special relationship with Israel, the growing importance of Middle Eastern oil, and the superpower contest for regional influence between the

---

"Peace is necessary and peace is inevitable, but it must be sought with delicacy, and with understanding of the fragility of the existence of a state that has had to struggle for every square meter of its territory and for every year of its existence."

Henry A. Kissinger,
March 1977

United States and the Soviet Union, have all led to an escalating role for the United States throughout the Middle East and specifically in regard to the Arab-Israeli conflict itself.

By 1977 it was widely recognized, in fact, that the United States had become, in the wake of the Yom Kipper War and through the initiatives of Secretary of State Henry A. Kissinger, the key outside participant in the conflict. Indeed, it was often said that the Arab-Israeli war had been partially transformed into a political war for support in Washington. But the Soviet Union remained actively involved, and it was feared the Soviets could pick up the pieces if U.S. peace efforts collapsed and America found itself in the middle of another Arab-Israeli war.

---

1 Shimon Shamir in A. L. Udovitch (ed.), *The Middle East: Oil, Conflict & Hope* (Lexington: 1976), pp. 229, 216. Shamir also has pointed out that "While the Arab-Israeli conflict is a salient cause of [Middle East] instability, it is definitely not the only conflict in the area which has led to armed confrontation in recent years. Other disputes that have evolved into sporadic wars include those between Arab and Kurd in Iraq, Greek and Turk in Cyprus, and Arab and black African in Sudan. Within the Arab world proper, civil wars of varying types have been fought in Lebanon, Yemen, and Dhofar. The Yemen civil war led to a massive Egyptian intervention which at one point threatened to involve Saudi Arabia and escalate into a regional confrontation. A Syrian invasion into Jordan was halted by a combination of Jordanian armor and United States-Israeli pressure, and only the dispatch of British and Arab League troops induced Iraq to reconsider its claim to Kuwait." (p. 195).

Sinai Accords: Postponement of Basic Issues

When Secretary of State Henry A. Kissinger negotiated an agreement in January 1974 between Egypt and Israel, it was hailed as a major breakthrough in Middle East diplomacy. The terms of the agreement were not very significant, but the fact that an agreement had been reached broke the tragic pattern of the Arab-Israeli conflict. A generation of wars, hatred and deep distrust had kept Arabs and Israelis from even entering negotiations. Kissinger succeeded in breaching these entrenched attitudes, because he focused negotiations on immediate and often minor issues.

He adopted the same approach to secure a second Israeli-Egyptian agreement in September 1975. Compared with the basic issues of recognition of Israel, the future of the Palestinians, permanent boundaries, the status of Jerusalem, and peace guarantees, the issues settled in the second Sinai troop disengagement accord were trivial. The tactic of avoiding basic issues succeeded again in producing an accord, but the success might have the opposite effect of the 1974 agreement. If the earlier talks showed that agreement was possible, the 1975 negotiations seemed to strengthen the impression that agreement was possible only as long as fundamental issues were not faced.

In one respect, however, the September accord did produce a fundamental change. It introduced Americans, for the first time, into the midst of the Arab-Israeli conflict.

Congress approved, somewhat hesitantly, the stationing of American technicians between Israeli and Arab armies in the Sinai Desert to monitor military activities. The Israeli government made clear that its ratification of the agreement hinged on such congressional approval. But the combination of avoiding issues and injecting Americans left in doubt the direction of the path toward a permanent peace. As journalist Edward Sheehan recounted, "Perhaps Dr. Kissinger's greatest achievement is to have bought time, to have prevented war, to have erected the foundation for the pursuit of real peace. But the method he chose was simply too slow, and if clung to may imperil peace for the great future."[2]

Initial Success

The temptation to continue the step-by-step approach to peace arose partly from its initial success. The 1974 military disengagement agreements that Kissinger negotiated between Egypt and Israel in January and between Syria and Israel in May defused the explosive situation that existed after the 1973 war. But they accomplished much more. Although limited to military arrangement, the meeting between Israeli and Egyptian officers in the tent at Kilometer 101 produced the first agreement between Israel and an Arab state in 25 years—since the armistice of 1949, ending the Israeli war of independence. Scarcely a month before the Egyptian-Israeli accord was signed, representatives of the warring governments for the first time ever met face-to-face in search of a political settlement.

The experience of negotiating and agreeing, Professor Nadav Safran of Harvard observed, challenged deep-seated Arab and Israeli attitudes toward the very notion of settlement and mediation. It not only "cracked...mental molds

2 Edward R. F. Sheehan, *The Arabs, Israelis and Kissinger* (Reader's Digest Press: 1976), p. 204.

Key Features of 1975 Pact

In Geneva on Sept. 4, 1975, Israel and Egypt formally signed a new Middle East accord, intended to provide another stepping-stone toward a final peace settlement between the Arab nations and Israel.

The key feature of the pact required the Israeli army to withdraw from the Sinai mountain passes of Gidi and Mitla. These areas were included in a new United Nations demilitarized zone. However, Israel still controlled 87 per cent of the Sinai Peninsula, seized from Egypt in the 1967 war. *(1967 map, p. 75; 1974, 1975 maps p. 70)*

Egyptian forces were allowed to advance to the eastern edge of the old U.N. zone established by the January 1974 accord. Egypt recovered possession of the Abu Rudeis oil fields along the Gulf of Suez, also held by Israel since 1967.

Egypt pledged that it would not resort to the threat or use of force or continue a military blockade against Israel in the straits of Bab el Mandeb linking the Red Sea to the Indian Ocean. The Egyptians also agreed to a provision stating that "nonmilitary cargoes destined to or coming from Israel shall be permitted through the Suez Canal," opening the waterway to Israel for the first time since 1956.

Both belligerents pledged that "the conflict between them in the Middle East shall not be resolved by military force, but by peaceful means," and that they "are determined to reach a final and just peace settlement by means of negotiations."

The agreement also limited military forces that each side was permitted to station adjacent to the U.N. buffer area, and called for the mandate of the U.N. peacekeeping mission to be renewed annually rather than for shorter periods.

As for the new U.S. role, a few hundred American civilian personnel began monitoring Egyptian and Israeli activities in the Sinai. The agreement resulted in two warning stations operated separately by Egypt and Israel and three other stations in the Mitla and Gidi Passes manned by American technicians.

that had made for rigidity, fatalism, and despair," but gave rise to "vested interest, intellectual or political, in depicting the other side's intentions and behavior in a favorable light," he wrote. Previously each side had placed the worst possible interpretation on every word, act or alleged thought of the other side—interpretations that became self-fulfilling prophecies. But now, Safran noted:

...The world [was] treated to the novel spectacle of Golda Meir and Moshe Dayan [Israeli defense minister] urging in the Knesset [Parliament] that there was no need to assume the worst about Sadat [Anwar Sadat, president of Egypt], arguing that he seemed to be genuinely desirous of peace and eager to turn his attention to the tasks of reconstruction and developing his country, and stressing that he had scrupulously observed the terms of the agreement in letter and spirit.

On the other side, one could hear Sadat reassure his people that they need not worry about the possibility of Israel's sitting tight and trying to freeze the situation created by the disengagement agreement; and one could see him hopping from one Arab capital to another in an effort to allay this and other doubts and justify the "gamble" he took on the grounds that the United States and Israel had changed.[3]

While the initial Egyptian-Israeli accord established the minimum basis of trust necessary to pursue more basic issues, the second accord still did not come to grips with them. Whether or not an opportunity was thus lost, the nature of the issues placed limits on step-by-step diplomacy. The Carter administration had to face this reality in 1977.

The Egyptians seek, above all, the return of territory; the Israelis want credible pledges of peace. Territory can be yielded in stages, but peace is difficult to slice into segments. Indeed, efforts to dilute a promise of peace may arouse the very suspicions the promise is supposed to allay.

The Israelis fear that their territorial security will be whittled away in exchange for verbal assurances whose calculated vagueness invites a renewal of hostilities on disadvantageous ground. The Egyptians fear that once they have clearly and unequivocally promised peace, the Israelis will feel less incentive to return still-occupied Egyptian territory. The other Arabs fear that Israel, with U.S. connivance, will make a separate peace with Egypt and thus leave the Syrians, Jordanians and Palestinians powerless to regain territory occupied by Israel since the 1967 war and bring about some fulfillment of Palestinian aspirations for self-determination.

The Geneva Conference

The approach to peace most often considered as an alternative to step-by-step diplomacy is a resumption of the Geneva conference. This forum, unlike Kissinger's "shuttle diplomacy," would consider those ultimate issues that have produced four wars in one generation. But the difficulties of step-by-step negotiations would be replaced by other hazards.

In Geneva, each side would be tempted to use the conference as a platform for appeals to world opinion. Declared positions would necessarily be the maximum demands, and retreats from these maximum positions would be made difficult by public commitment to them. For the sake of support at home, the participating Arab governments might compete in militancy, or at least take pains to avoid accusations by their allies of being soft on Israel. The presence of the Soviet Union and the conference might encourage Arab intransigence, since the Soviet Union is known to support the maximum Arab demands.

Even before it resumed, the Geneva conference might break down over the question of the Palestinians. The Israelis refuse to negotiate with the Palestine Liberation Organization, but the Arab states at a conference in Rabat, Morocco, Oct. 28, 1974, agreed to recognize the PLO as "the sole legitimate representative of the Palestinian people on any liberated Palestinian territory." *(See Palestinian Question, p. 109)*

Kissinger had urged the Israelis to avoid the pitfalls of Geneva by reaching agreement in his shuttle negotiations, but Israelis' feelings about Geneva have always been mixed. Professor Stanley Hoffman of Harvard observed that there is a "bizarre coalition" of Israelis favoring the Geneva approach—those who see a chance for a general settlement which "the step-by-step method squanders by running into too many obstacles," and those "who believe that Geneva would show once and for all that one cannot really reach peace with the Arabs."[4]

Hoffman urged the Israeli government to avoid the problems of both step-by-step and Geneva negotiations by taking a bold initiative. He would have Israel announce a plan under which, in return for peace treaties with adequate supervision and guarantees, the country would return substantially all the territory it occupied in the Six-day War in 1967. Hoffman recognized the risk that "Whatever are presented as maximum concessions and minimum demands become treated as minimum concessions and maximum demands." But, he added, "there are greater risks in the present stance: especially the risk that it leads nowhere or only to unmatched concessions by Israel."

But even had the government of Yitzhak Rabin in Israel wanted to take the risks involved in such an initiative, it probably lacked the strength. Rabin headed a shaky coalition and took office in 1974 with no real political base of his own. He had promised no withdrawal from the West Bank without calling a general election. And the Likud opposition party in March 1975 obtained 600,000 signatures—44 per cent of all who voted in the previous Knesset (parliamentary) election on December 31, 1973—on a petition opposing withdrawal from the West Bank even in exchange for peace. When the Likud acceded to power in 1977, it became even more difficult for the Israeli government to willingly give up the West Bank which it considered not "occupied" but "liberated."

Arab leaders, on the other hand, are unanimous in demanding complete withdrawal, and any who deviate from this position would be branded as traitors by the others and face great political dangers. Even by negotiating separately in 1975, Sadat aroused fears among other Arabs that Egypt would sell out the Palestinians in return for restoration of its own territory. After a visit to Damascus by Kissinger, en route home from his successful negotiation of the September accord, the Syrian government issued a statement condemning that accord as a "setback to the march of the Arab struggle."

While a Syrian-Israeli agreement for further military disengagement in Syria's Golan Heights was expected to be the next objective in Kissinger's peace efforts, the situation there left little room for compromise. To secure an agreement might require stationing American technicans, as in the Sinai accord, on this highly volatile frontier. Kissinger never was successful in achieving a second agreement between Syria and Israel.

The American role in the Middle East agreement was not limited to supplying technicians. The United States has provided large sums of money and military equipment. Israel received about $2.2-billion in military and economic assistance for fiscal 1976 and about $1.8-billion in fiscal 1977. In addition, the United States has begun providing Egypt nearly $1-billion annually in economic assistance. Lesser amounts have gone to Jordan and Syria.

The aid raised other questions besides economic cost. The United States held up consideration of Israeli arms requests before the September 1975 accord as a means of pressuring the Israelis to reach agreement with Egypt on the Sinai issues. Critics of this pressure contended that Israel could not afford to give up territory unless it felt militarily strong, and that psychologically it would be readier to compromise if it felt secure in U.S. support.

Though the Carter administration repeatedly pledged during the first half of 1977 not to use this form of pressure, observers have wondered what methods Carter will use to persuade both sides to make the requisite concessions and take the required risks.

3 Nadav Safran, "Engagement in the Middle East," *Foreign Affairs*, October 1974, pp. 48-49.
4 Stanley Hoffman, "A New Policy for Israel," *Foreign Affairs*, April 1975, p. 412.

Attitudes Since 1973 War

Sadat could afford to take his "gamble" on negotiating with Israel—even though Arabs had hitherto regarded such behavior as traitorous—because of the changes wrought by the 1973 war. Although Israeli forces were poised on the brink of military victory when Soviet and American pressures imposed a cease-fire, the Arabs had fought well, erasing the humiliation they had suffered in the 1967 war. Equally bolstering to the Arabs' self-respect was the impact of their oil embargo and price increases on the industrial West, before which generations of Arabs had felt impotent. The oil weapon also freed the Arabs from a fear that always discourages talks with an enemy—the fear of negotiating from a position of weakness,[5] a problem the Israelis now face for economic, political, and public opinion reasons even though they remain militarily supreme.

The Israeli poisition has changed dramatically since the October war. The initiative, unity and determination of the Arab fighting forces punctured the sense of security Israelis enjoyed after the easy triumph of 1976. The initial shock on the battlefield gave way to more lasting doubts. If even the smashing victory of 1967 could not bring peace and safety any closer, what faith could be placed in military superiority? By 1977, confusion and anxiety in Israel helped the Likud Party triumph over the party which had led Israel for nearly three decades.

A popular Israeli stereotype portrays the Arabs as people who "only understand the language of force," but the application of overwhelming force in 1967 widened the conflict, provoking greater Arab unity and enlarging the roles played by more distant Arab states, notably Saudi Arabia. King Faisal, by embargoing oil, abandoned his policy of keeping oil out of Arab-Israeli politics, and the Israelis faced a profound shift in the balance of power. Even if the Israeli army "won" the next war, countries dependent on Arab oil would face terrible pressure to "undo)' the victory. Fears that the next war might involve substantial civilian casualties and even bring about the use of nuclear weapons have made both sides exceedingly cautious.

The positions have been reversed to some extent since the 1973 war. By 1977, the Israeli government sought to limit negotiations, calling for a period of stabilization after another limited troop withdrawal. It hoped Arab unity would dissipate and the West would gradually wean itself from dependence on Arab oil, enabling Israel to negotiate later from a position of strength.

Rabin, in a plea reminiscent of Arab speeches after the 1967 war, called on his people to dig in for "seven lean years."[6] And Begin has made it plain he expects Israel's bargaining position to improve with time. It is Sadat and the other Arab parties now, who seeks to keep up the momentum of peace talks. The 1973 war, Sadat said in a 1975 May Day speech, broke the "wall of frustration and self-doubt" that had surrounded the Arab, enabling him to "speak with anybody and everybody without fears."

Intangibles in the Conflict

The problem, Kissinger explained to the House International Relations Committee March 25, 1975, after the collapse of his Middle East negotiations, "was balancing tangible positions on the ground against less tangible assurances which have symbolic meanings and importance." The power of symbols in the Arab-Israeli conflict has, from the very beginning, baffled and thwarted diplomacy.

In the first phase of the conflict the issue was, simply, Israel's existence, an issue which scarcely lends itself to compromise. It was in this period, historian Bernard Lewis observed, that the question of direct negotiation became crucial for both sides. "By entering into direct negotiations with Israel, the Arab states would be giving her a vital token of recognition," he wrote. "By refusing to negotiate they were maintaining their refusal to recognize her existence. For Arabs and Israelis alike the question of direct negotiation thus acquired a symbolic significance which it has retained for both sides ever since."[7]

Direct negotiations had symbolic significance, because the issue was Israel's existence. But from the beginning the existence of Israel has itself been a symbol. *(See Chapter on Israel, p. 1)* For nearly 2,000 years, Jews throughout the world have intoned the Passover slogan, "Next year in Jerusalem." For a scattered, persecuted, stateless people, the ritual could be little more than a fantasy, and its credibility scarcely grew with the passing ages.

But at the end of the 19th century, the Jews of Eastern Europe, oppressed by poverty and brutal pogroms, began to act on this fantastic hope. Modern Zionism grew from such unpromising ground, and within two generations the movement had carved out a sovereign state of its own. "If you will it, it is no dream," the Zionist leader Theodor Herzl had asserted, and soon the ancient dream of "next year in Jerusalem" was a reality. Most Israelis today insist that their sovereignty over the entire city of Jerusalem, whose Arab sector they conquered in the 1967 war, is "not negotiable." Soon after taking over as Prime Minister, Menahem Begin urged the United States to move the American embassy from Tel Aviv to Jerusalem.

Jerusalem symbolizes the Jews' ancient dream and in Jerusalem is a symbol of their recent nightmare. Yad Va'shem is the memorial center dedicated to recalling the Nazi holocaust and its six million Jewish victims. This unfathomable horror, as much as the Zionist ideal, is a foundation of the state of Israel, and, "In the rituals of government and diplomacy, Yad Va'shem is given a role parallel, and at times equal in its solemnity, to the role of national symbols that extol military glory, sovereignty, and independence."[8]

The trauma was real and could not fail to affect perceptions of reality, including the reality of being surrounded by Arabs condemning Zionism even if no longer specifically demanding the end of Israel. "The effects of the Nazi holocaust upon the national psychology," wrote Israeli author Amos Elon, "reached a new peak in the weeks preceding the Six Day War of 1967. Israelis, including many young people, were seized by abysmal fears; many were certain that another holocaust was being prepared for them by the rulers of Egypt, whose bloodthirsty statements were resounding hourly on the radio. Many impartial foreign observers have testified to the breadth of such feelings at the time, and how genuine the fears seemed."[9]

Ambivalence Toward the Arabs

Fears aroused by the Arabs may be complicated by less obvious sentiments. In an article in *Commentary*, Hillel

---

5 For background, see "Middle East Reappraisal," *Editorial Research Reports*, 1973 Vol. II, pp. 947-966, and "Arab Oil Money," *E.R.R.*, 1974 Vol. I, pp. 363-382.
6 Interview in the Tel Aviv daily *Harretz*, December 3, 1974.

7 Bernard Lewis, "The Palestinians and the PLO: A Historical Approach," *Commentary*, January 1975, p. 43.
8 Amos Elon, *The Israelis, Founders and Sons* (1971), p. 200.
9 Ibid., p. 216.

Golan Heights Disengagement, May 1974

LEBANON

Qiryat Shemona

U.N. ZONE

Jordan River

El Quneitra

SYRIA

GOLAN HEIGHTS (Occupied by Israel)

ISRAEL

Tiberias

Sea of Galilee

Miles
0   10

JORDAN

Israeli immigrants from Arab countries, but the plight of the Palestinian refugees suggests parallels that are difficult to ignore. Especially since the 1967 occupation of the West Bank and Gaza Strip, with their refugee camps, Israelis have been forced to recognize their Palestinian enemies as suffering exiles craving a return to their homeland. The fact that the Palestinian diaspora resulted from the creation of Israel does not make it easier for Israelis to face the Palestinian problem.

Problems of Palestinian Identity

One way to deal with the problem is to refuse to recognize the Palestinians as a separate nationality. Golda Meir, as premier, said in 1969 that there was no such nationality as Palestinian. And Menahem Begin, Israel's prime minister, has returned to this view—one Prime Minister Rabin had made somewhat more flexible by discussing the existence of a "Palestinian problem." It might seem inconsistent for Israelis, so anxious themselves for recognition, to deny it to another displaced people. But even the Palestinians were slow to regard themselves as a distinct people; in the past, many sought their destiny as part of a larger Arab nation. Only the repeated failures of the neighboring Arab states to restore them to their homeland drove the inhabitants of the former British Mandate of Palestine[11] to pursue a separate nationalism of their own.

The fact that Palestinian nationalism is new in its present form and also fragile creates psychological problems for negotiation quite apart from such substantive issues as territory and sovereignty. Throughout history, hatred of a common enemy has helped forge new national identities, and hostility to Israel is the most powerful force binding together the scattered Palestinians.[12] The expression of that hostility thus serves the creation of the Palestinian identity, but it scarcely encourages the Israelis to recognize it. The fragility of the Palestinian movement also discourages moderate Palestinians from disavowing the statements and actions of their own most militant factions.

Israeli unwillingness to recognize rival claims to the ancient land of Israel did not begin with Palestinian hostility. Amos Elon quoted the late David Ben Gurion, a founding father, as saying in 1917 that in the "historical and moral sense" Palestine was a country "without inhabitants." Elon went on to describe the attitudes of early Zionist leaders toward the native Arab population:

> Like most visionaries their eyes were hung with monumental blinkers.... They did not imagine that the Arabs who had been living there for centuries could possibly object to becoming a minority—a fully respected minority that would live in more comfort and wealth under the most liberal of regimes—through the advent of massive Jewish immigration from abroad.

"Later," Elon added, "a defensive mechanism of great complexity would be employed to protect this naive innocence from being contradicted by facts."[13]

Halkin, an American-born Israeli, recounted how scenes in occupied Arab communities prompted reflections on the way in which the Jew, while assimilating so much of European culture, preserved his identification with a lost land. The Jew clung to symbols which in biblical times were part of every day life. Halkin wrote:

> The flat, round bread he ate with every meal, dipping it in oil or vinegar when there was nothing else, is now eaten once a year in the form of the Passover matzah.
>
> The headcovering worn out-of-doors to protect himself from the blazing Palestinian sun, blackens, shrivels into round little *yarmulke* of the synagogue; his blowing, toga-like robe becomes the fringed *tallit* or prayershawl.
>
> Once or twice each year, on Rosh Hashanah or Tu B'shvat, he spends extravagant sums to buy exotic fruits of the Holy Land, dates and olives, almonds and figs....

But when the Europeanized Jew returns, after millenia, he finds "the Arab, the usurper, living there in his place, eating round, flat bread with every meal, covering his head with an elegant *kaffia* against the sun, wearing a long flat robe that ripples when he walks, eating figs and olives, almonds and dates for his daily fare, and living as a perfect matter of course the agricultural rhythms, the seedtimes and the harvests, of the Bible, Talmud, and prayerbooks. An upside-down world!"[10]

Whatever ambivalence such biblical scenes may stir in the Westernized Jew, another aspect of the Arab condition resonates with Jewish experience. The Arabs as a colonized people, like the Jews as a persecuted minority, have experienced powerlessness and humiliation at the hands of European civilization. Empathy with the Arabs finds only limited expression in Israel, and is not generally shared by

---

10 Hillel Halkin, "Driving toward Jerusalem: A Sentimental Journey through the West Bank," *Commentary*, January 1975, p. 51.

11 The word "Palestine" is of Roman origin, referring to the biblical land of the Philistines. The name fell into disuse for centuries and was revived by the British as an official title for an area which was mandated to their control by the League of Nations in 1920 after the breakup of the Turkish Ottoman Empire in World War I.

12 Spokesmen for both Israeli and Arab governments tend to use the figure three million when called on to state the size of the Palestinian population in the world. The United Nations Relief and Works Agency offers a lower estimate, 2.7 million, of whom 1.5 million are listed as refugees. About half of all Palestinians live under Israeli control.

13 Amos Elon, *op. cit.*, pp. 156-157.

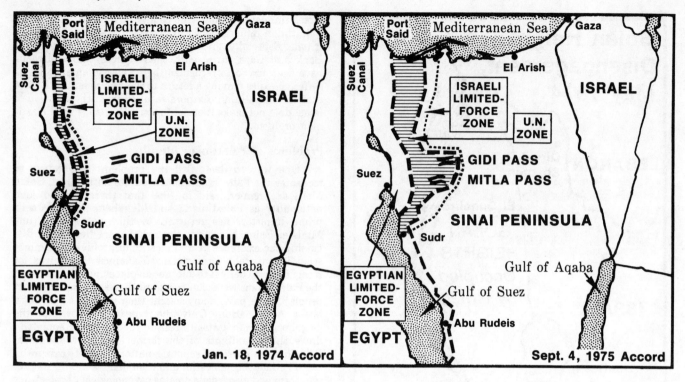

Nationalism Versus Arab and Jewish Traditions

If Zionist attention to Arab nationalism was slow in developing, it was at least in part because that nationalism was still nascent. The kind of identification with the nation-state which Westerners take for granted did not characterize the Arab world, where loyalties were divided among Islam, the Ottoman state and local or kinship groupings. These divided loyalties left the Arabs vulnerable to domination by homogeneous European nations. When the Arabs finally reacted with a nationalism of their own, it had to contend with the deepest strata of their own culture.

Bernard Lewis compared this inner conflict with the great struggle at the birth of Islam, when monotheism fought the idols of pagan Arabia. Now, however, the struggle is "against a new set of idols called states, races and nations; this time it is the idols that seem to be victorious. The introduction of nationalism, of collective self-worship, is the best founded and least mentioned of the many grievances of the Middle East against the West."[14]

The Jews, like the Arabs, were late in embracing the kind of nationalism that evolved in Europe. For centuries they sustained a double identity in which they were Germans or Russians or Americans or Moroccans as well as Jews, and prided themselves on their universality as well as their uniqueness. This divided identity left them, like the Arabs, vulnerable to European nationalism, especially when it went berserk in Nazi Germany. There is terrible irony in the fact that Jews and Arabs now kill each other in the name of the very nationalism which so long victimized both peoples.

The struggle of Israelis and Arabs to make nationalism dominant over traditional identities creates deep conflicts within each culture. The ambivalence is painful and often consciously denied; it is easier to confront an external enemy, projecting inner conflict onto hostile neighbors only too ready to reciprocate. Lebanese scholar Fouad Ajami wrote in *Foreign Policy* magazine: "Arabs and Israelis have

yet to accept themselves, let alone one another. People cannot be at peace with others unless they are at peace with themselves."[15]

Both Jews and Arabs have felt their historical uniqueness primarily through their religious achievements. As their religions now yield to nationalism, they cling all the more fervently to the traditional spirit of historical mission. The burden of the past thus assumes extraordinary power in the politics of the Middle East. Each people, Ajami wrote, is "haunted by its own ghosts, and tormented by deeply felt historical grievances that seem to justify its own violence and insensitivity."

Overcoming Enmity

Arab-Israeli wars are not really between governments but between peoples, and the enmity between these peoples spring from depths that diplomacy can scarcely touch. The Israelis, newcomers seeking acceptance in the region but finding themselves barricaded in a national ghetto, openly plead for non-governmental relations with the Arabs. Communication, commerce, travel and cooperation, they feel, can be both cause and effect of peace. They will feel safe, it has been said, when they can go shopping in Cairo.[16]

President Sadat, however, until 1977 always said the attainment of such calm relations must be the work of the "next generation." Beginning in 1977, Sadat and other Arab leaders began talking about the possibilities of "normalizing" relations with Israel within five years after the signing of a peace agreement.

Certain steps toward that goal of normalization can be taken even prior to a peace settlement and can help foster the minimum trust necessary to conduct negotiations. Sadat's decision to reopen the Suez Canal June 5, 1975, was in part a gesture to promote a climate of good faith for

14 Bernard Lewis, The Middle East and the West (1968), p. 70.
15 Fouad Ajami, "Middle East Ghosts," Foreign Policy, spring 1974, p. 94.
16 Remark attributed to Golda Meir.

further talks.[17] The likelihood of Egypt's renewing hostilities is reduced by the resumption of traffic through the canal and the return of people to the rebuilt cities along the banks that were virtually ghost towns for years after the 1967 war. In fact, the Israelis proposed that the canal be reopened and the towns revived as evidence of Egypt's good faith in its promises of peace.

During 1976, representatives of the PLO and of a small, but respected, Zionist party from Israel began holding talks in Paris. This was an unprecedented step that could lead to ideological changes on both sides, though there was little public discussion of these talks. And during 1977 a few Israelis openly traveled in Morocco and Tunisia, suggesting the beginnings of direct contact, however minimal.

Economic Pressures in Egypt and Israel

Egyptians, of course, have benefited economically from the renewed traffic through the canal. An Egyptian shift in priorities from confrontation to economic development has added credibility to its promises of peace and built vested interests in Egypt opposed to a new war.

Sadat already faces compelling economic pressures. In the euphoria after the 1973 war, his government raised hopes of more consumer goods, less inflation and an influx of foreign investments. But the continuing, costly confrontation with Israel, and Soviet refusals to postpone debt repayments, have thwarted these hopes. Inflation in Egypt was running at 30 per cent in mid-1975, and economic conditions provoked a series of violent protests, the most serious in January 1977. *(See chapter on Egypt, p. 24)*

Continuing domestic unrest has threatened Sadat's progress toward greater political freedom in Egypt. Reduced tension with Israel, on the other hand, would free desperately needed resources for economic development. Former Planning Minister Ismail Sabry Abdullah said that Egypt had spent $25-billion for military purposes since 1967 and had suffered an equal loss in war-related costs. Meanwhile, he said, Egypt received less than $900-million in financial aid from other Arab states.[18]

Israel, too, pays a heavy price for the lack of movement toward peace. The 1973 war wiped out the economic boom fostered by the 1967 triumph. Rabin spoke in 1974 of "the necessity to shock the public in order to return our consumer society to sanity." The inflation rate in Israel had reached about 35 per cent by 1977 and the Israeli pound had been devalued more than 100 per cent between the Yom Kippur War and 1977. Rabin stated about a year after that war that the "Israelis are ready to live with a 20 per cent lower living standard so as not to endanger themselves and their future." Yet he spoke of Israel's economic situation approaching a state in which it would be necessary to restrict growth in military strength.[19]

In both Israel and the Arab countries, a reduction in tension would permit a shift to economic priorities. The resulting internal development might, in turn, divert energies from belligerence. However, the absence of external pressures could also release internal conflicts in both camps.

The Priority of Fear

But a shift to internal priorities seems unlikely as long as Arabs and Israelis do not feel safe from each other. Egypt and Syria will not relax their military efforts as long as an enemy army occupies part of their territory; Israel will not risk a repeat of the 1973 war, which caught it off guard. Military spending is, in fact, escalating in most countries in the area.

Israeli President Katzir, noting that Israel spends more than one-third of its gross national product for security, told western journalists: "It would be a good thing if we could divert these resources to civilian purposes. However, a dead man cannot engage in agriculture or industry. Therefore we must care first and foremost for our existence, which is why we invest in security." In the same speech, Katzir said that "Israel has a nuclear potential." Asked if this is not a subject of concern, Katzir replied, "Why should this subject trouble us—let the world worry."[20] He later clarified his remark, indicating that he meant the general potential of scientists and the scientific-technological experience exist in Israel which could be put into effect if Israel wanted. Since Katzir spoke, Israel's nuclear potential has been estimated by the CIA and various reports at about a dozen small weapons.[21]

The Arabs, too, have a weapon the entire world has reason to fear—oil. If the Sinai agreement is not followed by further steps toward peace, the Arabs may soon feel that they have no alternative but another war. If another war occurs, another oil embargo can be expected. Saudi Arabia is the key oil country. And even though President Carter has expressed great appreciation for Saudi moderation and friendship the Saudis will be in a difficult position if the United States cannot deliver Israeli concessions. Crown Prince Fahd was quoted in May 1977 as saying, "We are capable of increasing oil production...but we have demands in return for that. First and foremost, we want the United States to throw all its weight into the process of reaching a just settlement of the Mid-East crisis based on Israeli withdrawal from all Arab territories occupied in the 1967 war, and the return of the Palestinians' rights to their homeland and a state of their own."[22]

On the Road to Geneva

Resumption of the Geneva conference, recessed in December 1973, had come to symbolize progress toward easing tensions and possibly a settlement in the Middle East. But there were questions in the summer of 1977 concerning when the conference could be reconvened, who would be there and what would be discussed.

Before the election of Menahem Begin in Israel the widespread assumption had been that it was necessary to work out, through quiet, traditional, diplomacy, general principles and agreements that could be discussed and finalized at Geneva. The problem of Palestinian representation, it was thought, would also have to be dealt with beforehand.

But soon after taking office, Begin called for the resumption of Geneva as early as October 1977, without conditions or prior agreements. President Sadat of Egypt showed some interest, but others, including the United States, were more cautious. There was considerable fear

---

17 *The Washington Post* commented editorially June 5 that Egypt had made a "pledge" to reopen the canal as "part of the first Egyptian-Israeli disengagement accord." However, Israeli Premier Rabin was reported in the Tel Aviv daily *Ha'aretz* on Dec. 3, 1974, as indicating that "there was no commitment that it [the canal] would be reopened before another move takes place."

18 Quoted in *The New York Times*, April 9, 1975.

19 Rabin was quoted in *Ha'aretz*, Dec. 3, 1974, and the *Jerusalem Post*, Dec. 4; Ya'acobi in *The Wall Street Journal*, April 25, 1975.

20 Quoted in the Tel Aviv daily *Ma'ariv*, Dec. 2, 1974.

21 *Time Magazine*, April 2, 1976.

22 *Saudi Gazette*, May 23, 1977.

that an aborted Geneva conference, or even one that broke-down after serious negotiations had begun, might propel the parties toward renewed warfare. There was also fear that a conference that was not properly prepared could degenerate into a forum for propaganda points as was often the case at the Security Council. In March 1977, for instance, Israeli representative at the U.N., Chaim Herzog, referring to an attempt by Egypt to debate the Middle East situation in the Security Council, charged that Egypt's plan "completely misfired and on the eve of Passover Israel administered another Passover plaque against the Egyptians."[23] This was precisely the kind of exchange that might add fuel to the

Middle East fire at Geneva, rather than help extinguish the memories of past wars.

C. L. Sulzberger, writing in *The New York Times* just before Begin's visit to Washington, offered his own prediction—one which only proved the future was very uncertain. "My own guess," Sulzberger wrote, "is that there will...be an Arab-Israeli negotiation in Geneva...beginning with wholly irreconcilable negotiating positions, and that the United States, while denying it, will put the heat on Israel by slowing arms deliveries. From there on everything is conjecture."[24]

23 Jewish Telegraphic Agency Daily Bulletin, 31 March 1977.

24 C. L. Sulzberger, "War, Peace and Generals," *The New York Times*, 13 July 1977.

ISRAEL AND THE ARABS: DECADES OF HOSTILITY

Since the establishment of Israel little more than a quarter of a century ago, Israel and its Arab neighbors have engaged in more or less continuing hostilities, punctuated by four major wars.

Having failed in 1947 to prevent the creation of a Jewish state in Palestine, the Arabs at first fought to destroy Israel and then to recover territories occupied by Israel in subsequent battles. Arab enmity has stiffened Israel's resolve to attain "secure and defensible borders," and this in turn has fostered Arab fears of Israeli "expansionism." The problem of the Palestinian Arabs, many of whom were made refugees during the 1948 and 1967 wars, has remained a continuing obstacle to peace.

The history of the Arab-Israeli conflict has been marked by a gradual movement toward Arab solidarity. Arab participants in the first war of 1948-49 distrusted each other almost as much as they distrusted the hated Zionists. By the time of the fourth war, in 1973, Arab unity had advanced to such a point that Saudi Arabia, the Persian Gulf states and several North African countries could join with Egypt, Syria and other "front line" Arab states (states bordering on Israel) in a common effort to recapture the territories lost to Israel in the 1967 war.

For its part, Israel has suffered growing isolation in the world community, most acutely since the 1973 war. Its refusal to give up Arab territories occupied in 1967 has cost the country many friends in the Third World, while the October 1973 Arab oil boycott forced Japan and most Western European governments to adopt a more pro-Arab stance. The Arabs' decision to use oil as a political weapon has also had an effect on U.S. policy in the Middle East. And by the mid-1970s, Palestinian nationalism had achieved considerably more sympathetic attention throughout the world.

The 1973 war gave the lie to the assumption that—despite U.S. backing for Israel and Soviet support for the Arabs—the two superpowers could avoid a confrontation during an Arab-Israeli war. The specter of superpower conflict arose as first the Soviet Union and then the United States felt compelled to resupply the belligerents with arms on a massive scale and later when the United States ordered its forces on worldwide alert following the threat of a possible Soviet troop movement into the Middle East.

By the time the first Arab-Israeli peace conference opened in Geneva in December 1973, it appeared that a permanent resolution of the Arab-Israeli conflict had become a real possibility at last. The 1973 war demonstrated the dangers of continued stalemate, while the Arabs' use of their oil weapon assured the interest of outside powers in promoting a peace settlement. But awesome issues remained to be resolved. While many Arab leaders might no longer demand the liquidation of the Jewish state, Arab insistence that Israel withdraw from all territories occupied in 1967 collides with Israeli insistence on secure and defensible boundaries. Even should an agreement be reached regarding the West Bank, Sinai and the Golan Heights territories, the emotional issue of Jerusalem still might prevent a full settlement of the Middle East conflict.

The fate of the Palestinians also remains in dispute. On November 24, 1976, the United Nations General Assembly in a 90 to 16 vote (30 abstentions) called for the creation of a Palestinian state—something Israel has refused to contemplate—and for Israeli withdrawal from all occupied territories by June 1, 1977. Responding, Israeli U.N. Ambassador Chaim Herzog proclaimed, "The General Assembly has been hijacked by a group of Arab extremists."

As Carter assumed the presidency in January 1977, a world-wide consensus seemed to be emerging that the coming few years might be the best opportunity—and quite possibly the last—for peaceful resolution of the Middle East conflict. Secretary of State Henry A. Kissinger left office proclaiming the Middle East to be at a "moment of unprecedented opportunity." U.N. Secretary-General Kurt Waldheim and Kissinger's successor, Cyrus Vance, have also expressed this view. Both toured Middle East capitals early in 1977 for first-hand assessments and during Vance's first press conference he emphasized that "it is critically important that progress be made this year" in the Middle East.

Still, the gulf separating Israel from the Arabs remains huge. Terence Smith returned from four years as Jerusalem correspondent for *The New York Times* noting that "I came home from those years deeply skeptical about the prospects for an early settlement in the Middle East. It is not that the political problem is beyond solution. That is basically a question of sovereignty versus security that two dispassionate lawyers could resolve.... But the *human* obstacles—the deep-seated mistrust on both sides, the fear of annihilation, the wounded national honor—these are the real stumbling blocks. They are the elements that have prevented a solution in the past and will continue to make one difficult to achieve in the future."[1]

War of Independence (1948-49)

The first Arab-Israeli war stemmed from Arab refusal to accept a United Nations plan to partition Palestine into separate Arab and Jewish states. The U.N. had been drawn into the Palestine dispute following the failure of Great Britain, which held a League of Nations mandate over Palestine, to hammer out a Palestine solution that would be acceptable to both Arabs and Jews.

In May 1947 the U.N. General Assembly established an 11-nation special commission to study the Palestine problem. A majority of the commission recommended the division of Palestine into a Jewish state, an Arab state and an internationalized Jerusalem. The Zionists endorsed the majority approach. A commission minority recommended a federal state with autonomous Arab and Jewish provinces and limits on Jewish immigration. Although the Arabs initially insisted on absolute Arab sovereignty over Palestine, they ultimately endorsed the minority plan.

---

1 See "Israel Journal: 1972-1976. Reflections on a Troubled People," *Saturday Review*, 5 February 1977, p. 8.

For a time it appeared that neither plan could command the necessary two-thirds support in the United Nations, but on Nov. 29, 1947, the General Assembly approved the partition plan by a narrow margin of 33 to 13 with 10 abstentions.

The Palestinian Arabs thereupon resorted to arms to prevent partition. In the civil war that followed, however, Palestinian Jews were able to win control over most of the territory allocated to them by the partition plan. Britain ended its mandate over Palestine on May 14, 1948, and the same day the Zionists proclaimed the establishment of the state of Israel. One day later, the armies of five Arab countries—Egypt, Transjordan, Iraq, Syria and Lebanon—invaded Palestine, and the first Arab-Israeli war began.

"The first Arab-Israeli war produced a shock from which the Arabs never truly have recovered—their defeat at the hands of the numerically inferior Jews," Harry B. Ellis wrote in a 1970 analysis prepared for the American Enterprise Institute, an independent research organization.

Although the Arab states involved had 40 times as many people as the infant Jewish state, the Arabs, torn by dynastic rivalries, never placed their armies under effective joint command and were unable to agree on common objectives. Meanwhile the outnumbered Jews profited from their sense of cohesion and their paramilitary experience fighting the Palestinian Arabs during the mandate period. Their war effort was augmented by the influx of men and equipment from abroad as the war went on.

The shooting war stopped on Jan. 7, 1949, and by Feb. 24 Egypt had separately signed an armistice agreement with Israel, followed by Lebanon in March, Jordan in April and Syria in July. Iraq refused to sign an armistice agreement and simply withdrew from Palestine.

When the fighting stopped, Israel held over 30 per cent more territory than had been assigned to the Jewish state under the U.N. partition plan. The Arab state envisaged by the U.N. plan never materialized; its territory was divided among Israel, Transjordan and Egypt. Israel gained about 2,500 square miles. Transjordan, which annexed the West Bank of the Jordan River and transformed itself into the state of Jordan, gained 2,200 square miles. Egypt took the Gaza Strip, about 135 square miles, which it held in the status of Egyptian-controlled territory. Jerusalem was divided between Israel and Jordan. U.N. mixed armistice commissions were established to police the frontiers, and several demilitarized zones were established between Israel and Egypt, Jordan and Syria.

The U.N. also created the United Nations Relief and Works Agency for Palestinian Refugees in the Near East (UNRWA) to assist Palestinian Arabs who had fled or been driven from their homes. It is thought that more than 700,-000 Palestinian Arabs—the exact number is disputed—who had lived in the area that came under Israeli control became refugees—38 per cent in the West Bank area, 26 per cent in the Gaza Strip, 14 per cent in Lebanon and 10 per cent both in Syria and Transjordan.

Although the refugee problem was widely regarded as a major obstacle to the conclusion of peace agreements between Israel and its Arab neighbors, Middle East specialist Nadav Safran, in *From War to War* (1969), held that other factors were more important in the refusal of Arab leaders to sign peace agreements. He quoted an interview with Azzam Pasha, first secretary general of the Arab League:

"We have a secret weapon which we can use better than guns and machine guns, and this is time. As long as we do not make peace with the Zionists, the war is not over; and as long as the war is not over there is neither victor nor vanquished. As soon as we recognize the existence of the state of Israel, we admit by this act that we are vanquished."

In addition to this psychological reluctance to accept defeat, Safran cited other considerations for the refusal of Arab leaders to make peace, including their fear of an outraged public opinion that had been encouraged to expect an easy victory and their apprehension that "peace would

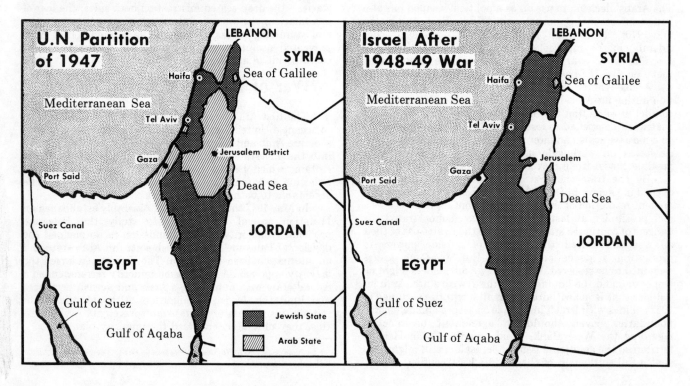

U.N. Partition of 1947

LEBANON
SYRIA
Haifa
Sea of Galilee
Mediterranean Sea
Tel Aviv
Gaza
Jerusalem District
Port Said
Dead Sea
Suez Canal
JORDAN
EGYPT
Gulf of Suez
Gulf of Aqaba

Jewish State
Arab State

Israel After 1948-49 War

LEBANON
SYRIA
Haifa
Sea of Galilee
Mediterranean Sea
Tel Aviv
Gaza
Jerusalem
Port Said
Dead Sea
Suez Canal
JORDAN
EGYPT
Gulf of Suez
Gulf of Aqaba

legitimize Israel's entry into the Middle Eastern political arena and allow it to maneuver freely among the rival Arab states and with interested outside powers in order to promote its suspected expansionist designs."

Although the refusal to sign peace agreements left open the possibility of future offensive action against Israel, Safran concluded that "there was not, at this stage, an active Arab commitment to a resumption of hostilities or to the total destruction of Israel."

But the absence of formal peace agreements resulted in continuing tension and bitterness, heightened by the Arab boycott of Israel and the barring of Israeli shipping from the Suez Canal and the Gulf of Aqaba, Israel's sea lane to Africa and Asia.

In May 1950 the United States, Britain and France issued the Tripartite Declaration in which they pledged themselves to limit arms shipments to the area and to oppose any attempt to alter the existing armistice lines by force. Yet the armistice lines were repeatedly violated by Arab commando raids into Israeli territory and retaliatory raids into Arab territory by Israel. The level of hostilities escalated, culminating in the second Arab-Israeli war of 1956.

Suez War (1956)

The immediate cause of the second Arab-Israeli war was Egyptian President Gamal Abdel Nasser's nationalization of the Suez Canal on July 26, 1956. Several events in the preceding year had emboldened Nasser to take this step: The British had withdrawn their 80,000 troops from the Suez Canal zone; the Soviet Union had agreed to supply large quantities of arms to Egypt on advantageous terms; and the United States, piqued by Nasser's opposition to the Baghdad Pact, had canceled its offer to help Egypt build the Aswan Dam.

Diplomatic efforts to settle the Suez Canal crisis foundered, and Britain and France—chief shareholders in the Suez Canal Company—determined to recapture the canal by force. They secretly enlisted Israel's participation in this effort.

On Oct. 26, 1956, the Israeli army invaded the Sinai Peninsula and in seven days had reached the Suez Canal. Egyptian troops were driven from the Gaza Strip and the Sinai. On Oct. 31, French and British air forces began bombing Egypt prior to invading the country. The United Nations speedily achieved a cease-fire and demanded the withdrawal of invading troops from Egypt.

Responding to intense international pressure from the United States and the Soviet Union, Britain and France withdrew the last of their forces from Egypt in December 1956. The last Israeli units were not withdrawn from the Sinai until March 1957, and then only under the threat that the United States would impose economic sanctions upon Israel if it failed to do so.

Meanwhile, a United Nations Emergency Force (UNEF) took up positions on Egyptian territory at the southern tip of the Sinai Peninsula and along the Gaza frontier. Israel gained free passage through the Gulf of Aqaba, and it warned that the removal of U.N. troops from Sharm el-Sheikh at the entrance of the gulf would constitute an act of war.

For a time Egypt also permitted Israeli cargoes on non-Israeli ships to transit the Suez Canal, but this concession was halted in 1959.

The 1956 war did not solve the Arab-Israeli territorial conflict and only temporarily altered the military balance in

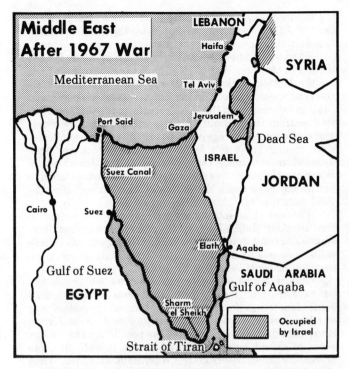

the area, since Russia immediately began replacing the military equipment lost by Egypt during the war. It did increase Arab hostility toward Israel, and as Nasser began to promote the concept of integral Arab unity this attitude intensified, since Israel was viewed as a physical barrier that split the Arab world.

The Arab-Israeli conflict became a "clash of destinies," in Safran's view, following the short-lived merger of Egypt and Syria in the United Arab Republic in February 1958. "Prior to the union with Syria, Egypt, along with the rest of the Arab League, stood for the application of the United Nations resolutions on partition and the return of the refugees, which admitted the right of Israel to exist; after the union, this line was abandoned for one that clearly intimated the liquidation of Israel under a variety of formulae, such as 'the restoration of the Arab rights in Palestine' or the 'liquidation of the Zionist aggression in Palestine,'" he wrote.

Although it had suffered a clear diplomatic defeat, Israel following the 1956 war "enjoyed ten years undisturbed by the border belligerency of its major and most dangerous foe, Egypt," Ben Halpern wrote in *The Idea of the Jewish State* (1969).

"The Negev, along the Sinai-Gaza frontier, was quiet, busy with civilian development. The continuing Suez blockade could be ignored because traffic flowed freely to and from Aqaba. While the UN Emergency Force was sketchily represented by troops in Gaza and Sharm el-Sheikh and by a light patrol along the Negev-Sinai line, no one—certainly not Israel—considered the Emergency Force a major factor in securing the border peace. Its chief function was to give the Egyptians an excuse for not reopening hostilities at a time when they felt unprepared," he wrote.

Six-Day War (1967)

The Six-Day War of June 5-10, 1967, erupted after months of mounting tension. But its immediate cause was the failure of diplomatic efforts to lift the blockade of the Gulf of Aqaba declared on May 23 by Egyptian President Nasser.

The gulf had been opened to Israeli shipping by Israel's victory in the Suez War of 1956, and it had been kept open by the United Nations Emergency Force stationed since then at the gulf's mouth on the Red Sea. Nasser's request on May 18 for removal of the U.N. force, from the Gaza Strip as well as from the gulf outpost, was accompanied by movement of substantial Egyptian forces into the Sinai Peninsula, raising Tel Aviv's fears of the long-threatened Arab attempt to terminate the existence of the Jewish state. When the United States and other Western nations failed to act promptly to break the blockade, as promised after the 1956 war, the third Arab-Israeli war was on.

The war substantially altered the power structure in the turbulent Middle East. Israeli planes in their first "preemptive" attack destroyed the bulk of the Egyptian air force while it was still on the ground. Israel's lightning move through the Sinai Peninsula broke the Egyptian blockade of the Gulf of Aqaba and once again put its soldiers on the banks of the Suez Canal. In the east, Israel's forces ousted Jordanian troops from the old section of Jerusalem and seized control of all Jordanian territory west of the Jordan River. The last move foreclosed the possibility that in the event of all-out war Israel could be cut in half; at some points, Jordan' territory on the West Bank extended to less than 9 miles from the sea. Finally, Israel captured the Golan Heights, the heavily fortified borderland hills from which Syria had for two decades harassed Israel's northeastern settlements.

Israel's smashing victory not only stunned the Arabs and their Soviet backers; it left Israel in a position of strength. In contrast to 1956, when Israeli forces were withdrawn in response to strong Washington-Moscow pressure, Tel Aviv at once announced that Israel would remain in the occupied territories until decisive progress toward a permanent settlement had been made.

Meanwhile, the Soviet Union broke off diplomatic relations with Israel and began to rearm the Egyptians. Nasser, charging that U.S. aircraft had contributed to Egypt's defeat, severed diplomatic relations with Washington. Six other Arab states followed suit. With the Soviet Union providing military assistance to the Arab governments, the United States moved in to fill the vacuum created by the French government's 1967 decision to end its role as chief supplier of armaments to Israel.

On Nov. 22, 1967, the U.N. Security Council unanimously approved a resolution (Security Council Resolution 242) aimed at bringing peace to the Middle East. The document called for withdrawal of Israeli forces from occupied Arab areas; an end to the state of belligerency between the Arab nations and Israel; acknowledgement of and respect for the sovereignty, territorial integrity and political independence of every nation in the area; the establishment of "secure and recognized boundaries"; a guarantee of freedom of navigation through international waterways in the area; and a just settlement of the refugee problem. *(Text, box, next column)*

Although U.N. efforts to end the Arab-Israeli conflict once again foundered, this resolution has remained the basis for all subsequent peace initiatives. Prior to the 1967 war, the Arabs had insisted that Israel return all lands in excess of the territory assigned to the Jewish state by the 1947 U.N. partition plan. Since the 1967 war, however, the Arabs have gradually modified their demands, and now insist only

U.N. Security Council Resolution 242, Nov. 22, 1967

The Security Council
Expressing its continuing concern with the grave situation in the Middle East,

Emphasizing the inadmissibility of the acquisition of territory by war and the need to work for a just and lasting peace in which every State in the area can live in security,

Emphasizing further that all Member States in their acceptance of the Charter of the United Nations have undertaken a commitment to act in accordance with Article 2 of the Charter,

1. *Affirms* that the fulfillment of Charter principles requires the establishment of a just and lasting peace in the Middle East which should include the application of both the following principles:
    (i) Withdrawal of Israeli armed forces from territories occupied in the recent conflict;
    (ii) Termination of all claims or states of belligerency and respect for and acknowledgement of the sovereignty, territorial integrity and political independence of every State in the area and their right to live in peace within secure and recognized boundaries free from threats or acts of force;

2. *Affirms further the necessity*
    (a) For guaranteeing freedom of navigation through international waterways in the area;
    (b) For achieving a just settlement of the refugee problem;
    (c) For guaranteeing the territorial inviolability and political independence of every State in the area, through measures including the establishment of demilitarized zones;

3. *Requests* the Secretary-General to designate a Special Representative to proceed to the Middle East to establish and maintain contacts with the States concerned in order to promote agreement and assist efforts to achieve a peaceful and accepted settlement in accordance with the provisions and principles in this resolution;

4. *Requests* the Secretary-General to report to the Security Council on the progress of the efforts of the Special Representative as soon as possible.

that Israel adhere to the principles of the 1967 Security Council resolution which they say calls on Israel to return to its pre-1967 borders. This resolution, referred to in 1973 resolution #338, became the basis for the Geneva Conference.

"War of Attrition"

Fighting was renewed in 1969 along the Suez Canal front after Egypt repudiated the cease-fire of 1967. This period, known as the "War of Attrition," was designed to wear the Israelis down and bring about territorial withdrawals. Although often violated, the cease-fire technically continued on the other fronts.

In June 1970, U.S. Secretary of State William P. Rogers proposed a cease-fire and a resumption of U.N. mediation efforts aimed at implementing the 1967 Security Council resolution to achieve a settlement based on withdrawal of Israeli forces from occupied territory and Arab recognition of Israel's right to exist within secure borders.

Egypt and Jordan, then Israel, agreed to invoke a cease-fire for 90 days, beginning Aug. 8, 1970, in their conditional acceptance of the Rogers peace formula. Once the agreement to seek a peace settlement was announced, however, public protest arose in many parts of the Middle East. Arab guerrilla groups and the government of Syria and Iraq rejected the peace initiative and denounced Nasser for accepting it. In Israel, six members of the hawkish Gahal minority party quit the cabinet of Premier Golda Meir when she announced the government's acceptance of the Rogers peace plan. Palestinian commandos dramatized their opposition to peace negotiations through a spectacular series of commercial aircraft hijackings.

Hopes for a peace settlement were dashed Sept. 7, 1970, with Israel's announcement that it was withdrawing from the peace talks. Its ambassador had met only once with the United Nations mediator, Gunnar V. Jarring. Israel's decision followed its repeated charges (only later validated) that Soviet missile batteries had been emplaced in the Suez Canal cease-fire zone in direct violation of the standstill agreement. (It was these missiles which made possible the Egyptian army's crossing of the canal in October 1973.)

In late 1971 and early 1972, the United States put forward a new proposal for indirect, American-mediated talks between Israel and Egypt on an interim peace settlement that included a troop pullback and reopening of the Suez Canal, but negotiations made little headway. Meanwhile, in what was an apparent effort to disassociate his regime from Soviet military support, Egypt's President Anwar Sadat ordered the departure of 20,000 Soviet military advisers from his country.

Israel's determination to retain Arab territory occupied in 1967 until a final peace agreement would be agreed to by the Arabs gradually began to weaken the country's position in the international community and finally led to the fourth Arab-Israeli war in 1973.

Yom Kippur War (1973)

Unprecedented Arab solidarity and the emergence of oil as an Arab political weapon marked the fourth Arab-Israeli war, which began Oct. 6, 1973. The war broke out on Yom Kippur, the holiest day of the Jewish calendar, when Egyptian and Syrian troops in a surprise move broke through Israel's weakly defended forward fortifications and advanced into the Sinai Peninsula and the Golan Heights.

By seizing the initiative, and with the use of sophisticated new Soviet weapons, the Arab forces were able temporarily to dictate the conditions of battle most favorable to themselves. Whereas Israel excelled at a "war of rapid movement and envelopment," Safran wrote in the January 1974 issue of *Foreign Affairs*, the Arabs "forced the enemy to fight a set battle, where the undoubted courage of their own fighting men and their numerical superiority in manpower and equipment could be used to best effect."

This "slugging type of war," Safran went on, "turned out to be extremely costly in men and especially in equipment to both sides," leading first the Soviet Union and then the United States to intervene as equipment suppliers to their client states.

U.N. Security Council Resolution 338, Oct. 22, 1973

The Security Council

1. *Calls* upon all parties to the present fighting to cease all firing and terminate all military activity immediately, no later than 12 hours after the moment of the adoption of this decision, in the positions they now occupy;

2. *Calls* upon the parties concerned to start immediately after the cease-fire the implementation of Security Council Resolution 242 (1967) in all of its parts;

3. *Decides* that, immediately and concurrently with the cease-fire, negotiations start between the parties concerned under appropriate auspices aimed at establishing a just and durable peace in the Middle East.

Despite the success of the initial Egyptian and Syrian strikes into Israeli-occupied territory, Israeli forces subsequently succeeded in breaking through the Egyptian lines to the western bank of the Suez Canal and advancing to within 20 miles of the Syrian capital of Damascus. The United States and the Soviet Union then joined to seek a cease-fire through the auspices of the United Nations, and the Security Council adopted a cease-fire resolution Oct. 22. A cease-fire was to go into effect "no later than 12 hours" after adoption of the resolution, and this was to be followed, for the first time, by negotiations "between the parties concerned."

Just as the truce appeared to be taking hold, however, it was threatened by a renewal of tension between the two superpowers. On Oct. 24, the Soviet Union proposed to the United States that the two nations join together to supervise the truce. The proposal was rejected by the United States, which backed the creation of a U.N. observer force without big-power participation. In the early morning hours of Oct. 25, U.S. armed forces were placed on worldwide alert in response to the possibility of a unilateral Soviet movement of troops into the Middle East to supervise the truce. The crisis was defused later that day when Moscow agreed to a Security Council resolution establishing an international peace-keeping force without big-power participation.

Israel and Egypt Nov. 11 signed a six-point cease-fire agreement worked out by Secretary of State Henry A. Kissinger. The agreement was signed at a United Nations tent at kilometer 101 on the Cairo-Suez road by Israeli Maj. Gen. Aharon Yariv and Egyptian Maj. Gen. Mohammed Abdel Ghany el-Gamasy; it was the first face-to-face encounter between Israeli and Egyptian negotiators. The meeting resulted in an exchange of prisoners of war and the lifting of the Israeli siege of the city of Suez and the Egyptian Third Army.

Geneva Conference and "Shuttle Diplomacy"

The first Arab-Israeli peace conference opened in Geneva Dec. 21. The participants were Israel, Egypt, Jordan, the United States, the Soviet Union and the United Nations. Syria boycotted the meeting. The first round of the peace conference ended the following day with an agreement to begin talks on separating Israeli and Egyptian forces along the Suez Canal. Egypt and Israel signed a troop disengagement accord Jan. 18, 1974, and the troop withdrawal was completed March 4. Meanwhile, efforts to negotiate a similar agreement between Israel and Syria continued. Syria and Israel signed a similar agreement May 31. In early 1975, Kissinger sought a second-stage disengagement in the Sinai Desert, but after 15 days of shuttling between Egypt and Israel he declared March 23 that his efforts had failed and returned to the United States.

After the breakdown in the talks, Egypt formally requested a resumption of the full Geneva Conference. But it became widely recognized that a propaganda battle at Geneva might degenerate into war and the conference was indefinitely postponed.

When Kissinger returned from his shuttle talks he and President Ford unofficially indicated their displeasure with Israel's negotiating position, and the United States began a "reassessment" of its policy in the Middle East. Meetings between Ford and Sadat in Salzburg, Austria, June 1 and 2, and a visit by Israeli Premier Yitzhak Rabin to Washington June 11 and 12 led to a resumption of step-by-step negotiations for troop disengagements, resulting in a limited agreement initialed by both sides Sept. 1. *(Middle East Diplomacy, p. 65)*

Yom Kippur War in Perspective

The latest war in the Middle East may be remembered as the war that broke the myths that three previous encounters between Israel and its Arab neighbors had built up. One casualty of the October 1973 fighting was the belief in Israeli invincibility which had prevailed since Israel's lightning victory during the Six-Day War in 1967.

It was clear when the latest fighting stopped that the Israelis were on the brink of another military triumph prevented only by U.S. and Soviet intervention. But Israel's failure to win a decisive victory, the success of the initial Egyptian and Syrian strikes into Israeli-occupied positions in the Sinai Peninsula and the Golan Heights and the high toll of Israeli casualties (more than 2,500 killed compared to less than 700 in the 1967 war) laid to rest the assumption that the Israelis were supermen who could not be beaten in battle. Despite their later setbacks, the Arabs proved that they could fight and fight well. "No matter what happens in the desert, there has been a victory that cannot be erased," Egyptian President Anwar Sadat declared in a speech before his country's National Assembly on Oct. 16, 1973. "According to any military standard, the Egyptian armed forces have realized a miracle. The wounded nation has restored its honor, the political map of the Middle East has changed."

The key to Arab successes on both the military and diplomatic fronts was an unprecedented display of Arab solidarity—a solidarity which disappeared after the second Egypt-Israel agreement and for the period in 1976 while the civil war raged in Lebanon, but which had by early 1977 reappeared. Sadat was the principal architect of this cooperation among states noted for their mutual suspicion. Since coming to power in 1970, after the death of Gamal Abdel Nassar, Sadat had been busy mending political fences with his fellow Arabs, especially with Syrian President Hafez Assad and Jordan's King Hussein. Though for about a year after the Sinai II agreement on September 1, 1975, Syria and Egypt became bitter rivals, Arab unity by 1977 re-emerged stronger than ever. Since the Riyadh and Cairo summits in October 1976 an Arab peace offensive has resulted in substantial pressures on Israel for major concessions if and when the Geneva Conference is resumed.

Sadat's rapprochement with the late King Faisal of Saudi Arabia and the sheikhs of the Persian Gulf states prior to the Yom Kippur War harnessed a weapon the Arabs had never before attempted to use—oil. The political and economic balance of the Middle East—and since of the entire world—was dramatically altered by the decision to embargo Arab oil exports to the United States in retaliation for massive U.S. aid to Israel. Through Kissinger's diplomacy the embargo was lifted March 18, 1974, by seven of the participating Arab countries (Syria and Libya lifted their embargoes later), but the oil weapon had already successfully demonstrated Arab power and remained a crucial element in Middle East diplomacy. Beyond the threat of future embargoes, the quadrupling of oil prices shifted the bulk of the world's financial reserves to the Arabian peninsula; and petrodollars have become a new and potent Arab asset. *(See Middle East Oil, p. 123)*

Although the 1973 outbreak of fighting seemed to catch both Israel and the world by surprise, Sadat had been saying for some time that he might be forced to resort to war. Besides the purely military objective of recapturing part of the territories lost to Israel in 1967, Sadat's primary motive for going to war was to refocus world attention on the Middle East. "Everyone has fallen asleep over the Mideast crisis," Sadat said in an interview with *Newsweek* senior editor Arnaud de Borchgrave in April 1973. "The time has come for a shock."

At that time, Sadat's main fear appears to have been that the spirit of detente between the United States and the Soviet Union would keep the big powers from pressuring Israel to return the occupied territories, and thus Israel's presence in these areas eventually could become legitimized. In his interview with de Borchgrave, Sadat recalled the Tet offensive in Vietnam, saying that while the Communists had suffered a military defeat, they had scored a decisive political and diplomatic victory. "It may be," de Borchgrave reported, "that Sadat is now willing to accept a military drubbing in order to weld Arab solidarity and step up pressure for an imposed international settlement."

When Arab unity was restored toward the end of 1976, a new Arab strategy began to emerge. By presenting a moderate image to the world, Sadat and most of the Arab leaders hoped to affect U.S. policy and create the conditions for resumption of the Geneva negotiations. Even the Palestine Liberation Organization (PLO) began to make gestures (however ambiguous) indicating a willingness to accept the existence of Israel if the bulk of the occupied territories would be turned over for the creation of a Palestinian state. Still, Arab sincerity as well as stability have been questioned, and the January 1977 riots in Egypt raised fears that the coalition of Arab moderates might at some point collapse.

Aware that the Carter administration's Middle East policies could largely determine Middle East events during the coming years ("The U.S. holds 99 per cent of the cards," Sadat has grown fond of repeating) a major effort was launched in 1977 to convince the United States that the Arabs no longer challenge Israel's existence but do question her occupation of Arab lands and her refusal to allow the fulfillment of "Palestinian rights." In February 1977, Sadat said in an interview in *Parade* magazine, "I want the American people to know...that never before have the prospects for peace been better. Not in the last 28 years—since Israel was created—have we had a better chance for a permanent settlement in the Middle East. We must not lose the chance."[2] But it remains unclear just what specific concessions the Arabs are willing to make to convince Israel that they are truly talking about a "permanent peace."

Israel's Growing Isolation

The Israelis have been forced by the 1973 war and subsequent developments to take a new look at their Arab neighbors. For one thing, the war demonstrated that the Arabs are closing the technological gap that has in the past protected the vastly outnumbered Israelis. Unlike the 1967 war—when, during the first hours of fighting, Israel destroyed most of the Egyptian air force while still on the ground—Israeli planes in 1973 encountered opposition from Egyptian and Syrian MIGs and from anti-aircraft missiles supplied by the Soviet Union.

"This time [in 1973] they marshaled all their resources, including oil, to achieve their purpose," said Moshe Dayan, then Israeli Defense Minister. "They took the international climate into account, the role the Russians would play, the importance of the detente between the Americans and the Russians. They realized that it was a changed world in 1973, and we have to realize it too."[3]

Another important change made evident by the 1973 war was Israel's growing isolation in the world community. The threat of an Arab oil boycott forced the Japanese and most West European governments to demonstrate a decidedly pro-Arab "tilt" during and after the October war. Western Europe consumes about one-fourth of the world's oil production and relies on the Arab states for nearly 70 per cent of its imports. In an attempt to appease the Arabs, the European Economic Community (Common Market), including the Netherlands and Denmark—generally considered to be pro-Israeli—issued a resolution on Nov. 6, 1973, calling on Israel "to end the territorial occupation which it has maintained since the conflict in 1967."

The Common Market declaration also pointed to the "recognition that, in the establishment of a just and lasting peace, account must be taken of the legitimate rights of the Palestinians." *(See box, p. 121)* Japan, almost entirely dependent on outside sources of oil—45 per cent of which comes from the Arab countries—also issued statements calling for an Israeli withdrawal. Numerous resolutions in the U.N. and other international bodies since the October War have indicated that developments of the 1970s tended to isolate Israel politically.

An especially hard blow to Israel was the loss of support in black Africa, where Israel had courted friendship since the 1950s through technical assistance programs. After the out-break of war in 1973 nine African states—Tanzania, Malagasy Republic, Central African Republic, Ethiopia, Rwanda, Cameroon, Dahomey, Upper Volta and Equatorial Guinea—broke off diplomatic relations with Israel. Togo and Zaire had previously severed their ties.

But most important, the new situation has affected U.S. policy. In order to protect growing relationships with many of the Arab states as well as to preserve America's special ties with Israel it is the United States which has come to encourage a general settlement. A more "even-handed" attitude has become detectable in U.S. policy, even though huge amounts of aid continue to flow and the U.S. commitment to Israel's survival remains firm and unbreakable.

Another War?

The prospects for a lasting peace remain uncertain. The fact that the Arabs were willing to begin direct peace negotiations with Israel at Geneva signaled a fundamental shift in their attitude. Previously, the humiliation of continual defeat, an unwillingness to negotiate from a position of weakness and a refusal to concede Israel's reality had kept the Arabs from such face-to-face talks. This is a shift also confirmed by some tangible gestures such as Eypt's rebuilding of its cities along the Suez Canal and President Sadat's reopening of the canal to Israeli cargoes.

Arab military performance in 1973 helped erase their sense of shame, and the power of oil and petrodollars has removed their fear of negotiating from a position of weakness. These factors, combined with the demonstration of Israel's military power after its initial setbacks in 1973, have created a new climate in which some Arab leaders have expressed the willingness to coexist with the Jewish state.

But the mood in Israel is wary and skeptical. The euphoria that followed the 1967 triumph gave way to despair and political chaos which helped bring the Likud party to power in 1977. Israelis recognize the power of the oil weapon and have an uneasy increasing need for U.S. military and economic aid. Israeli self-confidence has been partially undermined.

The dangers of further warfare and the potential benefits of peace may force both sides to recognize that their mutual interest may lie in a compromise settlement. Meanwhile, the dangers of another war loom ominously over all developments. Such a war could force the most serious U.S.-Soviet confrontation since 1962 and might herald the first use of nuclear weapons since World War II.[4]

2 See George Michaelson, "Peace Prospects Are Better Than Ever; An Interview With Egyptian President Sadat," *Parade*, February 6, 1977.
3 Quoted in *New York Times*, Nov. 27, 1973, in the second of a series of four articles on changing Arab attitudes.

4 See Robert J. Pranger, "Nuclear War Comes to the Middle East," *Worldview*, July-August 1977.

A DEEPENING ROLE AS MIDDLE EAST CONCILIATOR

In its withdrawal from Vietnam, the United States dramatically reduced its presence in the Far East; in the Middle East, however, the 1973 Arab-Israeli war has propelled American diplomacy into a decisive role.

The Nixon, Ford and Carter administrations have all been deeply active in attempting to defuse the explosive conflict. First it was Secretary of State Henry A. Kissinger's "shuttle diplomacy" which aimed to bring about small agreements which might improve the climate and lead to discussions on the major issues at a later date. Then President Jimmy Carter took over and began discussing a com-

---

"To let this opportunity [for peace] pass could mean disaster not only for the Middle East, but perhaps for the international political and economic order as well."

President Jimmy Carter
May 1977

---

prehensive settlement to be implemented over a period of years once agreed upon. Within a month of coming into office, Carter emphasized that 1977 "may be the most propitious time for a genuine settlement since the beginning of the Arab-Israeli conflict almost 30 years ago. To let this opportunity pass could mean disaster not only for the Middle East, but perhaps for the international political and economic order as well."

The impact of the oil embargo at the time of the 1973 war and of rising oil prices ever since has demonstrated that the daily lives of Americans could be affected profoundly by events thousands of miles away. Because another war almost certainly would bring another embargo, and because the United States is attempting to protect its friendships with Israel and with many of the Arab states, vital American interests spur the Middle East peacemaking efforts of the U.S. government. One of the most serious consequences of renewed Arab use of the oil weapon would be the danger of a split between the United States and its European allies, who depend on Arab sources for two-thirds of their petroleum and are not willing to see their economic lifeline cut because of what they consider at times excessive American support of Israel.

America's commitment to Israel, dating from decisive U.S. support for the United Nations plan that led to the creation of the Jewish state, is fundamental to American Middle East policy. This commitment originated in concern for the terrible plight of Jewish refugees from Hitler's genocide and has been sustained by considerable public support for a special friendship with Israel and by the politically active and influential Jewish minority in the United States.

Support for Israel, however, created strong anti-American feelings in Arab countries, opening many of them to Soviet influence. The arms with which Egypt and Syria attacked in 1973 had been supplied by the Soviet Union, and during the war shipments of Russian arms were countered by a massive airlift of U.S. weapons to Israel.

Paradoxically, however, Egyptian President Anwar Sadat regards the United States as the only country that can pressure Israel into returning Arab territory—because Israel depends on U.S. support. Anxious to reduce his own country's dependence on the Soviet Union, whose military advisers he had expelled in 1972, Sadat and then other Arab leaders staked their peace efforts on U.S. diplomacy. Henry Kissinger was the first to see the diplomatic opening immediately when the 1973 war broke out. Before leaving office he created a diplomatic axis running from Cairo and Riyadh to Washington. Relations with Syria were greatly improved as well while ties with King Hussein's Jordan remained close.

Fearful of both a new oil embargo and a revival of Russian influence in the Middle East if peace talks failed, the Ford administration pressured the Israeli government to agree to a compromise troop withdrawal from the Sinai Peninsula. The amount of pressure any administration can bring to bear, however, is limited by the strength of pro-Israeli pressures on Congress and by concern for the impact of Jewish votes and campaign funds. This concern became prevalent as the 1976 election approached and both President Ford and candidate Carter competed for Jewish support.

Still, President Ford told Sadat "with emphasis" on June 1, 1975, that "the United States will not tolerate stagnation in our efforts for a negotiated settlement—stagnation and a stalemate will not be tolerated."[1] Two years later, Jimmy Carter was telling a patiently waiting but increasingly anxious Sadat much the same thing.

U.S. Commitment to Israel

The United States—Israel's chief arms supplier and protector in international forums, target of a Middle East oil embargo and Arab hostility, and major barrier to Soviet influence in that region of the world—has made no formal treaty commitment to the defense of Israel. And the omission has been calculated.

The United States originally assumed the role of Israel's chief supporter with reluctance. Throughout the 1950s and early 1960s, Washington had shunned Israeli arms requests so as not to jeopardize either its friendships with Arab countries during the cold war or its oil interests in

---

1 Luncheon toast at meeting with Sadat in Salzburg, Austria, June 1, 1975. The Department of State Bulletin, June 30, 1975, p. 899.

the Middle East. But with the French decision in 1967 to cut off arms to Israel, U.S. policymakers felt they had no alternative but to step into the arms supplier role in order to counter Soviet assistance to Israel's enemies. *(Arms race, box, pp. 92, 93)*

No formal agreement was concluded, however. The signing of a defense treaty would have provided a rallying point for Arab hostility, thus placing U.S. friends in the Arab world in a very awkward position. An unwritten commitment was much easier to live with. Moreover, it was thought that a treaty might have encouraged intransigence on the part of Israel in any future negotiations over land acquired in the 1967 war.

Also, the Israelis generally have reacted negatively to suggestions of a mutual security treaty with the United States. The proposed security guarantees, Foreign Minister Moshe Dayan has said, implied that the United States envisioned borders that were "not worth anything." Guarantees could supplement defensible borders, he said, but not substitute for them. Nevertheless, it remains possible that some form of American guarantee for Israel's ultimate security might accompany a Middle East settlement if one can be achieved.[2]

In testimony before the House International Relations Committee on June 10, 1975, Secretary of State Kissinger replied in these words to a question about proposals for U.S. guarantees to Israel: "I believe that a final peace settlement in the area will require some sort of American—I don't know whether I want to use the word 'guarantee'—but some sort of American assurance as to the viability of the state of Israel." The presence of Americans at Sinai monitoring stations, a feature of the second Sinai disengagement agreement in September 1975, has sometimes been viewed as something that could be expanded into the Golan Heights and West Bank, for instance.

Past U.S. Policy

In 1967, when the United States stood on the brink of becoming Israel's chief benefactor, Senate Foreign Relations Committee Chairman J. W. Fulbright (D Ark. 1945-75) asked the State Department whether the United States had a national commitment to provide military or economic aid to Israel or any of the Arab states in the event of armed attack or internal subversion.

The State Department reply, written in early August 1967, two months after Israel had decisively defeated the Arabs and had occupied the Sinai Peninsula and the West Bank of the Jordan River, stated: "President Johnson and his three predecessors have stated the United States interest and concern in supporting the political independence and territorial integrity of the countries of the Near East. *This is a statement of policy and not a commitment to take particular actions in particular circumstances....* The use of armed force in the Middle East can have especially serious consequences for international peace extending far beyond that area. We have bent our efforts to avoid a renewal of conflict there. Thus, we have stated our position in an effort to use our influence in the cause of peace." [Emphasis supplied.]

Such references to U.S. support of "territorial integrity of the countries of the Near East" typify the expression of

---

2 Two studies are available on this important issue: Mark A. Bruzonsky, "A United States Guarantee for Israel?" Georgetown University Center for Strategic and International Studies (CSIS), April 1976; and N.A. Pelkovits, "Security Guarantees in a Middle East Settlement," Foreign Policy Papers No. 5, Beverly Hills and London: Sage Publications 1976.

Mondale on Middle East Peace

In a major address on Middle East policy in June 1977, Vice President Walter F. Mondale detailed the Carter administration's policies and strategies in a comprehensive way for the first time. He stated the following themes:

- "America has a special responsibility and a special opportunity to help bring about peace" in the Middle East;
- The United States has a "unique and profound relationship with the state of Israel";
- The United States also desires "to enjoy the friendship of much of the Arab world where we and our close allies have important interests";
- "Israel's survival is not a political question but rather stands as a moral imperative of our foreign policy";
- The Carter administration has made "progress in getting Arab leaders to recognize Israel's right to exist and to recognize—however reluctantly—that this commitment is essential to a genuine peace."
- For real peace Israel should be willing to "return to approximately the borders that existed prior to the war of 1967...and yet retain security lines or other arrangements that ensure Israel's safety";
- The Palestinians—if they demonstrate a willingness "to live in peace alongside Israel," should be given the opportunity "for a Palestinian homeland or entity—preferably in association with Jordan."
- "America can try to help establish the basis of trust necessary for peace. We can try to improve the atmosphere for communication. We can offer ideas, but we cannot, in the end, determine whether peace or war is the fate of the Middle East. That can only be decided by Israel and her Arab neighbors."

Delivered June 17, 1977, before the World Affairs Council of Northern California in San Francisco.

---

the U.S. commitment to Israel. By 1977, some analysts noted, the basic commitment also extended to Saudi Arabia and maybe even to Egypt, which has developed a very close relationship with the United States since 1974. The commitment to Israel, however, while not in the form of a treaty, has been reiterated by all recent administrations. The following statements are characteristic:

- A tripartite declaration by Great Britain, France and the United States May 25, 1950, providing that the three governments would act, within the United Nations armistice lines resulting from the 1948-1949 war between Israel and the Arabs.
- A reply by President Kennedy at a press conference May 8, 1963: "In the event of aggression or preparation of aggression [in the Near East], we would support appropriate measures in the United Nations, adopt other courses of action on our own to prevent or to put a stop to such aggression.... [This] has been the policy which the United States has followed for some time."
- An address by President Johnson June 19, 1967, at a foreign policy conference sponsored by the State Department: "Our country is committed—and we here reiterate that commitment today—to a peace [in the

United States Assistance to the Middle East, 1946-1976[1]

(U.S. fiscal year—millions of dollars)[2]

	1946-1952	1953-1961	1962-1967	1968-1972	1973	1974	1975	1976	Transition Quarter[7]	Total 1946-1976
ISRAEL										
Economic[3]	86.5	507.2	262.8	289.6	109.8	51.5	353.1	714.4	78.6	2,425.6
Loans	—	248.3	238.3	237.4	59.4	—	8.6	239.4	28.6	1,034.1
Grants	86.5	258.9	24.4	52.2	50.4	51.5	344.5	475.0	50.0	1,391.7
Military[4]	—	0.9	136.4	985.0	307.5	2,482.7	300.0	1,500.0	200.0	5,904.2
Credit Sales	—	0.9	136.4	985.0	307.5	982.7	200.0	750.0	100.0	3,454.2
Grants	—	—	—	—	—	1,500.0	100.0	750.0	100.0	2,450.0
Other[5]	135.0	57.5	33.7	124.4	21.1	47.3	62.4	104.7	12.6	595.2
ALGERIA										
Economic[3]	—	4.6	172.1	4.3	0.2	—	4.6	6.1	2.1	193.2
Loans	—	—	11.6	—	—	—	—	—	—	11.6
Grants	—	4.6	160.5	4.3	0.2	—	4.6	6.1	2.1	181.6
Other[5]	—	—	—	36.8	18.7	72.2	123.8	87.6	—	495.8
EGYPT										
Economic[3]	12.3	302.3	580.5	1.5	0.8	21.3	370.1	464.3	552.5	2,269.8
Loans	10.7	132.1	452.4	1.5	—	9.5	298.8	351.7	443.6	1,671.6
Grants	1.6	170.2	128.1	—	0.8	11.8	71.3	112.6	108.9	598.1
Other[5]	7.3	30.6	9.9	18.3	10.7	9.0	38.1	7.8	—	124.0
IRAN										
Economic[3]	42.3	548.2	173.3	8.7	1.3	1.4	1.7	1.0	0.1	760.0
Loans	25.8	197.0	80.4	—	—	—	—	—	—	295.8
Grants	16.5	351.2	93.1	8.7	1.3	1.4	1.7	1.0	0.1	464.2
Military[4] [6]	17.2	482.0	636.6	268.2	—	*	—	—	—	1,412.5
Loans	—	—	299.8	204.3	—	—	—	—	—	504.0
Grants	17.2	482.0	336.8	63.9	—	*	—	—	—	908.5
Other[5] [6]	—	70.0	134.3	437.7	281.9	290.6	5.3	40.0	—	1,180.5
IRAQ										
Economic[3]	1.4	21.6	22.5	3.2	0.2	—	—	—	—	45.5
Loans	0.9	—	13.5	—	—	—	—	—	—	14.4
Grants	0.5	21.6	9.0	3.2	0.2	—	—	—	—	31.1
Military[4]	—	49.4	0.6	—	—	—	—	—	—	50.0
Grants	—	49.4	0.6	—	—	—	—	—	—	50.0
Other[5]	—	—	11.6	0.1	0.1	—	—	—	—	11.8
JORDAN										
Economic[3]	5.2	275.4	285.7	93.6	71.1	64.5	99.3	61.9	86.6	1,047.7
Loans	—	4.7	16.7	6.1	15.9	15.6	25.0	18.6	19.0	116.1
Grants	5.2	270.7	269.0	104.7	55.2	48.9	74.3	43.3	67.6	931.5
Military[4]	—	16.2	55.9	126.2	54.8	45.7	104.6	138.3	—	551.5
Credit Sales	—	—	15.0	54.0	—	—	30.0	82.5	—	181.5
Grants	—	16.2	40.9	72.2	54.8	45.7	74.6	55.8	—	370.0
Other[5]	—	—	1.6	8.5	6.6	7.9	—	—	—	24.2
KUWAIT										
Other[5]	—	—	50.0	—	—	—	—	—	—	50.0
LEBANON										
Economic[3]	3.6	95.5	3.4	27.7	1.2	5.7	2.9	0.1	—	118.2
Loans	1.6	17.9	—	14.1	—	—	—	—	—	18.8
Grants	2.0	77.6	3.4	13.6	1.2	5.7	2.9	0.1	—	99.5
Military[4]	—	8.4	0.7	15.6	10.7	0.2	0.1	0.1	—	35.4
Credit Sales	—	—	—	10.0	10.0	—	—	—	—	20.0
Grants	—	8.4	0.7	5.6	0.7	0.2	0.1	0.1	—	15.4
Other[5]	—	2.4	—	14.2	—	5.8	60.2	—	—	81.1
LIBYA										
Economic[3]	1.8	173.1	41.4	4.3	—	—	—	—	—	212.5
Loans	—	8.5	—	—	—	—	—	—	—	7.0
Grants	1.8	164.6	41.4	4.3	—	—	—	—	—	205.5
Military[4]	—	3.3	11.8	1.6	—	—	—	—	—	17.6
Grants	—	3.3	11.8	1.6	—	—	—	—	—	17.6

SOURCE: Agency for International Development, *U.S. Overseas Loans and Grants and Assistance from International Organizations, July 1, 1945-September 30, 1976.*

	1946-1952	1953-1961	1962-1967	1968-1972	1973	1974	1975	1976	Transition Quarter[7]	Total 1946-1976
MOROCCO										
Economic[3]	0.3	290.4	266.9	228.9	29.6	20.0	23.7	45.1	4.0	892.4
Loans	—	192.9	131.7	124.9	18.0	—	8.0	24.8	—	490.7
Grants	0.3	97.5	135.3	104.0	11.6	20.0	15.7	20.3	4.0	401.7
Military[4]	—	2.4	54.4	45.9	9.9	3.6	14.9	30.8	0.2	169.5
Credit Sales	—	—	20.2	39.5	9.8	3.0	14.0	30.0	—	116.5
Grants	—	2.4	34.2	6.4	0.1	0.6	0.9	0.8	0.2	53.0
Other[5]	—	—	19.5	60.7	3.5	5.3	0.2	54.5	6.4	149.0
SAUDI ARABIA										
Economic[3]	4.7	27.9	—	—	—	—	—	—	—	31.8
Loans	4.3	—	—	—	—	—	—	—	—	4.3
Grants	0.4	27.9	—	—	—	—	—	—	—	27.5
Military[4]	—	63.1	185.8	46.2	0.2	0.2	—	—	—	295.9
Credit Sales	—	43.8	170.7	43.2	—	—	—	—	—	257.7
Grants	—	19.3	15.1	3.0	0.2	0.2	—	—	—	38.2
Other[5]	14.8	—	—	25.4	*	—	1.1	—	—	38.6
SUDAN										
Economic[3]	—	53.6	72.7	.8	15.3	5.0	8.2	1.7	0.1	127.9
Loans	—	10.0	39.8	0.2	13.1	2.8	—	—	—	46.0
Grants	—	43.6	32.8	.6	2.2	2.2	8.2	1.7	0.1	81.9
Military[4]	—	*	2.0	—	—	—	—	—	—	2.2
Credit Sales	—	—	1.5	—	—	—	—	—	—	1.5
Grants	—	*	0.5	—	—	—	—	—	—	0.7
Other[5]	—	—	—	7.9	15.6	18.1	—	2.7	—	38.1
SYRIA										
Economic[3]	0.4	44.9	43.9	1.0	0.2	—	104.6	34.9	78.7	279.7
Loans	—	23.6	29.4	—	—.	—	99.4	32.9	78.5	235.2
Grants	0.4	21.3	14.5	1.0	0.2	—	5.2	2.0	0.2	44.4
TUNISIA										
Economic[3]	0.2	240.9	279.8	225.5	17.0	10.3	13.2	11.7	0.9	773.2
Loans	—	49.0	161.1	127.3	9.4	—	—	2.3	—	336.8
Grants	0.2	192.0	118.8	98.2	7.6	10.3	13.2	9.4	0.9	436.4
Military[4]	—	5.3	20.0	19.5	2.2	4.3	7.2	15.6	10.1	88.1
Credit Sales	—	2.6	—	2.2	—	2.5	5.0	15.0	10.0	37.5
Grants	—	2.7	20.0	17.3	2.2	1.8	2.2	0.6	0.1	50.6
Other[5]	—	—	2.5	24.3	0.7	—	—	8.4	14.3	51.0
TURKEY										
Economic[3]	237.3	1,093.2	1,039.3	416.9	22.9	5.5	4.4	—	—	2,704.2
Loans	97.2	301.9	797.4	316.7	9.0	—	—	—	—	1,445.1
Grants	140.1	791.3	241.8	100.2	13.9	5.5	4.4	—	—	1,258.9
Military[4]	325.6	1,587.7	1,046.0	663.2	245.1	190.8	109.1	—	125.0	4,689.9
Credit Sales	—	—	—	15.0	20.0	75.0	75.0	—	125.0	310.0
Grants	325.6	1,587.7	1,046.0	648.2	225.1	115.8	34.1	—	—	4,379.9
Other[5]	32.3	32.0	7.1	52.1	67.1	30.5	26.2	70.2	2.8	316.8
YEMEN ARAB REPUBLIC										
Economic[3]	—	16.0	28.6	*	3.2	4.0	6.9	7.4	2.7	67.4
Grants	—	16.0	28.6	*	3.2	4.0	5.5	7.4	2.7	66.1
YEMEN, DEMOCRATIC REPUBLIC OF										
Economic[3]	—	—	—	2.8	0.1	1.6	—	—	—	4.5
Grants	—	—	—	2.8	0.1	1.6	—	—	—	4.5

[1] The table gives line categories under countries only where assistance was given. Egypt, for example, received no military aid during the 1946-1976 period.

[2] Figures may not add to totals due to rounding. Annual figures may not add up to cumulative totals because the fiscal year figures since FY 1955 represent new obligations entered into during those years on a gross basis; the cumulative figures for FY 1946-1976 are on a net basis, reflecting total obligations where funds obligated were not actually spent.

[3] Economic aid totals include official development assistance, Food for Peace programs, Peace Corps and miscellaneous programs.

[4] Military aid total includes the Foreign Assistance Act credit sales and grant programs, transfers from excess stocks, other grants and loans under special programs.

[5] Other includes Export-Import Bank loans, often made to the private sector within a country, Agriculture Department short-term credits, Overseas Private Investment Corporation direct loans and private trade agreements under PL-480, Title 1.

[6] Iran received Export-Import Bank military equipment loans as follows: $120-million in 1971, $100-million in 1972; $200-million in both 1973 and 1974.

[7] The figures listed in the column represent the dollar amounts allocated for the interim period between the old and new fiscal year budget, in this case July-Sept. 1976.

* Less than $500,000.

Middle East] that is based on five principles: first, the recognized right of national life; second, justice for the refugees; third, innocent maritime passage; fourth, limits on the wasteful and destructive arms race; and fifth, *political independence and territorial integrity for all."* [Emphasis supplied.]

● A statement by Secretary of State Henry A. Kissinger to reporters in Peking Nov. 12, 1973: "It has been a constant American policy, supported in every administration and carrying wide bipartisan support, that the existence of Israel will be supported by the United States. This has been our policy in the absence of any formal arrangement, and it has never been challenged, no matter which administration was in office." Again, on June 23, 1975, Kissinger stressed "our historical and moral commitment to the survival and well-being of Israel."

● At a press conference on May 13, 1977, President Carter restated the traditional pledge of uniquely close ties with the Jewish state. "We have a special relationship with Israel. It's absolutely crucial that no one in our country or around the world ever doubt that our No. 1 commitment in the Middle East is to protect the right of Israel to exist, to exist permanently, and to exist in peace. It's a special relationship."

Origins of Involvement

Vast oil reserves—and Britain's attempt to monopolize them at the end of World War I—first attracted the United States to the Middle East. Britain moved into the oil-rich region by securing a mandate from the League of Nations to Palestine and Mesopotamia (later to become Iraq). British companies, which produced less than 5 per cent of the world's oil, managed to corner more than half of the world's known reserves by 1919.

The United States, having fueled the Allied victory with large quantities of oil from U.S. reserves, protested British tactics and insisted on a share of the Middle East oil. The protests eventually paid off and, in 1928, several American companies joined with a European group to operate the Turkish (later the Iraq) Petroleum Company.

In addition to these early economic interests, the United States also had an influence on postwar peace settlements in the Middle East and Palestine policies. But a truly strong strategic interest did not emerge until the end of World War II, when the United States gradually began to fill the political role which the British and French were forced to relinquish in the Middle East. *(Background, box, p. 87)*

Soviet Challenge

America's first commitments to the Middle East were prompted by postwar Soviet expansionism. Greece, Turkey and Iran were the first states beyond the control of the Soviet army to come under this expansionist pressure.

The pressure on Iran began in early 1946 when the Russians refused to withdraw troops that had been stationed in northern Iran since 1941 under a wartime agreement with Great Britain whereby both powers acted to prevent Nazi influence in Iran. Under the agreement, both powers were supposed to withdraw their forces within six months of the end of hostilities. British and American troops—which had arrived after the U.S. entry into the war to help move supplies to the Soviet Union—were withdrawn, but Soviet troops remained after March 2, 1946,

the final date set for evacuation. In addition, the Russians demanded that Soviet experts help administer the Iranian government.

When Iran rejected this demand, the Russians engineered a revolt in the north by a Communist-controlled Tudeh Party and used their own forces to prevent the Iranian government's efforts to put down the revolt. A "puppet" Soviet regime was set up in the northern Iranian province of Azerbaijan with the objective of forcing the Iranian government in Tehran to recognize the new regime. Only after a protest by President Truman March 6 that the continued Soviet occupation violated wartime agreements did the Soviets begin their withdrawal from Iran, completing it on May 4, 1946.

Turkey came under Soviet pressure in the summer of 1945, when Moscow demanded cession of several Turkish districts on the Turkish-Russian frontier and revision of the 1936 Montreux Convention, which provided for exclusive Turkish supervision of the Dardanelles Straits between the Black Sea and the Mediterranean. The Soviets demanded joint Russian-Turkish administration of the straits, and the conclusion of a treaty with the Soviet Union similar to those between Russia and its East European satellites, and above all, the lease to the Soviet Union of naval and land bases in the straits for the "joint defense" of Turkey and the Soviet Union. These demands were renewed in a Soviet note to the United States and Britain in August 1946. President Truman replied by sending a naval force into the Mediterranean immediately upon receipt of the note. Twelve days later Britain and the United States replied to the Soviet Union in a joint note rejecting Moscow's demands. This was followed by an easing of Soviet pressure on Turkey.

In Greece, Communist guerrilla warfare, aided from Albania, Bulgaria and Yugoslavia, threatened to take over the country by early 1947. British troops and military assistance in 1945 had averted an earlier attempt by the Communists to gain control of Athens. By 1947, however, Britain was no longer able—as a result of wartime exhaustion—to continue to fulfill its traditional role, dating from the 19th century, of resisting Russian pressure in the eastern Mediterranean. The United States moved into the breach, as it had in Iran and Turkey, thereby initiating what became the pattern in the postwar era—namely, the United States replacing Britain as the protector of weaker states bordering on the Russian (or Soviet) empire.

Truman Doctine

Faced with the threatened Communist takeover in Greece and the probability that the collapse of Greece would lead to further Soviet pressure on Turkey, President Truman went before a joint session of Congress March 12, 1947, and spelled out the Truman Doctrine. The President asserted that the United States could only be secure in a world where freedom flourished. He said: "Totalitarian regimes imposed on free people, by direct or indirect aggression, undermine the foundations of international peace and hence the security of the United States."

President Truman stressed that it must be U.S. policy "to support free people who are resisting attempted subjugations by armed minorities or by outside pressure." To bolster the sagging Greek government and that of Turkey (again under Soviet pressure), the President urged Congress to authorize military as well as economic aid to the two countries, on a bilateral rather than a multilateral basis

Truman's Constituents

Israel has been intimately involved in American domestic politics ever since President Truman decided to work for its creation after World War II. Truman's decision ran counter to the advice of U.S. diplomats who served in the Middle East. When they warned that support for Israel would jeopardize American relations with the Arabs and damage wider American interests in the Middle East, Truman reportedly replied: "I'm sorry, gentlemen, but I have to answer to hundreds of thousands who are anxious for the success of Zionism; I do not have hundreds of thousands of Arabs among my constituents."

Israel is a major political issue among the estimated 5.9 million American Jews—who far outweigh Arab-Americans both in numbers and in financial resources—but how much impact the so-called "Jewish lobby" has had on American Middle East policy is a matter of conjecture. "U.S. Middle Eastern policy is not now and never has been, in spite of the Arab belief to the contrary formulated by 'Zionists,' foreign policy analyst James E. Griffiths wrote in 1969. "Yet the votes and financial resources of the American Jewish community exercise influence on it in favor of Israel for which there is no pro-Arab equivalent."

Internally divided and overwhelmed by pro-Israeli sentiment, supporters of the Arab cause in the United States concede that in the past they have been ineffective in turning public and congressional support to their side. But a pro-Arab lobby, dedicated to persuading Congress to reverse its "overcommitment to Israel," has begun to emerge from political obscurity. *(Israel lobby, p. 96; Arab lobby, p. 102)*

through the United Nations. Opponents argued that the Truman Doctrine undercut the U.N. and might provoke a clash with Russia, but a majority sided with the President. Along with the $400-million provided for Greece and Turkey went American civilian and military advisers.

(In an Oct. 17, 1951, protocol, the North Atlantic Treaty Organization extended mutual security guarantees to Greece, Turkey and the eastern Mediterranean Sea.)

Arab Reaction

The Arab states—which felt little threat from the Soviet Union and were still embittered by U.S. support for the 1917 Balfour Declaration calling for the establishment in Palestine of a national home for the Jewish people—did not like the Truman Doctrine. Arab hostility increased when, on Nov. 2, 1947, the United States, as well as the Soviet Union, voted with a two-thirds majority in the United Nations General Assembly to partition Palestine into Arab and Jewish states.

The United States and the Soviet Union were the first countries to extend diplomatic recognition to Israel when it declared independence minutes after British authority—mandated by the League of Nations and extended by the United Nations—expired at midnight May 14, 1948. Arabs saw U.S. actions as a betrayal of President Franklin D. Roosevelt's promise to King Ibn Saud of Saudi Arabia in 1945 that "no decision [will] be taken with respect to the basic situation in that country [Palestine] without full consultation with both Arabs and Jews."

The Arabs rejected the partition and went to war against Israel. The first Arab-Israeli war lasted from 1948 to 1949.

Arms Race

In an effort to bring stability to the Middle East, the United States joined with Britain and France in issuing the Tripartite Declaration May 25, 1950, in which the three powers attempted to reassure both Israel and the Arabs by declaring their opposition to the use of force throughout the Middle East.

The Tripartite Declaration notwithstanding, the United States had found itself drawn into a Middle East arms race of varying intensity since the mid-1950s. *(Box, pp. 92, 93)*

Baghdad Pact

In an effort to block Soviet pressure on the Middle East, the United States promoted the formation of a mutual defense treaty in February 1955 among Britain, Iran, Iraq, Pakistan and Turkey. U.S. officials participated in the defense and anti-subversion committees of the Baghdad Pact, and U.S. military and economic aid was provided to the members of the organization; but the United States did

"All the cards in this game are in the hands of the United States...because they provide Israel with everything and they are the only [one] who can exert pressure on Israel."

Egyptian President Anwar Sadat,
June 1975

not formally become a member. The main reason for this ambivalence was that Iraq and Egypt were rivals for leadership of the Arab world; and the United States, in the hope that Egypt could eventually be persuaded to join the pact, did not want to alienate Egypt by formally allying itself with Iraq. An alliance with Iraq would also create problems for U.S. relations with Israel.

The hope that Egypt could be induced into joining the pact was not achieved. Egyptian President Gamal Abdel Nasser denounced Iraq for allying with the Western powers, and he turned to the Soviet Union in the fall of 1955 for military equipment in an effort to gain a decisive military lead over Israel.

Moscow, which had reacted sharply to the formation of the Baghdad Pact, responded quickly to Nasser's request and became Egypt's major arms supplier. The Soviet Union thereby began to build a reputation throughout the Arab world as the main support the Arabs had among the great powers in their struggle with Israel.

Suez Crisis

Soviet arms shipments to Egypt in 1955 and 1956 convinced Israel to prepare for a preventive war against Egypt, before the military balance shifted in Cairo's favor. Israel's request for U.S. arms was rejected by President Eisenhower, who said March 7, 1956, he opposed U.S. arms for Israel because this would lead to an "Arab-Israeli arms race." When Nasser announced July 26 nationalization of the British-run Suez Canal and refused to guarantee the safety of Israeli shipping, the threat to Israeli, as well as to British and French, interests made the situation explosive.

Nasser's decision followed U.S. refusal July 19 to provide Egypt financing for the Aswan Dam project, a high dam on the upper Nile intended to furnish cheap electricity and increased irrigation. The United States had initially expressed interest in the project but turned it down as a result of Nasser's deepening ties with the Soviet Union. In nationalizing the Suez Canal, Nasser claimed that revenues would pay for the Aswan project.

The British government held 44 per cent of all shares in the Suez Canal Company; private French investors held 78 per cent of the remainder. Apart from these direct interests, both nations were heavily dependent on the canal as the shortest waterway to their oil supplies on the Persian Gulf.

"The search for a just and lasting peace in the Middle East is one of the highest priority items on the foreign policy agenda of our country."

Secretary of State Cyrus R. Vance
May 1977

Nationalization of the canal was intolerable in their view, because it meant that Nasser could bar their vessels at any time. The two governments immediately froze all Egyptian assets available and began talks on joint military action.

Israeli armed forces attacked Egypt Oct. 29, 1956, and completed occupation of the Sinai Peninsula to within 10 miles of the Suez Canal by Nov. 5. British and French aircraft attacked Egypt Oct. 31-Nov. 5, and an Anglo-French paratrooper force was dropped at the northern end of the Suez Canal Nov. 5. By Nov. 7, British and French forces had secured control of the canal.

Strong U.S. pressure in the United Nations and behind the scenes on Israel, Britain and France was instrumental in bringing the fighting to a quick halt and in forcing the three nations to withdraw their forces from Egypt by the end of the year. Although Americans sympathized with the invaders, they supported President Eisenhower's declaration Oct. 31 that the invasion "can scarcely be reconciled with the principles" of the United Nations. The U.S. condemnation of the invasion helped save Nasser from a disastrous defeat. The outcome of the conflict was a severe political and moral setback for Britain and France in the Middle East from which Nasser and the Soviet Union reaped the major benefits. The Suez crisis led to a major "power vacuum" in the Middle East, making the area prone to renewed Soviet penetration.

Eisenhower Doctrine

Following the Suez crisis, the Eisenhower administration feared that the Soviet Union might move into the Middle East vacuum resulting from the British and French diplomatic defeat. It was decided that the U.S. commitment to resist communism in the area had to be fortified.

On Jan. 5, 1957, President Eisenhower went before a joint session of Congress to urge support for a declaration which was promptly dubbed the Eisenhower Doctrine. H J Res 117 (Joint Resolution to Promote Peace and Stability in the Middle East) declared that "if the President determines the necessity...[the United States] is prepared to use armed forces to assist...any nation or groups of nations requesting assistance against armed aggression *from any country controlled by international communism."* [Emphasis supplied.]

The resolution did not draw a precise geographical line around the area to which it was intended to apply. The Senate and House reports on the resolution accepted the administration's view and defined the Middle East as the area bounded by Libya on the west, Pakistan on the east, Turkey in the north and the Sudan in the south. The Senate report said that no precise listing of nations was included in the resolution because this "would restrict the freedom of action of the United States in carrying out the purposes of the resolution."

The first test of the Eisenhower Doctrine and resolution came in 1958 following a coup in Iraq, which overthrew the pro-Western government, replacing it with a regime favorable to the Soviet Union and the United Arab Republic (UAR). The new Iraqi government immediately withdrew from the Baghdad Pact.

When the government of Lebanon came under similar pressure and its president requested U.S. assistance, Eisenhower ordered U.S. Marines from the 6th Fleet in the Mediterranean to land in Lebanon to protect the government. Citing the Eisenhower Doctrine, the President said July 15, 1958, that Lebanon's territorial integrity and independence were "vital to United States national interests and world peace." The UAR and the Soviet Union, Eisenhower charged, were trying to overthrow the constitutional government of Lebanon and "install by violence a government which subordinates the independence of Lebanon to...the United Arab Republic."

CENTO

Iraq's withdrawal from the Baghdad Pact convinced the Eisenhower administration that the three remaining "northern tier" members of the organization needed an additional pledge of U.S. support in resisting communism. With the Middle East resolution as a basis, the United States initiated negotiations with Turkey, Iran and Pakistan on defense arrangements, bringing the United States into closer cooperation with the Baghdad Pact, which was renamed the Central Treaty Organization (CENTO). Three identically worded executive agreements were signed in March 1959 between the United States and each of the three nations. Washington pledged in the agreements to come to the defense of the three countries in the event of Communist aggression or subversion.

The executive agreements with Iran, Pakistan and Turkey completed the United States' formal commitment to resist communism in the Middle East. Throughout the 1960s, the Kennedy and Johnson administrations repeated-

Partition of Palestine: U.S. Policy Developments

With the breakup of the Ottoman Empire during World War I, Palestine's fate was left up to Great Britain, France and Russia. Britain had pledged to support the independence of the Arab areas, in correspondence during 1915 between Sir Henry McMahon, high commissioner of Egypt, and Hussein ibn Ali, sherif of Mecca.

But, according to a secret agreement reached in 1916 by the British, French and Russian governments (known as the Sykes-Picot Agreement), the Arab areas were to be divided into British and French spheres of influence and Palestine was to be internationalized. Arabs later cited this agreement in charging they had been deceived by European imperialists.

Balfour Declaration. On Nov. 2, 1917, British Foreign Secretary Arthur Balfour pledged in a letter to Lord Rothschild, leader of the British Zionists, that Britain would support the establishment in Palestine of a "national home" for the Jewish people, on the clear understanding that "nothing shall be done which may prejudice the civil and religious rights of existing non-Jewish communities in Palestine...."

Reaction to the Balfour Declaration in the United States was positive. President Woodrow Wilson endorsed the statement before its publication, and Congress adopted a resolution approving the declaration in September 1922.

U.S. Role. The United States never declared war on the Ottoman Empire, an ally of Germany in World War I, but President Wilson strongly influenced the final peace settlement which set forth the basic boundaries of the Middle East states. His major contribution was the concept of interim League of Nations mandates which would eventually lead to independent states.

The United States sent a commission (King-Crane Commission) to the former Arab areas of the Ottoman Empire to determine their views on postwar settlements. The commission's final report in 1919—never formally accepted by the Paris Peace Conference or the U.S. government—called for a serious modification of the "extreme Zionist program" and advised against the creation of a Jewish state in Palestine.

Palestine Mandate. In July 1922 the League of Nations approved Britain's mandate to Palestine, which went into force Sept. 22, 1923. The mandate instrument included a preamble incorporating the Balfour Declaration and stressing the Jewish historical connection with Palestine. Britain was made responsible for placing the country under such "political, administrative and economic conditions as will secure the establishment of a Jewish National Home...."

In the mid-1940s the push to lift restrictions—set forth in a 1939 British government White Paper—on Jewish immigration into Palestine gained support, especially in the United States. (*Jewish migration to Palestine, pp. 4, 5*)

In August 1945, President Harry S Truman called for the free settlement of Palestine by Jews to the point consistent with the maintenance of civil peace. Later that month Truman suggested in a letter to British Prime Minister Clement R. Atlee that an additional 100,000 Jews be allowed to enter Palestine. In December, both houses of Congress adopted a resolution urging U.S. aid in opening Palestine to Jewish immigrants and in building a "democratic commonwealth."

Anglo-American Committee. In November 1945, Britain, anxious to have the United States share responsibility for its Palestine policy, joined with the United States in deciding to create a commission to examine the problem of European Jews and Palestine. In the meantime, Britain agreed to permit an additional 1,500 Jews to enter Palestine each month. A 75,000 limit had been set by the 1939 White Paper.

In April 1946 the Anglo-American Committee of Inquiry recommended the immediate admission of 100,000 Jews into Palestine, and continuation of the British mandate until establishment of a United Nations trusteeship. Truman immediately endorsed the immigration proposal, but Britain stipulated prior disbandment of underground Jewish forces in Palestine.

The President in October released a communication sent to the British government in which he appealed for "substantial immigration" into Palestine "at once" and expressed support for the Zionist plan for creation of a "viable Jewish state" in part of Palestine. A British government spokesman expressed regret that Truman's statement had been made public because it might jeopardize a settlement.

United Nations. Britain turned to the United Nations in early 1947, when a London conference of Arab and Zionist representatives failed to resolve the Palestinian question. The United Nations set up an inquiry committee which ultimately recommended that Palestine be divided into two separate Arab and Jewish states, with Jerusalem and vicinity to be an international zone under permanent U.N. trusteeship.

The United States and the Soviet Union agreed on the partitioning of Palestine, and on Nov. 29, 1947, the U.N. General Assembly voted to divide Palestine. Britain—setting May 15, 1948, as the date its mandate would terminate—refused to share responsibility with the U.N. Palestine Commission during the transitional period because the U.N. solution was not acceptable to both sides.

Civil war broke out shortly after the U.N. decision was made. In March 1948 the United States voiced opposition to the forcible partitioning of Palestine and called for suspension of the plan. The United States urged a special session of the General Assembly.

In April the Security Council adopted a U.S. resolution calling for a truce and a special session of the General Assembly. But it was too late to stop the division of Palestine. On May 14, the British high commissioner left Palestine, the state of Israel was proclaimed and the General Assembly voted to send a mediator to the Holy Land to seek a truce.

The United States granted Israel de facto recognition immediately. The Soviet Union recognized the new state three days later.

ly pledged to uphold the territorial integrity of Israel and the Arab nations. U.S. military assistance and arms sales delivered weapons to pro-Western governments in the area, and limited arms sales to Israel were initiated in 1962. *(Arms race box, p. 92, 93)*

1967 War

The perpetual Middle East tension flared June 5, 1967, into a third major Arab-Israeli war. During the Israeli-initiated, but Arab caused, Six-day War, Israel destroyed a substantial part of the armed forces of Egypt, Jordan and Syria. In addition, large amounts of Arab territory were captured—land which Israel has continued to occupy.

U.S. Role

A few hours after war broke out early June 5, Robert J. McCloskey, deputy assistant secretary of state for public affairs, declared that the U.S. position was "neutral in thought, word and deed." The McCloskey statement was criticized by members of Congress and others who pointed to U.S. ties with Israel. Later June 5, George Christian, presidential press secretary, said the McCloskey statement was "not a formal declaration of neutrality." And, at a late afternoon news conference at the White House, Secretary of State Dean Rusk June 5 said the term "neutral" in international law expressed the fact that the United States was not a belligerent. He said it was not "an expression of indifference."

Israeli planes and naval vessels June 8 attacked the U.S. Navy communications ship *Liberty* about 15 miles off the Gaza Strip.[3] American carrier-based aircraft went to the assistance of the military moves in the Mediterranean. The White House immediately notified Moscow of the developments on the first teletype communications system between Washington and Moscow. The system, which was installed in 1963 and was known as the "hot line," never before had been used in a crisis.

The U.S. Security Council June 6 unanimously adopted a resolution calling for a cease-fire. The truce went into effect June 10, although periodic clashes continued.

Peace Proposal

In his first major statement on U.S. Middle East policy following the outbreak of the war, President Johnson June 19 laid down a five-point formula for peace in the Middle East. *(See p. 81, 82)*

Johnson said the victorious Israeli troops "must be withdrawn." But he made it clear he would not press for a withdrawal to prewar lines in every respect.

Resolution 242

On Nov. 22, 1967, the U.N. Security Council unanimously approved a British resolution (Resolution 242) aimed at bringing peace to the Middle East. The resolution called for withdrawal of Israeli forces from occupied Arab areas; an end to the state of beligerency between the Arab nations and Israel; acknowledgement of and respect for the sovereignty, territorial integrity and political independence of every nation in the area; the establishment of secure and recognized national boundaries; a guarantee of freedom of

3 This attack was officially termed an accident, and Israel's apology was accepted by the United States government. Yet many questions have remained unanswered about the attack. A two-part article by Anthony Pearson in the May and June 1976 issues of *Penthouse* magazine suggests the attack was deliberate and speculates on possible Israeli motivations.

navigation through international waterways in the area; and just settlement of the refugee problem. *(Text, p. 76)*

There has been considerable disagreement over the precise meaning of Resolution 242. The Arabs have contended that the document required total Israeli withdrawal from the Sinai Peninsula, the Gaza Srip, the Golan Heights, the West Bank of the Jordan River and the eastern sector of Jerusalem. But the Israelis have insisted that the phrasing of the resolution—withdrawal "from territories" and not "from the territories"—did not require a total pullback from the 1967 cease-fire lines.

No War, No Peace

Resolution 242 provided the basis for subsequent U.S. peace proposals. In a departure from previous U.S. policy, the Nixon administration agreed early in 1969 to a series of bilateral talks with the Soviet Union as well as four-power talks which included Britain and France. The talks were carried on sporadically throughout the year, but little was achieved.

Rogers Peace Plan

The major elements of the U.S. diplomatic position were outlined by Secretary of State William P. Rogers on Dec. 9, 1969. Rogers called on Israel to withdraw from Arab territories occupied in the June 1967 war in return for Arab assurances of a binding commitment to a Middle East

---

"I would not hesitate if I saw clearly a fair and equitable solution to [the Middle East problem] to use the full strength of our own country and its persuasive powers to bring those nations to agreement."

President Jimmy Carter
May 1977

---

peace. He also put on record for the first time more detailed peace proposals made by the United States in October during bilateral talks with the Soviet Union. The proposals were rejected by Israel and scorned by the Arabs.

Meanwhile, the United States continued to support the efforts of United Nations envoy Gunnar V. Jarring to mediate a settlement between the Arabs and Israelis. On Jan. 25, 1970, Nixon reaffirmed U.S. support for Israel's insistence on direct peace negotiations with the Arabs. On Jan. 30, he asserted that the United States was "neither pro-Arab nor pro-Israel. We are pro-peace."

The Arab-Israeli conflict was potentially more dangerous than the Indochina war and could result in a direct U.S.-Soviet clash, Nixon warned in a televised interview July 1, 1970. He reiterated that the United States would not allow the military balance to shift against Israel.

The United States had "not excluded the possibility" of participating in a Middle East peacekeeping role, Rogers said Dec. 23, 1970, but he ruled out any joint U.S.-Soviet force as "totally impractical."

In late 1971 and early 1972, the United States put forward a new proposal for indirect, American-mediated talks between Israel and Egypt on an interim peace settle-

U.S.-Soviet Rivalry Over Naval Installations

Tit-for-tat superpower rivalry in the Indian Ocean advanced a notch July 28, 1975, when Congress approved a Pentagon proposal for constructing naval and air support facilities at Diego Garcia, a British-owned atoll 1,000 miles south of the Indian Peninsula. Action by both houses ended five years of bitter haggling with the White House, which argued that Diego Garcia was needed to match increased Soviet naval activity in the Indian Ocean and its outlying pockets—the Arabian Sea, the Persian Gulf, the Gulf of Aden and the Red Sea.

Diego Garcia will replace another port facility the U.S. Navy has been using in the area. In December 1971, the United States secured docking privileges at the island sheikdom of Bahrain in the Persian Gulf. But after the October 1973 war, Bahrain announced its intention of canceling this agreement, which it did on June 29, 1977. U.S. military activity on Diego Garcia was opposed by India and Australia on the grounds that it could lead to a U.S.-Soviet naval confrontation in the relatively peaceful Indian Ocean. But the Ford administration held that the United States must protect vital sea lanes through which oil tankers ply from the Persian Gulf to the West.

At an Aug. 28, 1974, press conference, President Ford defended the administration proposition, saying, "I don't view this [Diego Garcia] as any challenge to the Soviet Union. The Soviet Union already has three major naval operating bases in the Indian Ocean. This particular proposed construction, I think, is a wise policy, and it ought not to ignite any escalation of the problems in the Middle East."

A Defense Department spokesman identified the Soviet installations as Umm Qasr in Iraq, at the head of the Persian Gulf; at Aden in South Yemen, on the southwestern tip of the Arabian Peninsula, and at Berbera in Somalia, on the east coast of Africa. Aden and Berbera flank the Gulf of Aden, the passage leading from the Arabian Sea to the Red Sea, and thence to the Suez Canal and the Mediterranean beyond.

Tass, the official Soviet news agency, scoffed at Ford's assertions, claiming that he had been misinformed by his staff. Tass also quoted testimony of William E. Colby, director of the Central Intelligence Agency, before the Senate Armed Services Committee. Colby had described the Soviet presence in the Indian Ocean area as "relatively small."

Legislative History

The Nixon administration first requested new construction funds for Diego Garcia in fiscal 1970 legislation; but the Senate Appropriations Committee refused to fund the project, and the matter languished until 1974. A Senate-House conference report on the fiscal 1975 military construction authorization bill included $14.8-million for the Navy and $3.3-million for Air Force projects on the island. Obligation of these funds, however, was prohibited pending the President's certification of need. The fiscal 1976 bill allowed an additional $13.8-million for the Navy, contingent upon action taken by Congress on the President's message.

A resolution barring the Diego Garcia improvements was introduced by Senate Majority Leader Mike Mansfield (D Mont.). His resolution was rejected on July 28, 1975, by a 43-53 vote. In floor debate, Mansfield called the $28.6-million Navy request "only a down payment" for deploying an Indian Ocean patrol fleet that could cost $8-billion and require $800-million annually in operating expense.

Armed Services Committee Chairman John C. Stennis (D Miss.) dismissed speculation that Diego Garcia would lead to a three-ocean navy. He called the request "reasonable," noting that the closest refueling station was 3,500 miles away, in the Philippines. Dewey F. Bartlett (R Okla.), who had led a Senate inspection of the Soviet facilities at Berbera in early July, claimed the Russians were developing a "significant" naval and air station there that "exceeds any other facility" outside Soviet borders.

The same day as Senate action, the House turned down by voice vote an amendment by Rep. Robert L. Leggett (D Calif.) to remove $13.8-million for Diego Garcia construction. Leggett, who also inspected Berbera, described the Soviet base as "modest."

The congressional green light enabled the Pentagon to enlarge its existing communications station at Diego Garcia by construction of a 12,000-foot runway for military planes, anchorage for a six-ship carrier task force and storage capacity for 640,000 barrels of oil, enough to fuel the task force for 28 days.

Bahrain Problems

The naval facility at Bahrain had also experienced congressional problems. At first the Senate balked at the idea, and a resolution deleting a provision to pay Bahrain for the leasing of docking facilities was introduced. The administration request for the funds was embodied in an executive agreement, and the resolution called for the agreement to be submitted as a treaty for Senate ratification. However, the resolution was defeated on a 30-59 vote June 28, 1972. Sen. Hugh Scott (R Pa.), the minority leader, argued that the emir of Bahrain might tear up the agreement if it were in treaty form—on the basis that it might alienate other Arab states.

Reacting to U.S. support of Israel in the 1973 war, Bahrain rescinded the naval agreement on Oct. 20, 1973, accusing the United States of a "hostile stand against the Arab nations." After quiet negotiations, Defense Department sources revealed on Oct. 3, 1974, that Bahrain had shifted its policy and the United States would be allowed to retain the base. Reports indicated that Iran and Saudi Arabia put pressure on Bahrain, because they wanted a continued U.S. Navy presence in the Persian Gulf. But in June 1977 Bahrain finally did rescind the agreement.

The U.S. Middle East naval force consisted of one command destroyer and two additional destroyers, which rotated in and out of the Bahrain facility six months a year. The job of the mini-fleet was to show the flag in the Persian Gulf and on patrols along the south coast of the Arabian Peninsula and the east coast of Africa. A naval force will now operate from Diego Garcia.

ment that included a troop pullback and reopening of the Suez Canal; but negotiations made little headway.

Fourth War

The "no-war, no-peace" stalemate held until October 1973, when Arab frustrations over the deadlock triggered the fourth Arab-Israeli war. On Oct. 6, Egyptian forces crossed the Suez Canal and Syria attacked the Golan Heights.

Benefiting from the element of surprise, Arab forces initially inflicted great damage to Israeli defenses, but Israel was soon on the offensive. The intensely fought war led to the rapid depletion of combatants' arsenals and the airlifting of arms first by the Soviet Union and then by the United States.

To avoid total defeat and humiliation of the Arabs—which would very probably have barred any peaceful resolution of the conflict—the United States and the Soviet Union joined in pressing for a cease-fire. Following a Kissinger trip to Moscow, a joint U.S.-Soviet resolution calling for an immediate cease-fire and implementation of the 1967 U.S. Security Council Resolution 242 was presented to the Security Council Oct. 21. Egypt and Israel agreed, and the cease-fire was to go into effect the next day.

U.S. Troop Alert

But the fighting continued and Egyptian President Anwar Sadat—concerned for the fate of his army—called on both the United States and the Soviet Union to send in troops to enforce the cease-fire. A worldwide alert of U.S. armed forces was called in the early morning hours of Oct. 25, reportedly in response to the possibility of a unilateral Soviet movement of troops into the Middle East to supervise the truce.

Secretary of State Kissinger said that U.S. forces had been alerted because of the "ambiguity" of the situation, after it had been learned that Soviet units had been alerted. The emerging crisis was defused when later that day the U.N. Security Council agreed to form a peace force that would not include the troops of the five permanent members of the Security Council, thereby excluding U.S. and Soviet forces.

Geneva Conference

The United States assumed a leadership role in attempting to bring about a peace settlement in the aftermath of the 1973 war. Kissinger negotiated a six-point cease-fire agreement which was signed by Egyptian and Israeli military representatives November 11, 1973, at Kilometer 101 on the Cairo to Suez road.

In late December 1973, largely through Kissinger's continuing efforts, the Geneva Conference was convened in accordance with U.N. Security Council resolution #338. *(See p. 77)* It recessed after only two days of largely ceremonial and propagandistic exchanges, but it symbolized the kind of semi-direct negotiations which have since then been the focus of attention.

Disengagement Agreements

Subsequent Kissinger shuttles produced troop disengagement accords between Israel and Egypt in the Sinai Peninsula (Jan. 14, 1974) and between Israel and Syria in the Golan Heights (May 31, 1974).

Meanwhile, the United States began to mend its relations with the Arab states. During a Kissinger visit to Egypt, Kissinger and Sadat agreed to resume U.S.-Egyptian diplomatic relations—broken since 1967. The resumption of U.S.-Syrian relations was announced in Damascus June 16, 1974, during a visit by President Nixon, whose Middle East tour also took him to Egypt, Israel and Saudi Arabia.

In March 1975, Kissinger sought to secure an agreement for a second-stage disengagement in Sinai as part of his step-by-step approach to peace in the Middle East. But after 15 days of talks with Sadat and Israeli Premier

---

"Peace must bring a new relationship among the nations of the Middle East, a relationship that will not only put an end to the state of war which has persisted for the last quarter of a century, but will also permit the peoples of the Middle East to live together in harmony and safety."

Secretary of State Kissinger
Geneva Conference, Dec. 21, 1973.

Yitzhak Rabin, Kissinger returned on March 23 to the United States calling his efforts a failure. Kissinger and Ford made clear at the time that they held Israel primarily responsible for the breakdown in negotiations, although they subsequently avoided any public assessment of blame.

In response to the failure of the talks, the United States began a "reassessment" of its Middle East policies. Consideration of Israel's request for $2.5-billion in U.S. aid was suspended until completion of the reassessment, a thinly veiled form of pressure on the Israeli government to be more forthcoming in talks with Egypt.

The stalled negotiations began again as President Ford met with Sadat in Salzburg, Austria, and with Rabin in Washington during June 1975. The United States hoped that Egypt and Israel would reach an interim agreement and thus avoid the risks of failure in a resumed Geneva Conference. As a culmination of Kissinger's resumed shuttle diplomacy, representatives of the two sides initialed a limited agreement Sept. 1. But it was recognized, at best, as only a hopeful first step toward a permanent overall peace settlement. *(Middle East Diplomacy chapter, p. 65)*

Step-by-Step, "Shuttle Diplomacy"

In retrospect, step-by-step diplomacy only set the stage for the Carter administration's efforts for an overall settlement to the Middle East conflict. Professor John Stoessinger assessed Kissinger's tactics in this way:

"We cannot say with certitude whether the critics of the step-by-step approach...were justified in their assertions that Kissinger avoided the heart of the conflict by refusing to address himself to the Palestinian problem. What the record does indicate, however, is that Kissinger has managed to narrow the differences between Israel and the Arabs more successfully than any other mediator in the long history of that tragic conflict."[4]

Journalist Edward R. F. Sheehan, who once suggested to Kissinger that he should have attempted to resolve the conflict himself, not just take a few small steps toward a future settlement, reported Kissinger's response as follows: "What were the alternatives? The conflict in the Middle East has a history of decades. Only during the last two years have we produced progress. It's easy to say that what we've done is not enough, but the steps we've taken are the biggest steps so far. They were *the attainable*—given our prevailing domestic situation."[5]

By the "domestic situation," Kissinger apparently had in mind the pressures applied by Israel's friends in the Congress and elsewhere whenever the demands being placed on Israel were thought to be excessive or unbalanced vis-a-vis the concessions being asked of the Arabs.

When Kissinger and President Ford began the "reassessment" of American Middle East policy in light of the March 1975 breakdown in shuttle diplomacy and alleged Israeli intransigence, the government developed three options for future Middle East strategy. The preferred option to emerge was, according to Sheehan:

> "The United States should announce its conception of a final settlement in the Middle East, based on the 1967 frontiers of Israel with minor modifications, and containing strong guarantees for Israel's security. The Geneva conference would be reconvened; the Soviet Union should be encouraged to cooperate in the quest to resolve all outstanding questions (including the status of Jerusalem) which should be defined in appropriate components and addressed in separate subcommittees."[6]

To circumvent attempts by the Jewish lobby and Israel to frustrate such a policy, Sheehan reports, "Kissinger's advisers envisioned Ford going to the American people, explaining lucidly and at length on television the issues of war and peace in the Middle East, pleading the necessity for Israeli withdrawal in exchange for the strongest guarantees."[7]

But Israel's supporters lobbied vigorously against this policy. The Jewish lobby on Capitol Hill produced a letter signed by 76 senators on May 21, 1975, strongly endorsing Israel's demands for "defensible" borders and massive economic and military aid.

It was a clear warning to Ford and Kissinger not to consummate the shift they were contemplating in American policies and strategies. Theodore Draper wrote in the April 1975 issue of *Commentary* magazine (published by the American Jewish Committee): "The consequences of attempting to impose a one-sided settlement on Israel, covered up by a less-than-convincing guarantee, could be traumatic for both Israel and the U.S."

In view of these developments, the preferred option to come out of the "reassessment" was shelved in late May. Sheehan wrote that Kissinger decided that "at some future date, when the President was stronger, when his prospects were more auspicious, he might go to the people with a plan for peace based upon the first option."[8] Consequently,

Kissinger returned to shuttle diplomacy and by September brought about a second disengagement agreement between Israel and Egypt in Sinai.

This agreement resulted in considerable tension among the various Arab parties to the conflict. Sadat was strongly denounced by Syria's Assad and by the PLO for agreeing to what amounted, in some views, to a separate though temporary peace with Israel. This dissension within the Arab camp became an important factor in the Lebanese civil war which escalated the following year, becoming, in a sense, an Arab civil war. *(See Lebanon chapter, p. 46)*

Brookings Report

In December 1975 the Brookings Institution, one of Washington's most prestigious think tanks,[9] released a broadly supported study entitled "Toward Peace in the Middle East." This study was frequently cited during the coming months as an outline of the policies the United States should pursue in attempting to bring about a comprehensive Middle East settlement, step-by-step diplomacy having apparently reached its end.[10]

The Brookings report was signed by 16 well-known Middle East specialists and scholars including a number of prominent American Jews. It was also signed by Jimmy Carter's prospective National Security adviser, Zbigniew Brzezinski, and by Brzezinski's selection as Middle East expert at the National Security Council, William Quandt.

In brief, the Brookings report contained the following recommendations:

- Israeli withdrawal to the 1967 borders with minor, mutually agreed modifications;
- Recognition of "the principle of Palestinian self-determination";
- Resolution, probably at a resumed Geneva Conference, of all outstanding issues, including Jerusalem, thus leading to peace between all of the parties;
- Implementation of the agreement in stages over a number of years;
- Arab recognition of Israel, conclusion of a peace treaty, and normalization of relations;
- Some arrangement for multilateral and bilateral guarantees for Israel's security, with the United States probably playing a unique role.

Lull During the Presidential Campaign

The year 1976 was one of little movement in the Middle East and, in the United States, attention focused on the presidential elections. Both President Ford and candidate Carter competed for Jewish votes and support with ever-increasing pledges for Israel's welfare. "By the end of the campaign," one scholar noted in mid-1977, "it was difficult to conceive of a more pro-Israel candidate than Jimmy Carter."[11]

Still, the tension in U.S.-Israeli relations that appeared shortly after Carter took office as President was foreseeable throughout 1976.[12] A *Washington Post* editorial writer noted

4 John G. Stoessinger, *Henry Kissinger: The Anguish of Power* (Norton: 1976), p. 221.
5 Edward R. F. Sheehan, *The Arabs, Israelis and Kissinger* (Reader's Digest Press: 1976), p. 201.
6 *Ibid.*, p. 166.
7 Edward R. F. Sheehan, "Step by Step in the Middle East," *Foreign Policy*, Spring 1976, p. 58.
8 *Ibid.*, p. 176.

9 A review of the studies done on the Middle East by Washington's major think-tanks can be found in Mark A. Bruzonsky, "What Think-Tanks Think About the Middle East," *National Jewish Monthly*, December 1976.
10 The Brookings report became the central focus of an important series of hearings held by the Subcommittee on Near Eastern and South Asian Affairs of the Committee on Foreign Relations of the Senate during May, June and July 1976. The Brookings report along with considerable other useful information is printed in the report of these hearings entitled "Middle East Peace Prospects."
11 Stephen L. Spiegel, "Carter and Israel," *Commentary*, July 1977, p. 35.
12 For a review of developments during this period, see Mark A. Bruzonsky, "U.S. & Israel: The Coming Storm," *Worldview*, September 1976. Also see Georgiana G. Stevens, "1967-1977: Ameirca's Moment in the Middle East?" *The Middle East Journal*, Winter 1977.

Accelerating Arms Race in Middle East . . .

Four Arab-Israeli wars, along with interim clashes and inter-Arab feuding, have sparked one of the world's most intense arms races. Since the quadrupling of oil prices that began during the 1973 war, Middle Eastern countries have found the means to purchase vast new arsenals of the most modern weapons. Western governments, anxious to pay for the costly oil, have plunged into the Middle East arms market with gusto. None has pursued the sale of weapons in the region with more enthusiasm and success than the United States.

In 1974, the Middle East took 57 per cent of all arms exported. And the figure has steadily risen since then.

The Middle East accounts for the bulk of the steep rise in U.S. arms sales. From total sales of less than $1-billion in 1970, sales had jumped nearly ninefold by fiscal 1974, to a total of more than $8-billion. In the past few years sales have hovered around $9- $10-billion. During the years 1973 through 1976 Iran alone purchased $10.9-billion, Saudi Arabia $7.8-billion and Israel $4.4-billion—all from the United States. France and England also have cashed in on the booming arms trade. The Arab countries have also decided to attempt to produce some arms themselves. Money for Egyptian arms purchases and capital investments is being provided largely by Saudi Arabia, Kuwait and smaller Persian Gulf states.

The Egyptians began complaining that the Soviets were dragging their feet in resupplying them with weapons lost during the 1973 war. Then as relations between Egypt and the Kremlin worsened, President Sadat began looking to the West. He had hoped to find the United States willing to supply Egypt with defensive arms, but as of mid-1977 only a few cargo airplanes had been approved. Yet the Egyptian Ambassador to the United States, Ashraf Ghorbal, stated in a May 1977 interview that "If you are going to play...a very effective role, politically and economically, you must also help militarily."

U.S. Aid to Israel

U.S. military assistance to Israel entered a new phase after the 1973 war. Prior to that time, all U.S. assistance to Israel had been on either a cash sale or credit basis. But because of the magnitude of Israel's needs and the heavy toll the war had taken on the Israeli economy, the United States agreed to provide for the first time outright grants of military aid.

President Nixon Oct. 19, 1973, called on Congress to approve his request for $2.2-billion in military aid to Israel in order "to prevent the emergence of a substantial imbalance resulting from a large-scale resupply of Syria and Egypt by the Soviet Union." The U.S. decision to resupply Israel triggered the Arab oil embargo, but this did not deter the United States.

Of Israel's $2.5-billion aid request in 1975, $1.6-billion was to be for military assistance. The State Department held up the request to Congress during the "reassessment" period pending Middle East developments, but requested the aid in September in the wake of the new Sinai accord. Israel received in fiscal 1975 about $1.5-billion in military assistance aid and $740-million in economic assistance. In 1976 Israel received about the same amount of economic assistance but only $1-billion in military credits, half to be forgiven and the other half financed through special loans.

By 1975 Israel's armaments were substantially in excess of what she had when the Yom Kippur war had broken out just two years earlier, from 1,700 to 2,700 tanks and from 308 combat aircraft to 486. Both the Arab and Israeli sides were clearly competing in a massive arms race, one which continued in 1977 notwithstanding the constant talk of peace in the area.

Previous U.S. Policy

This overwhelming U.S. support for Israeli arms assistance was in marked contrast to U.S. policy of less than a decade earlier. The United States, anxious to line up Arab allies in the cold war and to preserve its oil interests in that area, had been reluctant to supply Israel with weapons in the 1950s and early 1960s.

Tripartite Declaration

The first—and last—great power agreement to limit the flow of weapons to the Middle East was the U.S.-British-French declaration of May 25, 1950. In response to the Israeli argument that an arms race would threaten Israeli military superiority, the three powers declared their opposition to the "development of an arms race between the Arab states and Israel." The declaration was silent on arms races among Arab states.

From 1950 to 1955, the declaration had the effect of keeping Arab-Israeli arms rivalry at a low pitch. But Britain, the major arms supplier during this period, found itself drawn into an arms race between Egypt and Iraq, long-standing rivals in the Arab world. To maintain its oil interests in Iraq and to assure Iraq's association with the West in the cold war, Britain furnished aircraft and training to create an Iraqi air force. To assure Egypt's acceptance of continued British control of the Suez Canal, Britain created an Egyptian air force. Israel, realizing that the Tripartite Declaration was breaking down, began its search for arms suppliers.

Israel's success in finding arms was displayed in a February 1955 retaliatory raid into the Gaza Strip in which Egyptian forces suffered a number of casualties. Egyptian President Gamal Abdel Nasser sought Western arms to restore the military balance but was refused. After this refusal—coupled with the establishment of a U.S.-British-sponsored mutual defense treaty linking Britain, Iraq, Turkey, Iran and Pakistan—Nasser turned to the Soviets and convinced them Egypt could be useful in their efforts to destroy Western positions in the Middle East. Soviet bloc weapons began flowing to Egypt in September 1975.

France Arms Israel

Following disclosure of the Czech-Egyptian arms agreement, Israel intensified its search for an arms

. . . Is Part of Larger Superpower Rivalry

supplier in the West, approaching the Eisenhower administration in September 1955. After six months of delay, Israel was informed—through testimony by Secretary of State John Foster Dulles before the Senate Foreign Relations Committee, that it should rely for security not on U.S. arms but on the "collective security" of the United Nations. Turning then to Britain, Israel was advised by Prime Minister Anthony Eden to make concessions to the Arabs to avoid war.

France viewed the Soviet-Egyptian arms agreement in the context of its colonial war with the Moslem rebels in Algeria. Concerned that Soviet arms would find their way from Cairo to Algiers, France first sought an agreement from Nasser not to assist the Algerian rebels. When Nasser refused, France decided to honor Israel's request for arms. Arming Israel was viewed as a way to pin down Soviet-supplied weapons in Egypt. But, in the mid-1960s, France began to re-examine its arms sales policy to the Middle East in the light of the conclusion of the Algerian war in 1962. France began discussing arms sales with Arab countries—Jordan, Lebanon and Egypt. The apparent shift in French sales policy led to increased Israeli pressure on the United States to relax its restrictions on arms sales to Israel.

Another factor pushing Israel toward the United States was the disclosure in December 1964 and subsequent termination of secret West German arms sales to Israel. When the secret sales became known, Egypt threatened to recognize East Germany. The Egyptians' threat caused Bonn to cease abruptly its arms flow to Israel.

U.S. Arms to Israel

The shift in French policy, the termination of German arms and the continued flow of Soviet arms to Egypt, Syria and Iraq forced the United States in the early 1960s to reconsider its position on supplying weapons to Israel. In 1962, Israel obtained U.S. surface-to-air Hawk antiaircraft missiles and in 1965, Patton tanks; in 1966 an agreement was reached for the first sale of U.S. combat aircraft to Israel. The agreement called for the sale of 48 A-4 Skyhawk fighter-bombers to Israel, delivered in 1968. Pressure built up during the 1968 presidential campaign to supply Israel with the most advanced aircraft. Both party platforms supported efforts to end the arms race in the Middle East; but, in the absence of arms control, both called for providing Israel with the latest in supersonic aircraft.

Congressional Mandate

Similar support was shown on Capitol Hill. The fiscal 1969 foreign aid authorization act (PL 90-554) contained a provision calling on the President to negotiate the sale of supersonic aircraft to Israel to provide it with a deterrent force capable of "preventing future Arab aggression." In October 1968, President Johnson announced that he had instructed Secretary of State Dean Rusk to open negotiations for the sale of jets to Israel. The sale of 50 Phantom F-4s was announced by the administration in late December.

In the fiscal 1970-1971 Foreign Military Sales Act (PL 91-672), Congress called for arms control negotiations in the Middle East and expressed support for the President's position that arms should be made available to Israel and other friendly states in order to meet threats to their security. Also in 1970, in the fiscal 1971 defense procurement authorization (PL 91-441), Congress expressed "grave concern" over "the deepening involvement of the Soviet Union in the Middle East and the clear and present danger to world peace resulting from such involvement which cannot be ignored by the United States." The legislation authorized the transfer of an unlimited amount of aircraft and supporting equipment to Israel by sales, credit sales or guaranty.

Policy Issues

The sudden flow of arms sales to the Middle East has aroused growing concern in Congress. Sen. Edward M. Kennedy (D Mass.) June 18, 1975, urged a six-month moratorium on arms sales to the Persian Gulf to prevent hooking "these nations on the heroin of modern arms."

Congressional critics have argued that the sales actually spur arms races rather than promote regional stability and that the United States is not able to control the use of the weapons once they have been sold. The critics also warned that the weapons can be used by totalitarian regimes to suppress legitimate interest groups within their own countries and that weapons sold to such countries as Saudi Arabia and Kuwait might one day be used against Israel.

Carter Administration

During the campaign, Jimmy Carter stressed: "We cannot be both the world's leading champion of peace and the world's leading supplier of weapons of war."

Soon after becoming President, Carter ordered a review of military sales policies. In May, Carter announced that "the United States will henceforth view arms transfers as an exceptional foreign policy implement.... The burden of persuasion will be on those who favor a particular arms sale rather than on those who oppose it." But the Carter policy in practice immediately faced challenges. Israel, for one thing, was not listed among those countries to receive favored nation treatment. This gave rise to a mini-confrontation between the administration and Israel's Capitol Hill supporters. The result was a special presidential letter affirming America's unique responsibilities to Israel.

By mid-1977 the Carter administration had approved $4.5-billion in worldwide arms sales. A number of co-production agreements with Israel were not approved, but Carter did request in June an extra $115-million in military equipment for Israel, plus an additional $250-million in July when Prime Minister Begin visited the United States. Sales to Saudi Arabia and Iran continued at an unprecedented level, small amounts of arms continued to flow to Jordan and some arms sales to Egypt were being considered.

during the campaign that "barring a shift by Israel—regardless of who wins the November election in the U.S.—a wracking American-Israeli showdown is near."[13] And the editor of the Jewish lobby's weekly newsletter, Wolf Blitzer, wrote in *The Jerusalem Post* that "The real crunch for Israel will probably come during 1977 if Ford is elected—it will be delayed by only a few months if a Democratic candidate wins."[14]

Carter's Middle East Policy

Jimmy Carter was quick to grasp the fundamentals of the Middle East stalemate. By March he had begun to outline a three-point Middle East settlement plan involving:
- Israeli withdrawal to approximately the 1967 borders;
- Creation of a "Palestinian homeland" probably in the West Bank and Gaza Strip areas;
- Establishment of real, lasting, permanent peace between Israel and her Arab neighbors.[15]

Carter was the first American President to discuss the concept of a "Palestinian homeland." It first came as a shock, for just two years earlier when asked if he had plans for dealing with the Palestinian problem, Henry Kissinger responded to a questioning journalist, "Do you want to start a revolution in the United States?"[16] But within a short time there was general recognition that some form of Palestinian self-determination must be granted if there was to be a chance of achieving any comprehensive settlement.

However, Carter refused to specify what he had in mind even after completing, by May, a first round of personal meetings with all the important leaders of Middle East countries. As of mid-1977 there were those who insisted Carter was aiming for a Palestinian state in the West Bank and Gaza Strip, perhaps loosely affiliated with Jordan; while others felt that what Carter had in mind was a Palestinian region federated with Jordan and excluding political control by the Palestine Liberation Organization. There was also some discussion of the U.S. offering Israel unspecified guarantees against future Arab attacks should Israel agree to withdraw from the occupied territories.

Menahem Begin's victory in Israel confused the picture even further. Not only had Begin's Likud party refused to contemplate the idea of a Palestinian homeland, but it was elected on a platform of never returning the West Bank and Gaza Strip areas to Arab sovereignty.

Begin came to the United States on July 19 for two days of talks with Carter and members of his unofficial Middle East team which included Brzezinski and Quandt at the National Security Council, Secretary of State Vance, Intelligence and Research Director Harold Saunders, and Assistant Secretary for Near Eastern Affairs Alfred Atherton at the State Department.

Although the atmosphere was cordial, it was clear during Begin's visit that the new American administration and the new Israeli government were very far apart on many important issues.

How these differences could be resolved seemed difficult to envisage during the summer of 1977, especially since American Middle East policy has always been greatly influenced by American domestic politics. Even before

Begin's American visit his personal envoy, Samuel Katz, had told American Jews, "We are confident that the Jewish community in America will stand out courageously and challenge its government if it becomes necessary."[17]

By mid-1977 Carter appeared to be very much aware of the criticism of his policies and of the possibility of a major battle with Congress and Israel's supporters should he continue to pursue them. Ford and Kissinger had faced a similar situation in the spring of 1975, but the Middle East pressures on the American government to find some way to defuse the time-bomb in the Middle East were much greater two years later.

An article in the American Jewish Committee's *Commentary* magazine in July 1977 noted that "On the basis of his administration's performance so far, there are grounds for fear and not too many grounds for hope."[18] And the liberal weekly, *The New Republic*, at the time of Begin's election, predicted that "We cannot rule out a monumental political crisis between Israel and the Carter administration, and between the administration and Israel's friends in Congress, with Americans forced to choose up sides in a bitter fight."[19]

But there was some reason for hope, as well. Carter had begun to define the basic outline for a comprehensive settlement, something Kissinger had not been able or willing to attempt. Though such a settlement might have to be implemented in steps, over a period of many years, the approach offered the possibility, at least, of eventual peace. But it was only a hope and only a possibility, and the dangers for the warring states in the Middle East and for the mediating superpower, the United States, were substantial.

Imposing a Settlement?

Though President Carter and Secretary of State Vance have repeatedly stressed that the United States would not and could not impose a settlement on both the Arabs and the Israelis, the possibility continued to be discussed in 1977. Former Under Secretary of State, George Ball, wrote in the April 1977 issue of the Council on Foreign Relation's journal, *Foreign Affairs*, that what Carter does about the Middle East will be the "acid test of political courage and decisiveness. If America should permit Israel to continue to reject inflexibly any suggestion of a return to earlier boundaries and the creation of a Palestinian state and to refuse even to negotiate about Jerusalem, we should be acquiescing in a policy hazardous not only to Israel but for America and the rest of the world. That would not be responsible conduct for a great power?"[20]

But Ball did not address himself to the problem of how an imposed settlement might also have to include imposition on a number of the Arab parties of a complete peace with normal relations with Israel, including credible American guarantees against any future warfare. It also appeared doubtful to many observers whether a decision to impose a settlement was domestically possible, in view of heavy pressures against the idea, or it was internationally realistic. But it was a measure of the seriousness of the situation that the discussion continued. One noted New York international affairs lawyer, Rita Hauser, stated in a

13 Stephen S. Rosenfeld quoted in Mark A. Bruzonsky, "Carter and the Middle East," *The Nation*, Dec. 11, 1976, p. 615.
14 *Ibid.*, p. 618.
15 Carter's initial handling of the Middle East problem is outlined in Mark A. Bruzonsky, "Mr. Carter Grasps That Nettle." *The Nation*, April 23, 1977.
16 Sheehan, op. cit., p. 167.

17 Joseph Lelyveld, "Katz in 'The Mountains' ", *The New York Times Magazine*, 10 July 1977, p. 63. Katz's attitude may have been in response to what was perceived as a Carter attempt to divide the American Jewish community from Israel.
18 Spiegel, op. cit., p. 40.
19 "Begin With Begin," *The New Republic*, 28 May 1977.
20 George Ball, "How to Save Israel in Spite of Herself," *Foreign Affairs*, April 1977, p. 471.

New York Times editorial that "Ball's thesis...has come to dominate the U.S. foreign policy establishment."[21]

"Taking a 'worst case' view" of possible future developments, Israeli scholar Shimon Shamir in 1976 offered "some of the critical problems that may ensue for the world community and particularly for the United States" should there be no progress toward ending the Middle East conflict. Among these, he cited: "A concentrated Soviet effort to establish firm control in parts of the Middle East; the emergence of a superpower confrontation from a local conflict; the proliferation of nuclear and other nonconventional weapons; interference with the flow of oil from the Middle East to the industrialized states; disruption of the international economy by Arab financial maneuvers; inter-Arab conflicts upsetting the status quo in the region; the collapse of pro-Western Arab regimes such as those of Jordan and Lebanon; an increase in the volume and effectiveness of terrorism (possibly with the development of more mobile and sophisticated weaponry); an accelerated arms race, which may put Western suppliers at a disadvantage; economic crisis and shortages in countries which possess limited resources, necessitating substantial economic aid; and perhaps even subversion and other interference by Communist China."[22]

With this list of dangers added to America's historical commitment to Israel and her growing relationship with the Arab world, the Middle East and its problems were sure to remain at the top of Washington's list of foreign policy priorities for the forseeable future.

---

21 Rita Houser, *The Washington Post*, 26 April 1977.

22 Shimon Shamir in A. L. Udovitch (ed.), *The Middle East: Oil, Conflict & Hope* (Lexington: 1976), pp. 229-230.

A POTENT, EFFECTIVE FORCE ON U.S. POLICY

In May 1975, 76 members of the United States Senate sent President Ford a special letter. "We urge you to make it clear, as we do, that the United States, acting in its own national interests, stands firmly with Israel in the search for peace in future negotiations, and that this promise is the basis of the current reassessment of U.S. policy in the Middle East."

The letter was the result of a major campaign by the American Israel Public Affairs Committee (AIPAC) urging the Senate to make "a significant contribution to the reassessment of American policy in the Middle East," as AIPAC's influential newsletter reported.[1]

The policy "reassessment" resulted from a breakdown in Secretary of State Henry A. Kissinger's "shuttle diplomacy" earlier that year, for which Kissinger and Ford publicly blamed Israel. As journalist Edward R. F. Sheehan later reported, "Kissinger's much trumpeted 'reassessment' of American policy in the Middle East was his revenge on Israeli recalcitrance." The letter of the 76 senators, Sheehan noted, "was a stunning triumph for the lobby, a capital rebuke for Kissinger in Congress." "It was," wrote Sheehan, "the Israeli lobby that dealt reassessment its *coup de grace.*"[2]

What is popularly thought of as the "Israeli lobby" is, beyond doubt, one of the most powerful influences on American Middle East Policy in the nation's capital. Still, it is often credited with an undeserved omnipotence. It is doubtful if Senator James Abourezk's (D S.D.) assertion that the lobby has the "ability to accomplish virtually any legislative feat involving military or economic assistance to Israel" is true. More accurate is the understanding that AIPAC has the ability to maximize pro-Israeli sentiments while often promoting what many consider anti-Arab policies. The billions of dollars in aid that Israel receives from the United States is an example of this maximizing influence. The 1977 legislation restricting compliance with the Arab economic boycott of Israel is an excellent example of this latter power.

Moreover, the "Israeli lobby" is not in reality a foreign agent required to register with the Justice Department and the Congress. It is a lobby financed and supported primarily by the American Jewish community. AIPAC's ability to stir the grass roots throughout the United States when support is needed is well known. While AIPAC and the Israeli government are in close and constant contact, it would be an exaggeration to say that Israel controls AIPAC. There have been instances where AIPAC has promoted policies or used tactics not fully supported by the Israelis.

In short, AIPAC is a unique institution mirroring the unique relationship between the American Jewish community and the state of Israel.

Effectiveness

During the past few years, AIPAC, which serves as the coordinator for the efforts of nearly all Jewish organizations, has been very active on a number of issues including:

● Amendments to the 1974 Trade Reform Act linking trade benefits to Communist emigration policies and limiting U.S. loans and credits to those countries.

● Requiring stringent restrictions on the sale of Hawk surface-to-air missile batteries to Jordan.

● A resolution calling for re-examination of U.S. membership in the United Nations if Israel were expelled.

● A bar on future U.S. support for the United Nations Educational, Scientific and Cultural Organization (UNESCO) because of its anti-Israel actions.

● Financial assistance for refugees from the Soviet Union and other Eastern European countries, with Israel receiving 80 per cent of the money.

● An agreement by the Ford administration not to supply weapons to Egypt during 1976 except for six C-130 transport planes.

● An amendment by Senator Abraham Ribicoff (D Conn.) to the Tax Reform Act of 1976 penalizing those American firms that complied with the Arab boycott of Israel by denying certain tax benefits on foreign source income; along with other anti-boycott legislation passed by the 95th Congress in 1977.

Congressional Anxiety

In a 1976 year-end summary, AIPAC reported that during the 94th Congress it had "vigorously countered" the "moves to 'reassess' America's relationship with Israel." "On the basis of this fine record of congressional support for closer U.S.-Israel cooperation, we are confident that the 95th Congress will demonstrate the same awareness of fundamental U.S. interests by supporting a secure, viable Israel,"[3] AIPAC noted in October 1976.

There is, however, a growing uneasiness in Congress regarding the Middle East predicament and it is reflected in a level of criticism of AIPAC unheard of just a few years ago. Most of this criticism is quiet and almost always "off-the-record." But the statements of General George S. Brown, Chairman of the Joint Chiefs of Staff, concerning the excessive influence of the "Jewish lobby" have received extensive press coverage. And journalists Russell Warren Howe

"What's good for American society is terribly important to the Jewish community."

—Hyman H. Bookbinder, Washington representative, American Jewish Committee

1 *Near East Report,* 28 May 1975, p. 93.

2 Edward R. F. Sheehan; *The Arabs, Israelis, and Kissinger* (Reader's Digest Press: 1976), pp. 165, 173, 176.

3 *Near East Report,* 27 October 1976, p. 179.

and Sarah Hays Trott uncovered substantial criticism when researching their book *The Power Peddlers, A Revealing Account of Foreign Lobbying in Washington.*[4]

Assessments of the strength of the Jewish lobby depend largely on the perspective of the assessor. Few, however, would agree with the president of Hadassah, the Jewish women's organization, who terms the Jewish lobby a "myth" created by journalists.

Senator Abourezk, of Lebanese origin and one of the few members of Congress espousing Arab interests, has said he is "envious" of the Jewish lobby. Abourezk told a Denver audience in March 1977 that "The Israeli lobby is the most powerful and pervasive foreign influence that exists in American politics." Others disagree. "There is a lot of mythology about the Jewish lobby," said Richard Perle, an aide to Senator Henry M. Jackson (D Wash.).

In any event, the lobby's effectiveness will apparently be tested by the Carter administration. President Carter, within a few months after entering office, outlined some clear differences between the United States and Israel concerning resolution of the Middle East stalemate. *(See U.S. Policy, p. 80)*

Even before Carter outlined his views, columnist Joseph C. Harsch wrote that there is "a desire and intention on Mr. Carter's part to regain the control over aid and support to Israel which President Eisenhower asserted and kept." "It will come down to a test of strength in Washington between the White House and the Israeli lobby," Harsch concluded. "The lobby has won most rounds since the days of Lyndon Johnson. Which will win this new round? It will be a fascinating test of Mr. Carter's political skill and strength."[5]

AIPAC

Organization

AIPAC's origins are in the American Zionist Council which originally promoted support for Israel's creation and welfare. The present name was adopted in 1954 and until 1975 the organization was headed by I. L. (Si) Kenen. By 1977, it functioned as an umbrella lobbying organization with a staff of more than 20, an annual budget of more than $500,000 and four registered lobbyists—an executive director, Morris J. Amitay, who replaced Kenen, legislative director Kenneth Wollack, and legislative liaisons June A. Rogul and F. Stephen McArthur.

AIPAC's money comes from individual donations ranging from $25 to $5,000, which are not tax-exempt. More than 15,000 members contribute to AIPAC. Presidents of most of the major Jewish organizations sit on its executive committee and the Washington representatives of more than a dozen of these organizations meet weekly at AIPAC's offices. AIPAC does not contribute to candidates or rate members of Congress.

Separately incorporated from AIPAC is Near East Research Inc., a nonprofit organization supported by subscriptions to its weekly newsletter, *Near East Report*, and individual donations. It conducts research on the Middle East and the Arab-Israeli conflict. Si Kenen, who in 1977 remained as honorary chairman of AIPAC, was its president and he contributed a column to *Near East Report*.

The newsletter, quartered with AIPAC near Capitol Hill, has an estimated readership of 30,000 and is legally

4 Doubleday, 1977. See Chapter Six, "The Mideast Conflict."
5 Quoted in Mazin Omar, "A Test for Carter's Mettle," *The Middle East* (London), April 1977, p.10.

separate from AIPAC. But Amitay is a contributing editor and his organization pays for mailing about 4,000 copies to every member of Congress, embassies, executive branch officials and United Nations delegations. The popular perception of the newsletter throughout Washington is that *Near East Report* is AIPAC's newsletter.

Operations

"The basic axiom of our work," said Amitay, "is the basic support for Israel that already exists. There is a broad base of support that shapes congressional feeling and the administration's. It makes our job a lot easier."

More and more AIPAC's job is to counter the lobbying of pro-Arab groups and what Wollack has called the "petro-diplomatic group," which have greater financial resources but less power, in the view of AIPAC. The organization basically is a one-issue lobby. "We are pro-Israel—I don't deny that," explained Amitay. "But we make our case on the basis of American interests. ...Our point of strength is not that we're so highly organized, but that so many people are committed to Israel. We have a lot of non-Jewish support. I'm very glad we don't have to rely just on Jews; if we did, we would be an ineffective group."

Amitay said the organization does "what all lobbies do"—informing Congress and the executive branch on the issues of importance to it—U.S.-Israeli relations, peace negotiations and military and economic assistance. It testifies for itself and on behalf of other Jewish groups, as it did on the proposed arms sale to Jordan. It provides members of Congress with information and encourages its members to communicate their views. "We keep in touch with other Jewish organizations," Amitay said, "but we're not a button-pressing organization" when it comes to mobilizing their members.

Amitay, since taking over in 1975, has greatly expanded AIPAC's Capitol Hill operations. Jews, Amitay explained, were "...concerned that after the Yom Kippur war [in 1973] that the United States' efforts to improve relations with Arab states not be at the expense of the security of Israel." Under Amitay AIPAC has moved into new, roomier headquarters a short distance from the Capitol.

Publicly, and for obvious reasons, AIPAC keeps its distance from the Israeli embassy in Washington. Amitay even asserts that "We maintain no formal relationship or substantive connection with the Israeli embassy." Many in Washington, take issue with the assertion of complete separation. However, the relationship appears to be close and often intimate—Amitay visits Israel regularly.

Evaluation

Amitay said, "I think we've had some effect." Capitol Hill staffers are more generous in their appraisal. "AIPAC is very effective," said a former staffer in the Senate minority leader's office. "They have a good grass roots operation, which is vital. It can deliver letters, calls to members from their home state. At any given moment, it can mobilize."

"It's effective with very little in the way of resources," said an aide to a Democratic senator.

A House source called AIPAC's involvement "helpful" in preventing the sale of anti-aircraft equipment to Jordan. Besides its Senate testimony, it mailed information to members of Congress on the possible impact on Israel of the sale. It also described this in mailings to Jewish groups across the country. One House aide estimated that it helped round up an additional 40 cosponsors to a resolution (H Con

Res 337) opposing the sale, which had more than 125 cosponsors.

Though widely acclaimed, AIPAC also has its detractors. One Washington professor often consulted by the Israeli Embassy on political matters acknowledged, when questioned about AIPAC, that "in the past two years I've heard more anti-lobby sentiment than in all the years before." And a former Foreign Service officer who worked on Middle East matters for a Democratic senator felt "AIPAC often does with a sledgehammer what should be done with a stilleto." "Many senators damn well resent the methods used," he said. Indeed, when Carl Marcy, former chief of staff of the Senate Foreign Relations Committee, who in 1977 was editor of the *Foreign Affairs Newsletter,* decided to investigate AIPAC's methods in enlisting the 76 senators who signed the May 1975 letter to President Ford, he was struck by the ability of pro-AIPAC groups to counter such an investigation. Marcy concluded that "after broaching the subject to several individuals and groups, we confess we were intimidated by such comments as the following: 'Good idea; do you have independent means to support your newsletter?' 'Don't be stupid, damned few people will think any presentation balanced no matter how hard you try.' 'What are you, nuts or something?' "

Other Groups

Other major Jewish groups play a less direct political role than AIPAC, leaving much of the explicit lobbying to it and devoting more of their time to issues besides Israel. Beyond their relatively small Washington operations, they have influence as sources of information to their members and the public through mailings and publications and as forums for political figures.

American Jewish Committee

"I'm concerned that there could be an impression that the Jewish community is interested only in Israel," said Hyman H. Bookbinder of the New York-based American Jewish Committee. "If we have to list our priorities, obviously that is the top one. In my own operation, it is my top priority, but we also pursue a variety of other issues." He listed the extension of the Voting Rights Act, voter registration by postcard, housing issues and civil rights as among those his two-person office follows. And they keep up with executive branch actions, and comment on regulations and proposals.

Behind that course is Bookbinder's belief that "we can't expect people to be interested in our issue and in the security of Israel unless we're interested in theirs. What's good for American society is terribly important to the Jewish community." A long history of scapegoating, he said, shows that "when there are social ills, Jews get it."

Additional Support

B'nai B'rith's Anti-Defamation League also has a Washington representative, David A. Brody, a registered lobbyist. The league, too, is particularly interested in domestic issues, not just Israel. Bookbinder and Brody, as well as individuals active in other major Jewish organizations, are on AIPAC's 70-member executive committee.

But the amount of lobbying that any of them do is limited by law because of their tax status; contributions are exempt. "We obviously work with each other, cooperate and share information," Bookbinder observed. "But every Jewish group bends over backwards to comply with the law. So we welcome an explicit operation like AIPAC."

An aide to a Republican senator insisted that Middle East issues are not lobbied only by Jewish groups, because broader questions than just Israel usually are involved. On the proposed arms sale to Jordan, for example, Americans for Democratic Action, the AFL-CIO and other unions worked against it, he said. Non-Jewish groups were active as well in barring U.S. payments to UNESCO in late 1974.

Among the other Jewish organizations which work closely with AIPAC in Washington are the American Jewish Congress, various religious groups, the Institute for Jewish Policy Planning and Research, the National Council of Jewish Women and Hadassah.

Since the Yom Kippur War, debate about Israeli positions and about U.S. Middle East policy has caused serious tensions even within the American Jewish community. A new organization, "Breira" which in Hebrew means "alternative," has acted as a catalyst for debate on both Israeli and American policies. The Jewish establishment has vigorously challenged Breira which at times has voiced an opinion separate from AIPAC on Capitol Hill. Many people believe that the existence of significant Jewish dissent from the usually hard-line policies promoted by AIPAC could have some influence on Carter administration Middle East policy and on congressional thinking.

Effectiveness

"When I have a question, I'll call one of those groups," said one aide who has worked for more than 10 years for several House members. "It's an educational situation, though, not overt lobbying like AIPAC, which keeps members constantly informed." An aide to a Democratic senator said he feels that these Jewish groups fill a need by being active in social programs, but, in his view, are "necessarily less effective" as an influence on Middle East policy because they are not one-issue-oriented.

Bookbinder believes the nationwide personal contacts the organizations' members have with influential members of Congress and executive branch officials are as important as any Washington operation.

Citizen Support

In the view both of people on Capitol Hill and of representatives of Jewish organizations, the so-called Jewish lobby gains much of its power from citizen activism—both Jewish and non-Jewish—and widespread public backing for the longstanding U.S. policy of support for an independent Israel.

Jewish Community

"The influence of the Jewish lobby is not a result of people walking the halls of Congress, but of people back home," said Perle. "What happens in Washington is really modest compared to what happens in the district. A newspaper story or an event in the Mideast can trigger an immediate response." Still, those familiar with AIPAC know how quickly a telephone or telegram campaign can be mobilized if AIPAC gives the signal.

According to 1973 census statistics, there are about 5.7 million Jews in this country, 2.8 per cent of the population. But with their higher political participation, they account for about 4 per cent of the vote nationally, wrote Stephen D.

AIPAC's Tough Executive Director

Morris J. Amitay, known around Washington as Morrie, became in 1975 AIPAC's second executive director in its 23-year existence. Amitay, 41, brought to his job a Jewish heritage, foreign affairs experience in the State Department and five years as a legislative assistant in Congress.

He replaced I. L. (Si) Kenen, who retired at age 70 but continued as president of Near East Research Inc. which publishes the weekly newsletter *Near East Report.* AIPAC's new and roomy office on North Capitol Street near Capitol Hill keeps its front door locked and scrutinizes visitors on a monitor for security reasons.

Kenen once told an interviewer that he rarely went to Capitol Hill to lobby, because support for Israeli  causes already was so strong. Amitay conducts a different operation, although support in Congress remains firm. "I'm trying to change things here from essentially a one-man operation to having many qualified people," Amitay said in an interview when he first assumed his position. "We hope for a greater presence on the Hill." AIPAC by 1977 had expanded to four registered lobbyists and a budget in excess of half a million dollars.

"Until a few years ago, this was a public affairs organization," Amitay continued. "Now it's a nuts-and-bolts operation, analyzing legislation, gathering information, keeping up with the issues." Though AIPAC has grown since Amitay assumed the directorship, the organization has remained tightly controlled. "I put a premium on working quickly, on a quick response. That's why I like being fairly small and un-bureaucratic," he has said.

Amitay is known as a tough, no-nonsense partisan. He is a highly efficient, though sometimes abrasive fighter who knows how to pull the right congressional levers. He is considered a much more controversial personality than his predecessor.

Journalists Russell Howe and Sarah Trott in their study of Washington lobbyists have focused on the personality difference between Amitay and Kenen. "Kenen stands in contrast to Amitay, the New Yorker whom he chose as his successor and who enjoys boasting of AIPAC power and the facility with which it procures 'confidential' documents from senatorial offices," the two authors have written. "Amitay seems to detect Hitlerian tendencies in all who disagree with Israel. Kenen, in contrast, attended the 1975 convention of the National Association of Arab Americans.... When pointed out in the audience..., he drew some laughs but did not appear to mind. Few could imagine Amitay submitting to this experience."[6]

6 The Power Peddlers: A Revealing Account of Foreign Lobbying in Washington (Doubleday: 1977), p. 280.

Isaacs in his book, *Jews and American Politics.* The electoral college system serves to further multiply that power, especially since Jews are concentrated in states with the greatest electoral votes.

An aide to Rep. Stephen J. Solarz (D N.Y.) noted that the "American Jewish community is very active and very well informed, especially in foreign policy matters, and it will make its views known." While this community is generally considered more liberal than the public at large on most issues, it draws some support from conservatives who have anti-Communist views and reservations about detente with the Soviet Union.

Bookbinder defined the Jewish lobby as "the totality of Jewish influence in America. If AIPAC went out of life tomorrow, it wouldn't mean the death of the Jewish lobby." He included the influence of academics, the business community, professionals, and socially and politically active individuals, including contributors to cultural and social causes as well as political campaigns.

"The idea of Jewish giving in return for support for Israel is very, very unfair," Bookbinder continued. "Compared to giving of others, like labor, corporations, environmentalists, the anti-war movement, Jewish giving is relatively unstructured and untargeted."

John Thorne, former press secretary to Senator Abourezk, indicated the Jewish lobby is "better organized and more effective than the oil lobby, because it's scattered all over the country. Its power is due to pressure from constituents."

Percy and Brown

One senator who has felt that pressure is Charles H. Percy (R Ill.), who has described himself as a long-time friend of Israel. After a trip to the Middle East, he said in early 1975 that Israel must be moderate in its actions, that it was being unrealistic in avoiding contact with the Palestine Liberation Organization and that it no longer could count on the automatic support of a large majority of the Senate.

Percy was deluged with some 20,000 pieces of mail, mostly negative, in response to those comments and his refusal to sign the May 1975 letter of support for Israel. *Near East Report* devoted more than a column to Percy's remarks.

Similarly, Gen. George S. Brown, chairman of the Joint Chiefs of Staff, gained almost instant notoriety when remarks he made at Duke University on Oct. 10, 1974, criticizing the power of the Jewish lobby became public a month later. Brown, speaking in the context of the possible hardships that might be caused in this country by another Arab oil embargo, said: "[People] might get tough-minded enough to get down the Jewish influence in this country and break that lobby. It is so strong, you wouldn't believe, now. We have Israelis coming to us for equipment. We say we can't possibly get the Congress to support a program like this and they say don't worry about the Congress. We will take care of the Congress. This is somebody from another country, but they can do it."

Brown was reprimanded by President Ford Nov. 13. Jewish groups and a few members of Congress called for his resignation. Brown issued an apology, referring again to the activism of the Jewish community: "...I do in fact appreciate the great support and the deep interest in the nature of our security problems and our defenses that the American Jewish community has steadily demonstrated...." Though Brown made other controversial statements—including that Israel is a "burden" to the United States—he kept his position.

Public Opinion

Coupled with Jewish activism is public support for the U.S. policy, in effect since before Israel's founding in 1948, that the existence of Israel is in the U.S. national interest. Opinion surveyor Louis Harris documented that support in a poll concluded in January 1975 and described in a *New York Times Magazine* article April 6. He found that 52 per cent of the public sympathized with Israel, up from 39 per cent in 1973 right after the Yom Kippur War, with only 7 per cent expressing sympathy for the Arab side. Sixty-six per cent of the public favored sending Israel whatever military hardware it needed, but no American troops. On what Harris called the "pivotal" question—whether the United States should stop military aid to Israel if this were necessary to get Arab oil—64 per cent of the public was opposed.

Congressional Support

"Traditionally, the bastion of pro-Israel sentiment has been the Congress," one scholar concluded in a 1976 study of "The Arab-Israeli Battle on Capitol Hill."[7] The *Jerusalem Post's* Washington correpondent, commenting on the new faces in the 95th Congress, agreed. "Pro-Israel support in Congress," Wolf Blitzer wrote in December 1976, "has traditionally been a crucial factor in balancing the more 'even-handed' slant in the administration, influenced by Arabists in the State Department."[8] A long-time pro-Israeli observer further notes that "Had it not been for Israel's support in Congress, things would have been quite gloomy.[9] Si Kenen has been quoted as saying that "Without the lobby, Israel would have gone down the drain,"[10] dramatizing the importance of AIPAC in influencing the Congress.

A combination of local activism and public opinion, as well as the deep feeling that U.S. interests in the Middle East are served by support for Israel, is reflected in Congress, whose membership matches closely the Jewish population at large.

The number of Jewish members increased to 27 in the 95th Congress from 23 in the 94th and only 14 in the 93rd. That is 5 per cent, just a little more than the voting strength of Jewish citizens. There are five Jews in the 100-member Senate—Jacob Javits (R N.Y.), Abraham Ribicoff (D Conn.), Richard Stone (D Fla.), Howard Metzenbaum (D Ohio) and Ed Zorinsky (D Neb.). There are 22 Jews in the House.

Explanation

Aides to key members of Congress argue that Jewish lobbying is not solely responsible for the support on Capitol Hill. A Senate aide who has worked on Jewish issues has said, "It is fundamentally misleading to talk about attitudes up here toward the Mideast and Israel in terms of Jewish lobbying. Seventy-five per cent of the Senate is pro-Israel for a whole lot of reasons."

7 Mary A. Barberis, "The Arab-Israeli Battle on Capitol Hill," *Virginia Quarterly Review,* Spring 1976, p. 204.
8 Wolf Blitzer, "Many Friends Among the New Faces in Congress," *Jerusalem Post Weekly,* 21 December 1976.
9 *Ibid.*
10 *The Power Peddlers, op. cit.,* p. 271.

An aide to a Republican senator said that the "interest and support on the Hill for Israel comes from the belief that it is vital to the U.S. national interest. American interests lie in the Middle East; extract Israel from the picture and things are not changed."

Staff Network

Augmenting the support for Israel is what has been called a "network" of staff aides who are interested in Jewish matters, particularly Israel.

They include such people as Richard Perle in Senator Jackson's office, Albert Lakeland in Javits' office and aides to other senators, among them Hubert H. Humphrey (D Minn.), Birch Bayh (D Ind.), Clifford P. Case (R N.J.), Abraham Ribicoff (D Conn.), Howard M. Metzenbaum (D Ohio) and Frank Church (D Idaho). Senator Richard Stone (D Fla.), one of Israel's strongest supporters, has assumed the chairmanship of the Subcommittee on Near Eastern and South Asian Affairs of the Senate Foreign Relations Committee. This important subcommittee was previously chaired by Senator George McGovern (D S.D.), whose staff man, Seth Tillman, was often thought of as leaning toward certain Arab positions. Under Stone, the subcommittee staff is headed by Steve Bryen, a sharp pro-Israeli advocate.

In the House, a partial list of the staff "network" would include aides to Speaker Thomas P. O'Neill (D Mass.), Sidney R. Yates (D Ill.), Charles A. Vanik (D Ohio) and Stephen Solarz (D N.Y.).

In his book, Isaacs quotes Amitay, a former aide to Ribicoff, as saying in 1973, "There are now a lot of guys at the working level up here who happen to be Jewish, who are willing to make a little bit of extra effort and to look at certain issues in terms of their Jewishness, and this is what has made this thing go very effectively in the last couple of years."

As head of AIPAC, Amitay takes a somewhat different view of the importance of any Hill network. "The talk of a Hill network is highly over-rated," he said. "It doesn't exist as such. It's not a question of the staff getting members of Congress to do things; it is the bosses who get the staff people interested."

Staffers who make up the so-called network describe it in terms of people knowing each other from working on common issues. They deny that they may constitute an "Israel lobby" in their own right. "Staff people are important, just because of the way senators operate," said one aide. "But they are important on any issue, not just Jewish ones. Strongly held views on Israel are due to constituents—no staff can create that kind of concern."

Assessments

The success of the Jewish lobby thus is attributed to a combination of U.S. national interest, widespread public support, community activism and an effective Washington operation. Former Senate Majority Leader Mike Mansfield (D Mont.) has said, "It's a strong lobby. But there are other strong lobbies, too." A long-time House aide observed, "Compared to some pressures, the Jewish lobby palls in comparison."

Bookbinder stresses that legislative successes are not the result of heavy-handed lobbying. "In the areas where we don't have substantial non-Jewish support, we don't make our case," he says. "Most of the Jewish community would like to have a federal welfare program or a multi-billion-dollar food assistance program, but we don't have that power.

"I'm concerned about a feeling that the Jewish lobby uses strong-arm, pressure-type tactics. That's nonsense.... There is a very small staff of professionals; we don't wine and dine members of Congress; we have no former congressmen on our payrolls; we don't give parties or have a direct political operation funding or rating people. It's just a bunch of fairly good, devoted people."

An aide to an influential Republican senator says he believes the claims of a powerful Jewish lobby are "way overrated and blown up" by State Department officials and Arab interests to suggest that Congress is "blinded and myopic" toward Israel. "Lobbies will win when they have the facts to support them," he observed.

The Israel lobby has facts—sometimes slanted—at its fingertips. AIPAC has both a director of research and a director of information and the instant availability of concise, hard information can often be translated into power and influence. *Near East Report's* indexed pamphlet *Myths and Facts: A Concise Record of the Arab-Israeli Conflict* is widely distributed and often the source of information on Capitol Hill.

Nonetheless, criticism of AIPAC has apparently begun to hurt. Some of AIPAC's supporters have called for a more low-key style but have expressed doubts that Amitay would be able to conform.

The coming years will be ones of great challenge for this key Washington lobby which is likely to be deeply enmeshed in the politics of American involvement in the Middle East.

OPENING DOORS THAT WERE PREVIOUSLY CLOSED

In the view of Middle East Arabs and their American kinfolk, they have spent a quarter century anguishing on the sidelines while the Zionist lobby has called the plays that have helped thrust the United States into the deadly game of Middle East politics—on Israel's team.

But, since the "Yom Kippur" war of October 1973, American supporters of the Arab cause have felt that at last they are getting into the game. Richard C. Shadyac, an Arab-American trial lawyer from Annandale, Va., who is a past president of the National Association of Arab Americans (NAAA), put it tersely: "The day of the Arab-American is here. The reason is oil."

Thomas Ruffin, former executive director of the NAAA, summed up the goal of the newly emerging Arab lobby. "We are not asking the United States government to take a pro-Arab, anti-Israeli position. We are asking that it weigh all aspects of the Middle East, that it adopt a balanced, even-handed approach in foreign policy."

Whether they consider it a tilt toward the Arabs or a bending back from Israel, American supporters of the Arab cause believe they witnessed movement on the part of the Ford administration, and they are hopeful about Carter. The shift began in the aftermath of the 1973 war, with its oil embargo and its oil price hikes. Until then, Arabists had viewed American policy, in tandem with that of the Soviet Union, as maintaining the status quo in the Middle East, meaning to them that the United States was doing nothing to remove Israel from occupied Arab lands and to bring about a "just settlement" of the Palestinian question.

But then Kissinger mediated initial troop disengagement accords between Egypt and Israel, and Syria and Israel. In the spring of 1975, he tried for further agreements among the belligerents. That mission came to naught, and an obviously annoyed President Ford commented, "If they [the Israelis] had been a bit more flexible...I think in the longer run it would have been the best insurance for peace." The White House announced a total reassessment of Middle East policy, covering "all aspects and all countries."

All this was an unfamiliar tune to Arab lobbyists, but it was music to their ears. Many Arab-Americans supported Gerald Ford for the presidency in 1976 on the basis of the shift in U.S. Middle East policy, especially in view of Carter's clearly pro-Israeli stands designed to attract the more numerous Jewish vote. Yet, by mid-1977, Carter's Middle East views were being applauded by many Arab-Americans. Joseph Baroody, incoming president of NAAA, may have captured the consensus when he stated soon after his election that "right now we're prepared to give Carter the benefit of the doubt. Until he gives us good reason to think he's not being even-handed, we'll assume he is being even-handed."

Complex Problem

Thus the stated objective of the Arab lobby is simple. Its leaders all speak of a "balanced, even-handed approach." But to carry that message beyond the White House to other American institutions, principally Congress, as well as the populace at large, presents a complex problem. Arab lobbyists must present their case effectively to the American establishment—its political, financial and communications power structure.

At any table where American sympathy and support are the stakes, Arabs hold that they sit down with a shorter stack of chips than do their Jewish counterparts. Baroody estimates that there are two million Americans of Arab origin or heritage in the United States, compared with six million Jewish-Americans. And the Arabs claim they spend far less on their cause in the United States than do their opponents, mainly because, they say, they have fewer funds and fewer sources of money.

But the real problem for the emerging Arab lobby has been mobilizing Arab-Americans to become active. Baroody indicates that "Arab-Americans have always been a group that quickly assimilated. There hasn't been an Arab-American consciousness, we never thought of ourselves as a group in that sense." But as with other ethnic affiliations, being Arab-American has come into style. The 1967 and 1973 Middle East wars stimulated Arab-American identification much the same as they did with American Jews. And the conflict of "dual loyalty" may have to be faced by Arab-Americans, just as by Jewish-Americans, should the Arab oil-producing states clamp another oil embargo on the United States.

The question arises as to how the Arabs can square away their pleading poverty with the piles of "petrodollars" Arab nations have amassed since the 1973 war. Hatem I. Hussaini, assistant director of the Arab Information Center in Washington, said, "Only recently have Arab nations begun to think about Arab information in this country. Arab governments think that diplomacy—visits by heads of states, prime ministers—is more important than publicity campaigns in the United States."

Dr. M. T. Medhi, founder of the Action Committee on American Arab Relations, believes that "The Arab governments are really not part of the 20th century. They have hundreds of problems of their own.... Hardly any of them have come to power as a result of elections and free campaigns. So they do not quite understand the need for spending money in America to change public opinion." Lebanese editor Clovis Maksoud who came to the United

"I hope we are becoming known as the Arab lobby...the Arab-American lobby that is."

—Joseph Baroody, president, National Association of Arab-Americans

States in 1975 as an Arab League emissary, recommended then that the Arab League information offices either be revamped or closed down in view of their ineffectiveness.

But there are signs of change. NAAA had by 1977 assumed a more effective voice in influencing U.S. policy and is considering registering as a lobbying organization just like AIPAC (the Jewish lobby, *see Israel Lobby, p. 96*). Saudi Arabia sent an information team to the United States in early 1977. The Egyptian press office has made impressive efforts under the lead of Press Minister Mohamed Hakki and Egyptian Ambassador Ashraf Ghorbal. One scholar who has studied the influence battle on Capitol Hill has even concluded that "if the NAAA succeeds in drawing together the different Arab groupings and molding them into a unified force, this may well present a future threat to the 'influence gap' that AIPAC and the Jewish-American community have created over the years."[1]

The NAAA's Ruffin also emphasized positive aspects of new Arab wealth rather than the negative side of the oil embargo and price increases. "Arab-Americans are much interested in forging a partnership of the Arab countries of the Middle East with the United States," he said. "We see that as a great economic and social benefit to our country. As Arab-Americans, we know the mentality of Arabs of the Middle East, who would like nothing better than to form such a relationship. Communism is abhorrent to those countries. The Arabs are sitting there with all those petrodollars, wanting to invest them in this country, and what happens? Congress places all kinds of roadblocks in their way. So they put them in France, West Germany and other countries."

NAAA's executive director, through July 1977, Michael Saba, who in early 1977 went on a 20-state lecture tour discussing Arab-American business relations, summed up, "Can't the United States have good relations with both sides, have two friends?"

Ugliness and Finery

The Arab lobby has to overcome a derogatory image of the Arab in the American mind, an image created and perpetuated by the American media, both Ruffin and Hussaini believe. "This whole thing of stereotypes is a big problem," Hussaini said. He added that cartoons portraying Arabs as ugly, mean and dirty have been devastating. Hussaini said another stereotype is not helpful either—the dashing Arab garbed in desert finery, the Arab of the movie, "The Sheik of Araby," or of the musical play, "The Desert Song." One Washington writer gave his assessment of the Arab lobby story when he concluded that "Arab-Americans suffer from an inferiority complex."[2]

Commenting on a 1975 Harris Survey that showed that a majority of Americans who know about Egyptian President Anwar Sadat look upon him favorably, Ruffin said this resulted from a favorable impression created by the American press. "At one time we faced total media bias," Ruffin said, explaining that Sadat has been the one Arab leader who has been able to knock a chink in the wall of prejudice. "After all, Sadat doesn't speak in a vacuum. He is speaking general Arab attitudes. A great portion of other Arab leaders hold the same views. They haven't had the same exposure as Sadat." *(Poll box, p. 105)*

1 Mary A. Barberis, "The Arab-Israeli Battle on Capitol Hill," *Virginia Quarterly Review*, Spring 1976, p. 223.
2 Benjamin Welles, "Arab Power," *Washingtonian Magazine*, December 1975, p. 120.

"We have to show our people that we can represent their interests."

—Michael Saba, former executive director, National Association of Arab-Americans

Arab-American Organization

A multiplicity of organizations promotes the Arab cause in the United States, but it is clear that two have had mounting importance since the 1973 war—the National Association of Arab-Americans, described by Saba as a strictly domestic product, and the Arab Information Centers (AIC), an adjunct of the League of Arab States and thus a foreign entity. Centers are located in five American cities.

Established in 1972 "to fill the absence of an effective political action group" on the national level, the NAAA staked itself out as the umbrella group for Arab-Americans unable to identify with pro-Arab activists who leaned "too much to the left."

NAAA is attempting to act as the major Arab lobby according to its new president Joseph Baroody. "Our basic support is among affiliates. There are 1100-1200 Arab-American organizations in this country. Virtually all of them are cultural, religious, or charitable, but many of them have affiliated with us for political reasons."

Baroody, a Washington consultant with Wagner & Baroody often representing business interests with government agencies, wants his organization "to represent what we think are the best interests of the United States in the Middle East." "This organization has said for a long time," he said, "that we don't have any quarrel with the existence of Israel." But Baroody insists that support for Israel's existence does not necessarily mean support for many of Israel's policies or giving in to the pressures of the Israeli-Jewish lobby.

Michael Saba, NAAA's young and personable executive director, took over in August 1976 after a stint in the Peace Corps in Malaysia and graduate school in anthropology at the University of Illinois. He has emphasized relations with what he terms "affinity groups" and grass-root lobbying. There has been some criticism of the organization's efforts for the past few years as not being sufficiently political, for not countering the influence of such organizations as AIPAC. Saba has replied that "We look to AIPAC as an organization that's done incredible things in this country and has mounted programs very significant in terms of American foreign policy. As a potential counter we need first to get our members more involved. The Jews relate more to the issues of the Middle East. We have to show our people that we can represent their interests."

NAAA moved into a suite of offices in Washington's Watergate building in 1975. There are three paid staff members and a varying number of volunteers, often university students on work-study programs. Should NAAA decide to do more lobbying, as now appears likely, additional staff would be needed. In addition, the NAAA has volunteer offices in a dozen other cities.

All members of the NAAA are Americans. "We don't accept money from any Arab country, though, the good Lord knows, there have been times when we could have used it," a former executive director said. The exact budget figure for NAAA is not released. Most of the money, however, comes from dues from over 4,000 members—ranging from $2 to $100 annually with a "very limited number" contributing $1,000 to $5,000. While Arab governments do not contribute directly to NAAA, it does appear they help out by sponsoring such things as sessions at the annual convention and by taking out advertisements in the annual journal.

The NAAA is not registered as a lobby with the clerk of the House of Representatives, but Baroody favors such a step. Its dues are not tax deductible, nor is the organization tax exempt. "We are open in what we do. We keep no secrets. We are just not up there on Capitol Hill that much," Ruffin said. But this may change since NAAA is considering adding staff persons for the lobbying assignment.

NAAA headquarters does not pay honoraria to members of Congress for speeches to Arab-American groups, but sometimes arranges for appearances of members before local NAAA chapters, which have the option of paying fees if they so desire.

"Our clout has to be on the basis of seeking justice and objectivity in the American approach to the Middle East. We are giving Congress a viable alternative which up to a year or two ago it never had," Ruffin noted a few years ago. He cited trips to the Middle East by Senators George McGovern (D S.D.) and Charles H. Percy (R Ill.), which he said resulted in the two "coming away with a desire for a more even-handed approach to the situation."

There have been many other such trips during the past few years and many more voices for even-handedness in the opinion of NAAA leaders. NAAA also points with pride to the seven Arab-Americans in the 95th Congress, up from one just a few years ago. President Baroody adds that "We are currently encouraging political activity by Arab-Americans throughout the country. If we hear of somebody somewhere out in the country who wants to run for the city council or whatever we'll encourage them and as individuals we'll support any of them."

White House Meeting

In Ruffin's estimation, a meeting that 12 NAAA leaders had with President Ford at the White House June 26, 1975, has been the high point of the organization's five-year existence. He called the session "very historic," noting that it

"The Israeli lobby is the most powerful and pervasive foreign influence that exists in American politics."

—Sen. James Abourezk (D S.D.)

was only the second time such a meeting with an American President had taken place. (A small group of Arab-Americans met with President Lyndon B. Johnson at the time of the 1967 Arab-Israeli war.) The NAAA arranged the Ford meeting through Kissinger, who, Ruffin said, was interested in the NAAA's presenting its views to the Chief Executive.

A four-page position paper was presented to Ford by the then-NAAA president, Edmond N. Howar of Washington, D.C. Howar, of Palestinian descent, is president of an investment real estate and building firm. He told Ford that the NAAA supported the President's initiatives in the Middle East, especially the administration's total reassessment of foreign policy in the region. However, Howar said the organization was concerned that certain fundamental problems not be overlooked. Ford was presented with these six points: a demand for Israeli withdrawal from territories occupied in the June 1967 war; protection of the rights of Palestinians; U.S. recognition of the Palestine Liberation Organization; special status for Jerusalem; a complaint about Israeli military incursions into Lebanon, and an expression of concern that congressional attitudes "are keeping more petrodollars from being invested in the United States."

The delegation gathered around the great table in the White House Cabinet Room. Flanking Ford were Kissinger and William J. Baroody Jr., the brother of Joseph Baroody, who is now NAAA's president.

The Baroodys are the sons of William Baroody Sr., whom *The Washington Post* called in an Aug. 17, 1975, article in *Potomac*, its Sunday supplement, the "ruling patriarch of Washington's first family of political conservatism." William Baroody Jr. was Ford's assistant for public liaison.

In speaking to Ford, as it does elsewhere, the NAAA stressed its American roots and its claim that its lobbying efforts are for objectives in the best interest of the United States. Ruffin said, "We don't speak on behalf of any Arab country or any Arab leader. In relation to U.S. policy, first and foremost we are Americans. We would not be pro-Arab if that hurt the United States. The reason why we are against the pro-Israel policy is that it hurts the economic and social well-being of the United States."

Within weeks of Jimmy Carter's taking over the White House, NAAA requested a meeting with Ford's successor. By mid-1977 delegations from NAAA had met with the President's assistant, Midge Costanza, and with National Security Council Middle East specialist William Quandt. Meetings with Secretary of State Vance and other State Department officials were being arranged and a meeting with President Carter was expected for later in the year.

NAAA's desire to try to influence Congress more than in the past was reflected in a "Message to the Congress" from NAAA shortly after rioting and demonstrations erupted in the occupied territories held by Israel in early 1976. "Now we must address ourselves to the real problem of the Middle East,...Palestinians in and out of Israel," the message insisted. "Will this nation remain a bulwark against oppression and tyranny?" was the question NAAA posed to the traditionally pro-Israeli Congress.

Fifth NAAA Convention

In May 1977, the NAAA held its fifth annual convention in Washington's Shoreham Americana Hotel, drawing a few hundred delegates to a busy schedule of speeches, panel discussions and festivities. The NAAA maintains

Poll Results

Recent public opinion polls have generally shown that the Arabs still have a long way to go before they would substantially shift traditional American support from Israel to the Arab cause.

However, according to an April 1975 Gallup Poll, there has been a notable erosion of strong pro-Israeli support to an "evenhanded" position. In that poll only 37 per cent sympathized with Israel, compared with 54 per cent in December 1973, shortly after the October 1973 war. But, in shifting away from one-sided support for Israel, Americans moved into neutral, not pro-Arab positions. Sympathy for the Arab states registered only 8 per cent in both polls.

Harris

In an article in *The New York Times Magazine* April 6, 1975, Louis Harris of the Harris Survey concluded that American support for Israel is bedded in rather solid ground. He wrote that events in the previous year and a half had produced three conceptions in many American minds, especially among American Jews: "(1) Jewish power could well be eroding in the United States, accordingly weakening the lifeline to Israel; (2) traditional Israeli intransigence would no longer be tolerated; (3) with the decline of sympathy for Israel and the growth of Arab power, a spate of anti-Semitism could well break out in the Western world." Harris wrote, "The trouble with almost all of this is that it simply doesn't square with prevailing public opinion.... Instead of slipping away, support for Israel is now at a record peak." This view was based on January 1975 findings showing 52 per cent of Americans siding with Israel and only 7 per cent with the Arab states.

Foreign Policy Association

Opinion among those citizens most concerned with foreign policy, however, tends to be noticeably more "evenhanded." The Foreign Policy Association polled participants in its Great Decisions Program in early 1977. It found 23 per cent supporting a tilt toward Israel and 16 per cent a tilt toward the Arabs while 68 per cent favored continuing "evenhandedness" and 67 per cent approved the recommendations outlined in the Brookings Institution Report *(see U.S. Policy chapter, p. 80)*. The great majority, 89 per cent, favored "encouraging both sides to reach a settlement in direct negotiations, with 85 per cent approving a U.S. role as "mediator" and 78 per cent supporting reconvening of the Geneva Conference.

The election of Menahem Begin as prime minister of Israel has raised the possibility of some further erosion in support of Israel and an even greater movement to the "evenhanded" category.

close communications with the Arab diplomatic colony in Washington, as evidenced by the fact that several Arab envoys addressed the four-day convention or hosted banquets, breakfasts or other social functions.

Outgoing President Minor George reminded the convention delegates that "at all times we are an American organization pursuing American interests." He then went on to condemn the "economic and intellectual terrorism that has plagued us during the past 30 years" regarding Middle East policy.

Convention delegates also heard the new Deputy Assistant Secretary of State for Near Eastern and South Asian Affairs, Nicholas Veliotas, assure them that the Carter administration was aiming for a comprehensive settlement of the Arab-Israeli conflict and for resumption of the Geneva Conference later in 1977. "We believe, sincerely," Veliotas told the convention, "that the Arabs are ready to accept Israel in the Middle East, that our Arab friends are not just seeking tactical advantage for the next round of hostilities. There's certainly no reason to believe that the Arabs would be any readier 5, 10 or 15 years from now. On the contrary, a good case can be made that the trend will be away from moderation, away from compromise, if we don't make some progress in the near future."

The convention closed with a "Palestine Day." Dr. Hisham Sharabi, an NAAA board member who teaches at Georgetown University and edits the *Journal of Palestine Studies,* and Dr. Hatem Hussaini spoke of Palestinian aspirations.

A few days after the convention ended *The Washington Post* took notice of NAAA's growing importance with a story, which began in some editions on page one, discussing the organization's growth and influence.

In 1977 there were seven members of Congress who are of Arab-American descent. Senator James Abourezk (D S.D.) has announced he will not seek reelection. In the House the members are: James Abdnor (R S.D.), Adam Benjamin Jr. (D Ind.), Abraham Kazen Jr. (D Texas), Toby Moffett (D Conn.), Mary Rose Oakar (D Ohio), and Nick J. Rahall (D W.Va.).

Pushing the Cause

NAAA has set for itself the goal of awakening the sense of Arabic heritage in two million Arab-Americans beyond the confines of Middle East food and music. There has been considerable success so far, but even NAAA's leaders freely admit there's a long long way to go before there is an effective Arab lobby in Washington that can match the efforts of American Jews.

In April 1977 representatives of NAAA visited the Middle East, meeting with various Arab officials including Yasir Arafat of the PLO. Championing the Palestinian cause has become one of NAAA's important priorities. "I don't think there's any way to come up with a settlement in the Middle East without dealing face-to-face with the Palestinians," President Baroody has said.

As for the strength of the still new organization, "We've reached something of a cross-roads, either we grow more important and powerful or we fade away," he continued. It's unlikely NAAA will soon fade away.

AAUG

Another organization, the Association of Arab-American University Graduates (AAUG), has become more visible during the past few years in addition to NAAA. Not especially active in Washington, AAUG has not attracted the attention of journalists and politicians. But it has become an influence. "AAUG—they're more the academics and intellectuals," Baroody noted. "They identify more with internal Arab politics," Saba said. Comparing the two organizations, John Richardson, president of American

Near East Refugee Aid, believes that "AAUG is in sort of a bind of its own making. Many of its key spokes-people are immigrants, and many are Palestinians. It sees itself as primarily a Third-World organization. Whereas NAAA is made up of Americans of Arab background (mostly of Syrian and Lebanese background), AAUG is made up of Arabs, many of them hostile to the policies of the United States. NAAA people want desperately to be assimilated. Many people in NAAA are anxious about challenging American policy. Many Arab-Americans don't feel sufficiently secure to take a counter position. AAUG-types see themselves naturally in a position of antagonism with the United States."

Arab Information Centers

Hussaini, of the Arab Information Center in Washington, admitted in an interview that as a propaganda organization the centers are a long way from Madison Avenue-style public relations. He noted that between the 1967 and 1973 wars in the Middle East, the Arab Information Centers, as an arm of the Arab League, faced some lean times, occasioned by the rupture of diplomatic relations between many Arab nations and the United States.

The league provides at least $500,000 a year for AIC operations, which Hussaini described as "small, when you consider the size of this continent." He claimed that one Jewish organization alone, B'nai B'rith, spends $7-million annually on promoting the Israeli cause. The main Arab Information Center is in New York City, where 10 staffers are employed; the Washington office has five persons, and the Chicago, Dallas and San Francisco centers are manned by one person each.

Speakers and films are provided by the AIC for interested American university, church and civic groups. In 1975, the AIC began publication of a biweekly newsletter, *The Arab Report*, which it sends to members of Congress, journalists, television commentators and public opinion-makers. Also published is *The Palestine Digest*, principally a collection of news articles, editorials and speeches sympathetic to the Arab cause.

For years, Hussaini said, the Arab Information Center has worked with this set of facts: "The Congress is in general pro-Israel. The Congress in general is not interested in the Arab point of view." The 1973-74 oil crisis changed these attitudes only a bit, he thinks. Now the AIC gets a few more queries from individual members of Congress regarding the Arab side of the Israeli conflict, on oil policy or other Middle East matters.

Hussaini maintains that the AIC is not a lobby, because it does not have permanent spokesmen working the halls of Congress. "We are not as sophisticated, as well financed as the Israeli lobby," he said.

Hussaini is a Palestinian whose family left Jerusalem at the time of partition in 1948. He was educated at the American University in Cairo and the University of Massachusetts and taught government at Smith College before joining the AIC. He lives in the United States on a residence visa. Hussaini is a member of the Palestine National Council (the legislative body of the PLO) and was at the meeting in Cairo in early 1977. He maintains strong ties with some elements of the PLO and has been involved in recent PLO efforts, as yet unsuccessful, to open a Washington information office.

Arab Spokesmen

One of the Arab Information Center's major propaganda efforts in 1974 and 1975 was the sponsorship of two extensive American tours by Clovis Maksoud, special envoy of the Arab League. Maksoud, who has a reputation as a dynamic speaker and a prime conveyor of the Arab viewpoint, is a Lebanese lawyer educated at the American University in Beirut, George Washington University in Washington, D.C., and Oxford University in England. He was senior editor of *Al Ahram*, Cairo's leading Arab newspaper, from 1961 to 1971, and in 1975 he was a chief editor of another prominent Arab newspaper, *An-Nahar*, of Beirut.

The two American tours took Maksoud to 60 cities in 32 states, where he has talked to members of Congress, foreign policy and public opinion makers, civic and church organizations. He addressed such prestigious forums as the Commonwealth Club of San Francisco, the Los Angeles World Affairs Council and the National Press Club.

Maksoud said he could not claim any switching toward the Arab cause in the United States, but added: "We have broken the monopoly of interest the Zionist lobby has held for the American people."

Maksoud in 1977 was teaching at Georgetown University in connection with the Center for Contemporary Arab Affairs. But he is still often called upon to speak on behalf of the Arab cause.

Early in 1977 Ghassan Tueni, publisher of *An-Nahar*, visited the U.S. as a special representative of the Arab League. His tour around the country, lasting about two months, was not arranged by the Arab Information Center—Tueni's preference. Tueni met members of Congress, spoke at such places as the Council on Foreign Relations, and conducted a low-key effort to promote the Arab "peace offensive"—hoping to convince the Americans to put pressures on their Israeli friends for the kind of settlement the Arabs were apparently offering.

Palestinian Assistance Groups

Another component of pro-Arab activity in the United States encompasses organizations working to provide assistance to Palestinian refugees. Prominent among them is American Near East Refugee Aid Inc. (ANERA), headed by Chairman of the Board John H. Davis, a former commissioner general of the United Nations Relief Works Agency for Palestine, and President John P. Richardson.

At the outset of an interview, Richardson volunteered that he is of British and Norwegian descent, although he made it evident that he champions the Arab cause. However, he asserted that ANERA is a people-to-people program, providing aid to Palestinian refugees from private donations. "It is puzzling to me why ANERA is included in broad-brushed strokes on lobbying," he said. "If lobbying means having views, then I guess we are included...but that we are a lobbying organization is strictly false."

ANERA came into existence after the June 1967 war. Among the activities the organization has supported are the Industrial Islamic Orphanage in Jerusalem, the Palestine Hospital, Gaza College, the YWCA of East Jerusalem, Arab Women's Society of Jerusalem and the Association for the Resurgence of Palestinian Camps.

Over the past few years, ANERA has raised a few million dollars in the United States, of which 82 per cent, mostly in cash grants but a small part in pharmaceuticals, has gone to the Palestinian organizations ANERA patronizes, Richardson said. The other 18 per cent has been spent on administration and travel.

Richardson feels that information about the Arab world has changed views in the United States, promoting "even-handedness." "It's not the Arab lobby that really is changing things," he insists. "The Arab lobby isn't the main force changing things. What's changing things is the working press which is writing about the Arab story—Arab wealth, Arab political development."

Richardson has often been called upon to testify before congressional committees on the plight of the Palestinians.

Oil Contributions

Richardson makes speeches and is in close contact with individuals, foundations and corporations. Gulf Oil contributed $2.2-million to ANERA after the 1973 war, a donation that has not been approached by other American corporations. "The only sector that has been at all responsive is the oil companies, and for that we are appreciative," Richardson said.

What Arab lobbyists see as corporate niggardliness is also a sore point with Richardson, who said, "There is a tremendous surge of commercial activity in the Middle East. American companies are going for big money over there. We are trying to raise money for a good cause, that is for people, for Palestinian refugees. My hope is that they [U.S. firms] will make a small investment in people while they are making a hell of a lot of money out of the Arab world."

Arab Boycott

Defending the Arab boycott of American firms trading with Israel appears to pose no problem for Arab spokesmen. In early 1975, the Arab Boycott Office, meeting in Cairo, prepared a new blacklist. This led to a charge by the Anti-Defamation League of B'nai B'rith that some American companies were violating U.S. civil rights laws by discriminating against Jews. President Ford branded any such discrimination as "repugnant to American principles," adding that the boycott "has no place in the free practice of commerce as it has flourished in this country."

During the campaign, President Carter was outspoken in his denunciations of the boycott. But in practice, it has been the Congress where anti-boycott sentiments have been strongest. Both the Ford and Carter administrations urged caution and restraint in the type of anti-boycott legislation considered.

In separate interviews, Ruffin, Hussaini and Richardson indicated they thought that trade boycotts were a time-honored American institution, dating back to boycotting of British products by American colonials before the revolution. They noted that the United States has blacklisted Cuban products since the advent of Fidel Castro in 1959.

Hussaini said the news media have depicted the boycott "as something against Jews, and that is not it at all. It is simply this: Since Israel and the Arab countries are in a state of war, it is logical that companies—whether they are owned by Christians, Jews or Moslems—that contribute to the war economy of Israel be boycotted. Companies that send shoes to Israel are not boycotted."

Jews do the same thing, the Arab spokesman maintained, saying that at the time it was revealed that Iran might buy into Pan American Airlines, many Jews in New York called the airline and said, "We will not travel with you." Hussaini said, "They thought Iran was an Arab country." American Jews also boycotted tourism in Mexico for a time after Mexico's vote at the U.N. to condemn Zionism as racism.

During 1976 and 1977 the boycott battle heated up. Jewish organizations and many members of Congress insisted that only the primary boycott was valid—that is the direct boycott by Arab countries of Israeli products. Secondary and tertiary boycotts—of companies who have dealings with Israel and of companies who have dealings with other companies who have dealings with Israel—were repeatedly challenged. And, of course, actual discrimination against Jews who want to do business in the Arab world or are connected with companies who want to do business has been loudly condemned.

The Arabs have countered with the insistence that there is no discrimination against Jews as Jews. In fact, Jewish businessmen have been going to all Arab countries, including Saudi Arabia, it is pointed out. But known Zionist partisans are another matter, many Arabs believe. After all, they say, Zionism is an expansionist force threatening the Arab world and preventing the Palestinians from achieving their rights to self-determination. The real issue, they insist, is a peace settlement, a settlement which will then end the state of war which is the real reason for the boycott.

The political and propaganda struggle continues even though compromise legislation has made its way through the 95th Congress placing restraints on compliance with secondary and tertiary aspects of the boycott. Indications are that the legislation is considered acceptable to all the parties, which means that it contains ambiguities and loopholes that could allow more time for a political settlement to defuse the controversy.

Other Tactics

Additional facets of Arab lobby activity in the United States include educational and cultural organizations. The Chicago-based Federation of Arab Organizations encompasses small social and cultural Arab clubs throughout the nation. The American Friends of the Middle East, headquartered in Washington, D.C., provides educational counseling to help place students from the Middle East in U.S. universities. Also in Washington, the Middle East Resource Center and the Middle East Research and Information Project (MERIP) provide information generally considered pro-Arab.

In a story on the Arab lobby in the June 30, 1975, *New York Times*, several former Cabinet officials were named as well-financed and high-powered participants in the Arab effort. The *Times* said that records on the registration of foreign agents on file at the Justice Department show that the Washington law firm headed by Clark Clifford, Defense Secretary under President Johnson, had a contract with the Algerian government calling for $150,000 a year in fees, and that Richard G. Kleindienst, Attorney General under President Nixon, had a $121,000-a-year contract for legal services to the Algerian ministry of industry and energy.

Other former Cabinet officers or their law firms, named by the *Times*, who have worked for Arab or Iranian interests included John B. Connally, former Treasury Secretary under Nixon, and William P. Rogers, Nixon's former Secretary of State.

Another story, this one on the cover of *Parade* magazine,[3] the supplement to Sunday newspapers

3 Robert Walters, "Big-Name Americans Who Work for Foreign Countries," *Parade*, 20 June 1976.

throughout the country, said that a "highly confidential document...outlines plans for the Arab world to spend as much as $15-million annually on a far-reaching propaganda program in the United States." "That secret proposal," the article continued, "calls for a massive effort to swing American public opinion about the seemingly endless Middle East conflict away from the Israelis and toward the Arabs." *Parade* investigated the situation and reached the conclusion that the "Arab nations have mobilized a vast network of influential lawyers, Washington lobbyists, public relations experts, political consultants and a host of other highly paid specialists to implement the plan." "The Arabs," according to *Parade's* investigators, "have spent far more money than any other interest group to support the activities of foreign agents in this country." Among those singled out were former Senators J. William Fulbright and Charles Goodell as well as lawyers Clark Clifford, Richard Kleindienst, William Rogers, and Frederick Dutton.

A PALESTINIAN HOMELAND: IN SEARCH OF A SOLUTION

When newly inaugurated President Jimmy Carter spoke out in favor of a "Palestinian homeland" at a town meeting in Clinton, Massachusetts, on March 16, 1977, he opened up a debate about one of the most controversial aspects of the entire Middle East imbroglio.

By mid-1977 Carter had not defined just what he had in mind. He had emphasized that he envisioned the homeland to have links with Jordan, but he had not discussed the role of the Palestine Liberation Organization (PLO), the boundaries of the homeland (though it was widely assumed he had the Israeli-occupied West Bank and Gaza Strip in mind), or the political nature such a homeland would assume. There were those who thought that Carter had in mind a quasi-sovereign state—demilitarized, economically linked with other states in the area (possibly including Israel), but able to call itself independent. There were others who insisted that Carter not only would not accept an "independent Palestinian state" but might eventually agree with Israeli Prime Minister Menahem Begin that a single Palestinian-Jordanian state was also a reasonable solution to the Palestinian question. The Israelis have in mind a single state to the east of the Jordan river with which Palestinian Arabs in the West Bank and Gaza Strip areas might identify—without these areas actually reverting to Arab sovereignty.

When Begin came to meet with Carter in Washington on July 19 and 20, one of the major issues of disagreement between Israel and the United States was the future of the West Bank and Gaza Strip where approximately one million Palestinian Arabs are living. While Carter contemplated these areas as a Palestinian homeland, Begin claimed them as "liberated territories" rightfully belonging to the state of Israel and absolutely essential for Israeli security.

Concerning the role of the PLO, with which the United States has had cautious dealings of an indirect nature for a number of years, the Israelis were vehemently opposed to any PLO participation in peace negotiations at Geneva or elsewhere. How to allow for Palestinian representation at Geneva, as the Arabs were demanding, while circumventing Israel's unwillingness to deal with the PLO, was a major stumbling block in the planning for eventual resumption of negotiations at Geneva in hopes of finding some path toward peace in the Middle East.

Origins of the Controversy

"A land without a people for a people without a land," was the way one of the early Zionist leaders described Palestine. But the territory the Jews had left nearly 2,000 years earlier was not vacant when their descendants returned; for centuries it had been inhabited by Arabs. Now the Arabs of Palestine are "a people without a land."[1]

Some of them (about a half million) live in the state of Israel, created when the United Nations voted to partition Palestine in 1947. (A civil war ensued in which the nascent Jewish state held off the Palestinian Arabs and the ar-

mies of the surrounding Arab states.) Others (about one million) live in the West Bank and Gaza Strip, parts of Palestine occupied by Israel since the 1967 war. The rest, another million and a half, are scattered among neighboring Arab states. About half of the three million Palestinian Arabs are registered as refugees; many of them were made refugees for the second time in 1967. The victims of repeated Arab defeats, living in bitterness and often in poverty, lacking a territory they can call their own and a state to represent them among nations, this hapless people nonetheless has seized the attention of the entire world. One hundred five governments voted, against precedent, to let the Palestinians' leader address the United Nations General Assembly November 1974, and the Great Powers of the world worry about the threat the Palestinians pose to international peace. How has such a small and unfortunate people gained such attention?

Palestinian organizations trace their growing influence to their reliance on violence, and many observers would agree. "It is sadly true," an American diplomat in the Middle East observed, "that you seem to have to hurt someone to get any attention, and the Palestinians didn't get any attention for 20 years.[2]

But the record of terrorism by which the Palestinian organizations forced the world to consider their demands may exclude them from negotiations to deal with those demands. The Israelis refuse to talk with this "coalition of murderers,"[3] and the Egyptians fear that Palestinian intransigence may prevent a peace they need to regain their own occupied territory and to rebuild their economy. As the Arab states coalesce behind Egyptian, and more recently Saudi Arabian, leadership for renewed peace talks, the Palestinians fear that their Arab allies may sacrifice their goals in a compromise with Israel. The possibility of a compromise peace, which could result in a small Palestinian state, divides the Palestinians, whose leaders already may be losing some of their newly won influence.

Recognition of the PLO

The United Nations General Assembly voted Oct. 14, 1974, to recognize the Palestine Liberation Organization (PLO), a federation of Palestinian groups, as "the representative of the Palestinian people." The vote was 105 to 4, with 20 abstentions; only Israel, the United States, the Dominican Republic and Bolivia voted against the resolution. On Nov. 13, 1974, Yasir Arafat, leader of the

---

1. The history of the past century in Palestine is well presented in the following two recent books: Abaron Cohen, Israel and the Arab World (Beacon Press, abridged edition, 1976); and J. C. Hurewitz, The Struggle for Palestine (Schocken: 1976). For background information on the Arabs of Palestine see the following: Quandt, Jabber and Lesch, The Politics of Palestinian Nationalism (University of California Press, 1973); Michael Curtis (ed.), The Palestinians (Transaction: 1975); Fawaz Turki, The Disinherited, Journal of a Palestinian Exile (Monthly Review: 1974); Thomas Kiernan, Yasir Arafat (1976); Frank H. Epp, The Palestinians (Herald Press: 1976); Moshe Ma'oz (ed.) Palestinian Arab Politics (Jerusalem Academic Press: 1975); Sabri Jiryis, The Arabs in Israel (Montly Review: 1976), and Jureidini and Hazen, The Palestinian Movement in Politics (Lexington: 1976).

2. Quoted by Al McConagha, Minneapolis Tribune,, June 1, 1975.

3. The phrase used by Israeli Ambassador to the U.S., Simcha Dinitz, in an interview with Worldview magazine, July-August 1977.

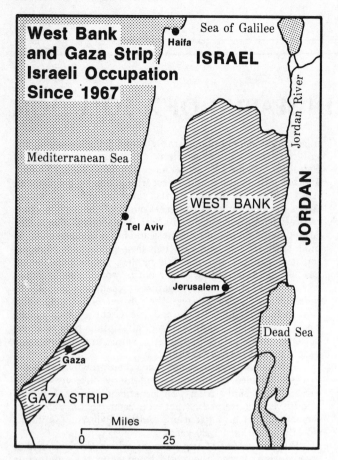

West Bank and Gaza Strip Israeli Occupation Since 1967

Sea of Galilee

Haifa

ISRAEL

Jordan River

Mediterranean Sea

WEST BANK

JORDAN

Tel Aviv

Jerusalem

Dead Sea

Gaza

GAZA STRIP

Miles

0   25

PLO and head of its largest guerrilla group, Al Fatah, addressed the assembly. It was the first occasion when a nongovernmental organization participated in one of the assembly debates. "I have come bearing an olive branch and a freedom fighter's gun," Arafat said. "Do not let the olive branch fall from my hand." *(Speech excerpts, box, p. 113)*

"I think one of the integral parts of an ultimate settlement has to be the recognition of the Palestinians as a people, as a nation, with a place to live and a right to choose their own leaders."

Jimmy Carter
November 1975[4]

Arafat's speech followed a unanimous agreement by 20 Arab heads of state, meeting in Rabat, Morocco, Oct. 26-28, to recognize the PLO as "the sole legitimate representative of the Palestinian people on any liberated Palestinian territory." Jordan's King Hussein, whose government had ruled the West Bank territory from the 1948 partition of Palestine until it was occupied by Israel in the 1967 war, abandoned his bitter opposition to the PLO claims and accepted the agreement "without any reservations." Since

---

4. Stated at The National Democratic Issues Conference, Louisville, Kentucky, Nov. 23, 1975.

then Hussein has maintained considerable contact with Arab leaders on the West Bank, but has publicly maintained that the Palestinians now speak for themselves in international diplomacy.

The removal of the moderate Hussein from negotiations regarding the West Bank dismayed Western countries hoping for a compromise peace in which Israel would yield territories occupied in 1967. Because Israel refuses to deal with the PLO, the prospects for peace talks appeared grim after the Rabat conference. Israeli Premier Yitzhak Rabin, in an interview with *U.S. News and World Report* after talks with President Ford in Washington June 11 and 12, 1975, reiterated the Israeli refusal to negotiate with the PLO, let alone accept its rule of the West Bank, because its goal, he said, is "the elimination of Israel." Rabin said that a "Jordanian-Palestinian state" is the solution of the Palestine issue."[5]

When Menahem Begin became Israeli prime minister in June 1977, he not only ruled out any thought of a Palestinian state, he also indicated that Israel would permanently retain the West Bank and Gaza Strip areas in which such a state could be created. "Such a state can never come into being" Begin said on his July 15 arrival in New York, prior to his meeting with President Carter.

Post-Rabat developments indicate that the setback to Hussein may have been temporary. The PLO failed to form a government in exile and to prepare for negotiations with Israel, as intended by the conference. Although its official position toward Israel has softened somewhat since it achieved its new status, the PLO still refuses to publicly acknowledge a goal of coexistence and permanent peace with Israel. To do so would probably split the organization and give extremist groups an opportunity to win the support of the many rank-and-file Palestinians unwilling to abandon their dream of eventually returning to all of Palestine as their homeland. A March 1977 meeting of the Palestine National Council (the most representative policy-making body of the PLO) in Cairo did not result in a willingness by the PLO to give up the Palestine National Covenant—as some analysts had expected—which calls for Israel's elimination as a Jewish state.[6]

But the new willingness of most Arab states to recognize Israel's permanence within its 1967 boundaries makes the more militant groups in the PLO an embarrassment to their intended negotiations. As a result, Hussein's influence was again in the ascendency as momentum toward peace talks picked up in mid-1975, before the Lebanese civil war, and again in 1977, after the end of the Lebanese conflict. During the civil war the PLO had suffered considerable losses at the hands of the Christian forces combined with the Syrians, who intervened in mid-1976 to prevent a leftist-Palestinian victory.

In May, 1975, Anwar Sadat paid the first visit by an Egyptian leader to Jordan, and Syrian President Hafez Assad arrived June 10 for three days of talks with Hussein. The meeting stood in dramatic contrast to years of bitter feuding between Jordan and Syria, which in 1970 sent tanks

---

5. U.S. News and World Report, June 23, 1975, p. 30.
6. The Palestine National Covenant, sometimes referred to as the Palestinian Charter, remains the basic ideological statement of the PLO. It was first written in 1964 and was revised in 1968. Though new statements and resolutions havse altered many crucial PLO positions, the Covenant remains the symbol of the unwillingness of the PLO to contemplate any form of coexistence with Israel. Soon after becoming Secretary of State, Cyrus Vance stated that only when the PLO rids itself of the Covenant, thus making clear its willingness to recognize Israel, might the U.S. consider dealing with the PLO. For the text of the Palestine National Covenant see Appendix II in Fabian and Schiff (eds.), *Israelis Speak* (Carnegie Endowment for International Peace: 1977), or John Norton Moore (ed.), *The Arab-Israeli Conflict* (Princeton Univ. Press: 1974), Vol. III, p. 698.

across the border in an effort to keep Hussein from smashing the Palestinian guerrillas who were challenging his regime. This time the two leaders agreed to coordinate their defenses, and Assad promised to protect Hussein from PLO infiltration, thus permitting the Jordanian king to concentrate more of his forces along his border with Israel. Relations between Syria and Jordan have improved since 1975 and Egypt's Sadat and King Hussein have frequently consulted on Middle East diplomacy, most noticeably in July 1977 when they declared the necessity for an "explicit link" between the envisioned Palestinian state and Jordan.

Changes Within the PLO

Growing unity among the governments seeking to negotiate with Israel has brought pressure on the PLO to moderate its stand, but the process alienates the radical groups within the organization. The extremists, however, are being wooed by Libyan strongman Muammar al-Qaddafi, who still vows the destruction of Israel and by Iraqi leader

---

"There should be provision for Palestinian self-determination, subject to Palestinian acceptance of the sovereignty and integrity of Israel within agreed boundaries. This might take the form either of an independent Palestinian state...or of a Palestinian entity voluntarily federated with Jordan but exercising extensive political autonomy."

The Brookings Report[7]

Saddam Husayn who insists his country "will never recognize the right of Israel to live as a separate Zionist state."[8]

As part of an intensified campaign to thwart peace negotiations, Qaddafi in June 1975 invited to Tripoli two radical Palestinians—George Habash, who organized a series of airplane hijackings in 1970, and Ajmad Jabril, whose exploits include plotting the April 11, 1974, massacre in Qiryat Shemona, in which 18 Israelis, including eight children, were killed by terrorists infiltrating from Lebanon. Egyptian intelligence sources told *Newsweek* that Qaddafi offered Habash $16-million to kill Sadat, a proposal allegedly accepted with the boast: "If we can penetrate the heart of Tel Aviv, you can imagine how much easier it will be in Cairo or Alexandria."[9]

The popular Front for the Liberation of Palestine, a Marxist group which crystalized under Habash's leadership in 1967, withdrew from the PLO executive council in September 1974, accusing it of "deviation from the revolutiionary course" and alleging that the organization had made contacts with the United States aimed at peace negotiations. In February 1975, Habash offered to bring his group back into the PLO if Arafat would reject any settlement involving recognition of Israel. Arafat has not agreed

to recognize Israel, but his public position has become considerably more moderate. In May 1977, just before the Israeli election that brought Menahem Begin to power, Arafat told syndicated columnist Georgie Anne Geyer that recognition of Israel was possible and Arafat was prepared to discuss guaranteeing Israel's security. The PLO was "embarking on a new program of international legitimacy," Arafat told Geyer."[10]

Arafat's goal of a "secular democratic state" for all of Palestine by definition excludes a distinctly Jewish state and raised deep fears in Israelis about their future status and even their presence amidst an Arab majority. But in his United Nations speech, Arafat repeatedly referred to this goal as a "dream," and he has indicated PLO readiness to accept a Palestinian state in the occupied territories as an interim arrangement.

David Holden of the *Sunday Times* of London observed that "For the present, probably, no Palestinian leader could survive without paying lip service to that vision of paradise regained, any more than Israeli politicians can get by without espousing the sacred 'law of return' and its vision of the ultimate ingathering of Jews throughout the world."[11] Holden suggested that before Arafat can relegate that dream to the indefinite future (much as the West Germans have done with their dream of a reunited Germany) and accept the partition of Palestine, he needs assurance that the Israelis also will move from established positions toward compromise. The Israelis, however, feel that any compromise they offer would be only a first step toward the dissolution of their state, and the loss of the security it provides, unless they first secure explicit pledges recognizing its permanence. With Menahem Begin's Likud Party now in power, any dealings with the PLO under any circumstances have become most unlikely.

Struggle Over Homeland

The Palestinians were the chief victims of Zionist insistence on the establishment of a Jewish state in part of Palestine and Arab insistence on destroying the new state. Who shoulders the major blame for the Palestinian plight—the Zionists, the Arab countries of Egypt, Jordan, Lebanon, Syria and Iraq, or the Palestinians themselves—depends to a large extent on the allegiance of those affixing blame.

Thus, the pro-Zionist publication *Near East Report* argued on July 17, 1974, that "those familiar with the history of 1948 [the year Israel declared its statehood] know that the Palestinians could have had their national homeland and that the Israelis accepted the establishment of a Palestinian state and at that time were ready to live in peace with it. But the neighboring Arab states invaded Palestine in order to crush the Jewish state and divide it up. The Palestinians were the victims of the Arab war against Israel."

A totally different account came from Sami Hadawi, a Palestinian refugee. Hadawi wrote that as soon as the United Nations adopted its resolution of Nov. 29, 1947, calling for the partition of Palestine into Jewish and Arab states, "Zionist underground forces came out into the open and began to attack Arab towns and villages, driving out all

7. See United States Policy chapter, p. 80.
8. Interview in *U.S. News and World Report*, 16 May 1977, p. 96.

9. *Newsweek*, 30 June 1975.
10. Georgie Anne Geyer, *The Washington Post*, 28 May 1977, p. A13.
11. David Holden, "Which Arafat," *New York Times Magazine*, March 23, 1975.

the non-Jewish inhabitants and massacring those who stood by their homes...."[12]

Similarly, Zionists and Palestinians blame each other for the massacres that followed partitioning and for an Arab exodus then and also after the establishment of the Jewish state on May 14, 1948.[13] Israel contends that Arab governments broadcast appeals to the Palestinians to leave until Arab armies liberated their homelands—a contention that Arabs label a "myth."[14] Regardless of the truth or falsity of the matter, it is generally acknowledged that Israelis razed property abandoned by Palestinian Arabs to discourage their return.

Whatever the basic reasons for the departure, some 525,000 to 900,000 Palestinians—there is no agreement as to exactly how many—had fled their homeland by the end of 1948. Most of them left with no material possessions and were dependent on the goodwill of Arab countries where they sought refuge—or on the United Nations—for food, clothing and shelter.

Soon after independence, the Knesset (Parliament) passed the "law of return," which provided that "every Jew has a right to immigrate to Israel." Don Peretz noted in his famous study, *Israel and the Palestine Arabs* (1958), that the abandoned property of the refugees "was one of the greatest contributions toward making Israel a viable state." He reported that of the 370 new Jewish settlements established between 1948 and the beginning of 1953, 350 were on absentee [owned] property, and nearly a third of the new immigrants (250,000) settled in urban areas abandoned by the Arabs."

Treatment of Israeli Arabs

The Israeli Declaration of Independence called on the "sons of the Arab people dwelling in Israel to keep the peace and play their part in building the State on the basis of full and equal citizenship and due representation in all its institutions, provisional and permanent." How well the 160,-000 Palestinians who remained in the country were treated remains a subject of dispute."[15]

A pro-Palestinian study in 1973 was highly critical of the military administration under which Israeli Arabs lived until 1966. Military rule, the study asserted, deprived the Palestinians in Israel "of their human and political rights." It "provided for administrative detention, even for unspecified reasons, for an unlimited period without trial or indictment, for the banishment of a person from the country, for confiscation of the property of those who violate military laws and for the imposition of total or partial curfew."[16] Jacob M. Landau offered another viewpoint in his book, *The Arabs in Israel* (1968). "These restrictions applied to Jews as well as to Arabs, and, in fact, only a few Arabs and Jews were closely affected," he wrote. "And even they could apply to Israel's Supreme Court...for redress."[17]

Just as there are conflicting and generally subjective reports on the treatment of Israeli Arabs, there is also wide

disagreement about how the Arabs who came under Israeli occupation after the June 1967 war have fared. Hisham Sharabi, who edits the *Journal of Palestine Studies* and teaches at Georgetown University in Washington, D.C., reports that their treatment has become increasingly repressive. And the Israeli League for Human and Civil Rights, led by Israel Shahak, professor of organic chemistry at the Hebrew University of Jerusalem, a former Zionist and survivor of the Nazi concentration camp at Bergen-Belsen, has been critical of the government's policy of destroying the property of suspects and expelling them.

> "...[T]here is no more room for agreements without an attempt to tackle the central problem of the Middle East conflict, the relationship between Israel and the Palestinians, which almost certainly means, in practice, the PLO."
> Zbigniew Brzezinski, et al.,
> Summer 1975[18]

The *London Times*, in June 1977, released a major study of torture in the occupied territories. It concluded that "Torture of Arab prisoners is so widespread and systematic that it cannot be dismissed as 'rogue cops' exceeding orders."[19] Israeli officials condemned the report and issued a lengthy, detailed rebuttal which subsequently appeared in the *London Times*.

Israeli officials typically say that while a few injustices may have been committed, the government's policy in this potentially explosive situation has been liberal, fair and enlightened. If Israeli policy were as repressive as its critics claim, they add, the Arabs under Israeli rule would have shown far more sympathy for the aims and terrorist activities of the guerrillas. That there has been relatively little collaboration between the PLO and the Palestinians in Israel proves, according to these spokesmen, that the vast majority of Arabs are willing to coexist and cooperate with the Jewish state.[20]

Hanna Nassir, president of Bir Zeit College, on Israel's West Bank, says that Israeli claims that the PLO lacks support in the occupied territories are self-serving. Nassir and four compatriots were deported to Lebanon Nov. 21, 1974. This followed student demonstrations and a shopkeepers' strike in Ramallah expressing support for the PLO during U.N. debate on the Palestinian question.

Nassir "categorically and emphatically" denies reports that he organized the demonstrations and instigated the strike, calls his deportation illegal, even by Israeli law, and a violation of the Geneva Convention, and points out that "there was no charge, no trial and no court decision." He says he was summoned to the military governor's office, where, after midnight, police handcuffed and blindfolded him and drove him with the other detainees to the Lebanese border. "There cannot be any leadership in the occupied territories," Nassir says. "That is why the people of the

12. Sami Hadawi, *Palestine: Loss of a Heritage* (1963), p. 2.

13. See David Pryce-Jones' *The Face of Defeat: Palestinian Refugees and Guerrillas* (1972), p. 15. See also "Arab Guerrillas," *Editorial Research Reports*, 1969 Vol. I, pp. 316-323.

14. Fawaz Turki, for example, wrote in *The Disinherited: Journal of a Palestinian Exile* (1972): "An examination of the monitoring records in the West revealed no such appeals to the population of Palestine from neighboring states; rather it was revealed that the Palestinians were exhorted *not to leave.*" [His emphasis.]

15. Almost 60 per cent of the Israeli Arabs today live in Galilee, particularly in the all-Arab towns of Nazareth and Shafa Amr. More than 20 per cent live in the Little Triangle, the region east of the coastal plain and north of Tel Aviv. Seven per cent live in the Haifa area and in the Negev. The rest are scattered throughout the country.

16. Al-Hakam Darwaza. "The Palestine Question," a study prepared for the Khartoum Conference on African-Arab Problems of October 1975, pp. 82-83.

17. For a critical assessment by a former Israeli lawyer who now assesses the Israeli scene for the PLO in Beirut, see Sabri Jiryis, *The Arabs in Israel* (Monthly Review Press: 1976).

18. "Peace in An International Framework," *Foreign Policy*, Summer 1975, p. 10. Brzezinski, who is President Carter's adviser on national security affairs, was in 1975 professor of government at Columbia University and director of the Trilateral Commission.

19. "Israel and Torture," *Sunday Times* (London), June 19, 1977.

20. See "Israeli Society After 25 Years," *Editorial Research Reports*, 1973 Vol. I, pp. 315-322.

'I Am a Rebel, and Freedom Is My Cause'

Following are excerpts from the address by Yasir Arafat, chairman of the Palestine Liberation Organization, to the United Nation's General Assembly Nov. 13, 1974:

If we return now to the historical roots of our cause, we do so because present at this very moment in our midst are those who, while they occupy our homes, as their cattle graze in our pastures, and as their hands pluck the fruit of our trees, claim at the same time that we are disembodied spirits, fictions without presence, without traditions or future. We speak of our roots also because until recently some people have regarded—and continued to regard—ours as merely a problem of refugees. They have portrayed the Middle East question as little more than a border dispute between the Arab states and the Zionist enclave. They have imagined that our people claims rights not rightfully its own and fights neither with logic nor valid motive, with a simple wish only to disturb the peace and to terrorize wantonly. For there are amongst you—and here I intend the United States of America and others like it—those who supply our enemy freely with planes and bombs and with every variety of murderous weapon. They take hostile positions against us, deliberately distorting the true essence of things. All this is done not only at our expense, but at the expense of the American people, and of the friendship we continue to hope can be cemented between us and this great people, whose history of struggle for the sake of freedom we honor and salute....

Mr. President, the roots of the Palestinian question reach back into the closing years of the 19th century, to that period we call the era of colonialism as we know it today. This is precisely the period during which Zionism as a scheme was born; its aim was the conquest of Palestinian land by European immigrants, just as settlers colonized, and indeed raided, most of Africa....

The difference between the revolutionary and the terrorist lies in the reason for which each fights. For whoever stands by a just cause and fights for the freedom and liberation of his land from the invaders, the settlers and the colonialists cannot possibly be called terrorist. Otherwise the American people in their struggle for liberation from the British colonialists would have been terrorists; the European resistance against the Nazis would be terrorism; the struggle of the Asian, African and Latin American peoples would also be terrorism. This, Mr. President, is actually a just and right struggle consecrated by the U.N. Charter and by the Universal Declaration of Human Rights. As to those who fight against the just causes, those who wage war to occupy, colonize and oppress other people, *those* are the terrorists....

I am a rebel, and freedom is my cause. I know well that many of you present here today once stood in exactly the same adversary position I now occupy, and from which I must fight. You were once obligated by your struggle to convert dreams into reality. Therefore you must now share my dream. I think this is

exactly why I can ask you now to help, as together we bring out our dream into a bright reality, our common dream for a peaceful future in Palestine's sacred land....

In my formal capacity as chairman of the Palestine Liberation Organization and as leader of the Palestinian revolution, I

proclaim before you that when we speak of our common hopes for the Palestine of tomorrow we include in our perspective *all Jews* now living in Palestine who choose to live with us there in peace and without discrimination.

In my formal capacity as chairman of the Palestine Liberation Organization and leader of the Palestinian revolution, I call upon Jews one by one to turn away from the illusory promises made to them by Zionist ideology and Israeli leadership. Those offer Jews perpetual bloodshed, endless war and continuous thralldom.

We invite them to emerge from their moral isolation into a more open realm of free choice, far from their present leadership's effort to implant in them a Massada complex.

We offer them the most generous solution that we might live together in a framework of just peace, in our democratic Palestine....

Later Dismay

Arafat later expressed dismay that his speech, which he considered a conciliatory move toward compromise, was viewed as militant in the Western press. In an interview with Eric Rouleau, published in *Le Monde* Jan. 7, 1975, he said:

...the Zionist propagandists have emphasized one phrase from my speech taken out of context to give support to the despicable theory according to which we are really trying to throw the Jews into the sea.

Yes, I did say that I dreamt—I emphasize the word— of a unified and democratic Palestine, But is it a crime to dream? Is there a law against imagining the change which would come about in years to come?

Do I have to force myself to forget the house where I was born in Jerusalem a few yards from the Wailing Wall, a house whose destruction Mrs. Golda Meir ordered under the occupation regime? Do I have less right to be there than this Russian lady, naturalized American, who has come to install herself in my ancestral land?

West Bank have accepted the PLO as its leadership. They have more freedom."[21]

Palestinians charge that since 1967, more than 1,500 of them have been expelled without charges or trial from the West Bank and Gaza, including many doctors, teachers, mayors and other professional people capable of leadership.

The local elections which took place in the West Bank in April 1976 confirmed two things: first, that support for the PLO was quite strong as many candidates for mayor and other offices who were known supporters of the PLO were elected; and second, that the Israeli occupation provided for

a substantial measure of free expression even on delicate political matters.

Just a month earlier, Israel experienced an escalation of the periodic rioting and demonstrations from the occupied territories. In this demonstration, Israeli Arabs from the Galilee area protested a government land expropriation scheme. Violent clashes with police occurred in more than a dozen Israeli Arab villages. Six Arabs were killed, and over 70 others were injured.

A few days after the rather startling show of militancy revealed in the West Bank elections, Israeli annexationist groups staged a two-day march through the West Bank to proclaim "the inalienable right of every Jew to every part of the land of Israel."

21. Interview with *Congressional Quarterly*, June 25, 1975.

The Arab League: An Effort at Cooperation

Officially known as The League of Arab States, it was at a 1974 meeting of this organization in Rabat, Morocco, that the Palestine Liberation Organization was elevated to the position of "sole legitimate representative of the Palestine people."

In 1976, the PLO officially became the twenty-first member of the Arab League.

Origin and development. A long-standing project that reached fruition late in World War II, the League was founded primarily on Egyptian initiative following a promise of British support for any Arab organization that commanded general approval. In its earlier years, the organization focused mainly on economic, cultural, and social cooperation. On April 13, 1950, a Convention on Joint Defense and Economic Cooperation was concluded that, among other provisions, obligated the members in case of attack "immediately to take, individually and collectively, all steps available, including the use of armed force, to repel the aggression and restore security and peace." In 1976, the Palestine Liberation Organization (PLO), which had participated as an observer at all League conferences since September 1964, was admitted to full membership.

Structure. The principal political organ of the League is the Council, which is composed of representatives from each member state. It meets twice a year, normally at the foreign ministers' level. Each member has one vote in the Council; decisions bind only those states which accept them. The Council's main functions are to supervise the execution of agreements between members, to mediate disputes between members, and to coordinate defense measures in the event of attack.

The treaty provided for the creation of permanent committees dealing with matters of concern to the League. In 1977, there were ten such committees which meet at least once a year: Political, Cultural, Economic, Social, Military, Legal Affairs, Information, Health, Communications, and Arab Human Rights.

Three additional bodies were established by the 1950 Convention: a Joint Defense Council to function in matters of collective security and to coordinate available military resources; a Permanent Military Commission, composed of representatives of the general staffs, to draw up plans for joint defense; and an Economic Council, composed of the ministers of economic affairs, to coordinate the development of Arab economies. A Council of Arab Information Ministers has existed since 1964. The General Secretariat is responsible for internal administration and the execution of the Council's decisions. It administers several bureaus: the Bureau for Boycotting Israel (headquartered in Damascus, Syria), the Arab Narcotics Bureau and the Arab Bureau of Criminal Police.

Activities. Spurred by an active secretary general, the political activities of the League have intensified. For example, the secretary general has participated in negotiations concerning settlement of the Middle East crisis. He has repeatedly sought to promote peace in Lebanon, including activity in connection with the dispatch of a truce observation force to the area. His continuing role as a mediator is illustrated by his unsuccessful attempt to settle disputes between Morocco, Mauritania, and Algeria in March 1976, and by his offer to participate in a joint League-OAU commission investigating Sudan's charge that it was invaded by Libya in July 1976.

The League has also been active in economic affairs. For example, the Conference of Economic and Foreign Ministers agreed at a meeting at Rabat, Morocco, in April 1976 to establish an Abu Dhabi-based fund to provide credits for Arab states suffering balance-of-payments deficits. The objectives of the fund are to stabilize Arab exchange rates, to insure the convertibility of Arab currency, and to work toward creation of a single Arab monetary unit.

Members: Algeria, Bahrain, Egypt, Iraq, Jordan, Kuwait, Lebanon, Libya, Mauritania, Morocco, Oman, Palestine Liberation Organization, Qatar, Saudi Arabia, Somalia, Sudan, Syria, Tunisia, United Arab Emirates, Yemen Arab Republic, Yemen People's Democratic Republic.

Source: *Political Handbook of the World: 1977*
(McGraw-Hill: 1977), p. 489.

Even before Menahem Begin took over the Israeli government, it was very clear that rising Arab and Jewish militancy over the West Bank would be a major stumbling block in any attempt to find a comprehensive formula for a peace settlement.

West Bank Economy

Israeli occupation unquestionably has improved economic conditions. "Wherever one goes on the West Bank," John M. Goshko wrote June 1, 1975, in *The Washington Post*, "the eye constantly is struck by those symbols—late model cars on the roads, forests of TV antennas sprouting over every town, running water in remote villages—that are the physical evidence of growing affluence. From what was essentially an area on relief only a decade ago, the West Bank today stands on the threshold of becoming a consumer society."[22]

The territory has no natural resources, and half its population is under age 14—conditions which make its economic prospects bleak. But West Bank agriculture has benefited from expanded markets and Israeli technical assistance. A tenfold increase in the number of tractors in the area since 1967—from 120 to 1,200—is one indicator of the change under occupation. Agricultural production has increased 12 per cent annually in real terms, and the West Bank's gross national product has risen by an average of 18 per cent annually since 1967.[23]

About one-third of the West Bank labor force (40,000 out of 120,000 total) now work in Israel. "Palestinian workers in Israel have become a vital element in Israel's economy," according to John Richardson, president of American Near East Refugee Aid. "They are 45 per cent cheaper to employ than Israelis, and they are available in

22. John M. Goshko, "The West Bank: Going it Alone," *The Washington Post*, June 1, 1975, p. C 1.

23. *Ibid.*, p. C 5.

large numbers to do the menial work that Israelis are learning to disdain." Richardson sees the integration of the West Bank and Gaza into the Israeli economy as deliberate policy and cites an Israeli Defense Ministry document:

> The areas are a supplementary market for Israeli goods and service on the one hand and a source of factors of production, especially unskilled labor, for the Israeli economy on the other.[24]

In 1977, the Carnegie Endowment for International Peace released a major study discussing what happened to the West Bank and Gaza Strip economies since 1967. In the book's preface, the President of Carnegie, Thomas L. Hughes, noted that the author, Brian Van Arkadie, "has framed his appraisal in terms that could be meaningful as a baseline for thinking about the future of the West Bank and the Gaza Strip." Van Arkadie, in summary, found that "In 1967, the Israeli government did not conceive or attempt to implement any systematic, large-scale plan to alter the economic structure of the West Bank and the Gaza Strip. Nor was there an Israeli master-plan for changing the external economic relationships of the two occupied territories. For the West Bank and the Gaza Strip economies, however, what happened after 1967 was more complex and no less profound than if such a master-plan had actually existed.[25]

Rise of Terrorists

Israel's refusal to repatriate more than a handful of Palestinian refugees and the Arab countries' hesitation about resettling them[26] led to the establishment of the United Nations Relief and Works Agency (UNRWA) to care for the refugees. Describing the plight of those in the refugee camps in the mid-1950s, Don Peretz wrote: "Many living in leaky, torn tents were middle-class urbanites who had owned modest but adequate houses in their native land.... The self-reliance and individual initiative of former tradesmen and farmers were drowned in the boredom and frustration which the camps bred...."

This resentment and frustration contributed to the growth of terrorist groups and the eventual formation of the Palestine Liberation Organization. In the early 1950s, some of the Arab governments trained and subsidized groups of Palestinian guerrillas. Egypt, for example, set up battalions of "Palestinian Fedayeen" under the direct command of Egyptian officers in the Gaza Strip. But according to Fawaz Turki, "the first clandestine organization that was a truly Palestinian expression" was Al Fatah ("conquest"), formed in the early 1950s by Palestinian students at Stuttgart University in West Germany.

One of these students was Yasir Arafat, alias Abu Ammur, who left Jerusalem with his family during the 1948 war and settled in Gaza. Arafat had become convinced that the Palestinians must look to themselves, not the Arab governments, for the recovery of their homeland. Within a few years, Al Fatah and other Palestinian groups were established in Europe and the Middle East to coordinate Palestinian liberation strategy. Members of the groups received military training in Algeria and by the early 1960s were conducting terrorist raids in Israel.

The Palestine Liberation Organization was established by Arab heads of state at the first National Palestinian Congress—held in the Jordanian sector of Jerusalem in May 1964—before the sector was seized by Israel in the 1967 war. The delegates declared that "the Palestinian problem will never be resolved except in Palestine and by the force of arms."

The PLO did not actually emerge as an autonomous faction in the Arab world, however, until Israel's victory in the Six-day War. Yehoshafat Harkabi maintains that the defeat of Arab armies raised the stature of the Palestinians, because only they were carrying on the fight. The Palestinians, he wrote, "were transformed from an inferior factor into standard-bearers of Arab nationalism and a source of pride."[27] The PLO became more militant. The incident which more than any other gave the PLO stature in the Arab world was an Israeli attack on an Al Fatah camp in the Jordanian village of Karameh on March 21, 1968. In a 12-hour battle, about 300 guerrillas held off Israeli attackers, inflicted heavy losses and forced them to retreat. Young Palestinians rushed to join the movement.

The increased confidence and militance of the PLO were evident in the Palestine National Covenant, revised by the National Congress in Cairo in July 1968. It holds, among other things, that "armed struggle is the only way to liberate Palestine" (Article 9) and that "the partitioning of Palestine in 1947 and the establishment of the state of Israel are entirely illegal" (Article 19). In the new state the PLO proposed to create, only "The Jews who had normally resided in Palestine until the beginning of the Zionist invasion will be considered Palestinians" (Article 6).[28]

Although the first Zionist immigrants arrived in the 1880s, Palestinians have used various dates at different times as the cutoff point in determining which Jews could remain in a Palestinian state. Sometimes 1917, the date of Britain's Balfour Declaration supporting a Jewish "national home" in Palestine, is used; at other times 1947, the date of partition, has been invoked. Palestinians now state that all Jews presently in Israel could remain, but past inconsistency on this point does not reassure Israelis, who in any case have no intention of living in a non-Jewish Palestinian state.

Guerrilla-National Uneasiness

One of the ironies of the Palestinian predicament is that PLO sympathizers often display as much animosity toward their supposed benefactors and friends, the Arab states, as they do toward the Zionist enemy. And the Arab states directly involved—Egypt, Syria, Jordan and Lebanon—have, while paying lip service to the Palestinian cause, tried to restrain the commandos. Jordan and Lebanon and most recently and devastatingly, Syria, have even used their armed forces against them.

The Palestinian presence has created a dilemma for Arab governments. A complete assimilation of the refugees would probably have destroyed the delicate political and religious balance in Lebanon and caused a dislocation of the Egyptian and Syrian economies. To allow the guerrillas to use their territory as a base from which to attack Israel invites quick and strong retaliation on and around the refugee camps. And to permit PLO terrorist attacks would make Israel even more reluctant to return the territories occupied

---

24. John Richardson "Special Issue on the West Bank and Gaza," *The Link*, published by Americans for Middle East Understanding, spring 1975, pp. 2, 5.

25. Brian Van Arkadie, *Benefits and Burdens: A Report on the West Bank and Gaza Strip Economies since 1967* (Carnegie Endowment for International Peace, 1977).

26. The one exception was Jordan, the poorest of the Arab countries, which accorded the Palestinians citizenship.

27. Yehoshafat Harkabi, "The Position of the Palestinians in the Israeli-Arab Conflict and Their National Covenant," *New York University Journal of International Law & Politics*, spring 1970, p. 218.

28. See footnote 6.

Defining Palestine, Palestinians

The word "Palestine" is of Roman origin, referring to the biblical land of the Philistines. The name fell into disuse for centuries and was revived by the British as an official title for an area mandated to their control by the League of Nations in 1920 after the breakup of the Turkish Ottoman Empire in World War I.

The British mandate also applied to Trans-Jordan (now Jordan), although it did not lie within the area designated "Palestine." Trans-Jordan lay entirely east of the River Jordan, and Palestine lay entirely to the west.

However, because the mandate applied to both regions, there is the argument that "Palestinian" applied to persons east as well as west of the River Jordan. There is a further argument that it applies not just to Arabs—as is common practice today—but also to Jews and Christians in the former mandated area.

Palestine as a legal entity ceased to exist in 1948, when Britain, unable to keep control, relinquished its mandate and Israel declared its independence. The previous year, Palestine had been partitioned by the United Nations into Arab and Jewish sectors. Israel enlarged its partitioned areas in a war of independence in 1948-49. Other parts of Palestine fell under the control of Jordan and Egypt.

---

in 1967. Moreover, most of the Palestinians have not desired assimilation in the countries in which they found themselves. The concept of "the return" to Palestine is deep and widespread among the Palestinians. Palestinian children are taught that their homeland is this or that village in Palestine—a homeland which most have never even seen.

The strongest action against the guerrillas until the civil war in Lebanon was taken by King Hussein of Jordan in 1970 after years of seeing their raids from Jordan bring Israeli reprisals against his country. In trying to stop the raids, he incurred the wrath of extreme leftist guerrillas, who threatened his life and promised to "revolutionize" his kingdom. Hussein's troops clashed with the guerrillas in September 1970 and succeeded in closing all but a few of their bases.[29] The next July, the Jordanian army attacked the remaining outposts. Some of the guerrillas were captured and imprisoned, while others fled to Israel or Syria. Under pressure from the other Arab governments, Hussein later declared an amnesty for all terrorists except those who had been convicted of murder or espionage. This did little to lessen PLO demands for his overthrow and the destruction of the "reactionary" Hashemite kingdom.

Lebanon, too, even before the civil war in 1975, took military action against the guerrillas. After armed clashes in 1969, the PLO was restricted to bases in southern Lebanon under the supervision of the Lebanese army. This arrangement came to be known as the Cairo Agreement. Then, in May 1973, after Israeli retaliation in Lebanon for commando attacks in Israel, the army again fought with the PLO. The fighting lasted two weeks and resulted in more than 350 deaths and 700 injuries. As was the case with Jor-

---

29 During the fighting, a secret terrorist arm, Black September, was set up. It claimed responsibility for, among other things, the murder of Jordanian Prime Minister Wasfi al-Tal in Cairo Nov. 29, 1971, and the massacre of Israeli athletes at the Olympic Games in Munich in September 1972. For the story of Hussein's expulsion of the PLO and the growth of Black September see John K. Cooley, *Green March Black September* (Frank Cass, London: 1973).

---

dan, Syria threatened to send military aid to the Palestinians while Israel was ready to assist Lebanon. Syrian troops did cross the border, but their quick withdrawal prevented a widening of the conflict. Further armed skirmishes between Lebanese rightists and Palestinians broke out in 1974 and 1975—and helped precipitate the civil war. The rightists began demanding that the government take control of the refugee camps. The PLO was equally insistent that the Palestinians retain control. By mid-June 1975, a series of efforts to form a new government had failed, and though PLO leader Arafat was said to want to avoid bringing on another civil war, he failed. *(See chapter on Lebanon, p. 46)*

The PLO fears that the Arab governments might make deals with Israel at the Palestinians' expense were exacerbated when the U.N. Security Council adopted Resolution 242 on Nov. 22, 1967. The resolution called for, among other things, "the withdrawal of Israel from territories occupied" in exchange for recognition of Israel by the Arab countries. Jordan and Egypt endorsed 242, provided that it was interpreted to mean all, not part, of the territories Israel seized. Syria did not accept the resolution until after the October 1973 war. The PLO has repeatedly denounced the resolution on the ground that it refers to refugees, not Palestinians, and puts forth no specific plan for a "just settlement."

But after the 1973 war, the PLO gained considerable esteem. Gradually the "refugee" problem became recognized as one involving peoplehood and a national identity. In October 1974 the leaders of the Arab states clearly took the leadership of the Palestinian Arabs from King Hussein and Jordan and gave it to Yassir Arafat and the PLO. By 1976 the PLO had become the twenty-first full member of the Arab League. By 1977, more than 100 countries had granted the PLO some form of recognition.

Proposed Solutions

Proposed solutions to the Palestinian problem are as numerous and varied as the interests of the parties involved. Therefore, it is hardly surprising that none of the major proposals has been acceptable to all parties. These include:

- A separate Palestinian state on the West Bank and in Gaza.
- Palestinian-Jordanian federation or a looser form of confederation.
- Assimilation of the refugees in the Arab countries where they are now living.
- Their repatriation followed by the establishment of a secular Palestinian state incorporating Israel.

A Palestinian State

Probably the most-publicized proposal is for a separate state on the West Bank and in Gaza. The Soviet Union, the Arab states and even the majority in the PLO now support the idea of creating a separate Palestinian state in territory now occupied by Israel. The United States has not yet taken an official position. But President Carter's endorsement of a "Palestinian homeland" linked in some way to Jordan has in a *de facto* way altered previous American policy which only referred to a Palestinian "refugee problem" in accord with U.N. Resolution 242. Israel's Labor Party government strongly opposed the idea of such an independent Palestinian state on the ground that radical, irredentist Palestinians would control it and use it as a base from which to launch terrorist attacks against Israel and work for Israel's destruction. Menaham Begin's right-wing Likud government—which came to power in June 1977—refuses to accept the very concept of Palestinian nationalism (except strictly within the context of Jordanian nationalism) and insists that the West Bank area itself is the historical heritage of the Jewish people and should eventually be incorporated into Israel.[30]

For the United States, Israel and Jordan, an independent Palestinian state on the West Bank and in Gaza would pose considerable difficulties and dangers. For American policymakers, there is the fear that it might turn to the Soviet Union for military and political support and destabilize rather than stabilize the region. Israelis are aware that even if PLO leaders were denied any significant voice in the Palestinian government, there would be pressure on that government to seize more of the territory that was formerly Palestine. Israel is also worried that the heavy concentration of Israeli Arabs living in the northern Galilee region of the Jewish state might identify with the new Palestinian state.

Some regard a Palestinian state on the West Bank and in Gaza as impractical. Tunisian Foreign Minister Habib Chatti wrote in *The Washington Star-News* on March 24, 1974: "The only workable solution to the Palestinian problem—and one which we are sure their leaders would accept—is the creation of a new Palestinian state. But the West Bank of the Jordan River and the Gaza Strip...would not suffice for such a state. The Palestinians would need more than these overcrowded bits of territory—and additional land would have to come from Israel and Hussein's Jordan."

The issue of the economic as well as political viability of a potential West Bank-Gaza Strip Palestinian state has been increasingly discussed since 1974.[31]

Though Jordan's King Hussein reluctantly acquiesced in the Rabat Conference legitimization of the PLO on "liberated Palestinian territory," he remains deeply concerned about the fate of the West Bank and troubled about the possibility of the area coming under a political regime which would challenge his Hashemite throne. Some observers thought he agreed to the Rabat decision because he expected the PLO to fail in its aim and thus give him further opportunities to reassert his claims to the West Bank.

By July 1977 the King was again asserting that the Jordanians and Palestinians "are one people and one family" and that any Palestinian state would have to have an "explicit link" with Jordan. His attitude toward a separate Palestinian state was publicly ambiguous but privately negative.

Though a majority of the world's nations have endorsed the idea of self-determination for the Palestinians, which also seems to be the basic concept behind President Carter's "Palestinian homeland" phrase[32] the political nature of such a possible Palestinian state, its location and the political leadership that would rule, were all open matters as of August 1977.[33] Menaham Begin's "Peace

U.N. General Assembly Endorses Palestinian State

On November 22, 1974, a few days after Yassir Arafat addressed the General Assembly, that body passed a resolution, #3236, by a vote of 89 to 7 with 37 abstentions. The resolution included the following operative clauses:

"1. Reaffirms the inalienable rights of the Palestinian people in Palestine, including:

(a) The right to self-determination without external interference;

(b) The right to national independence and sovereignty;

2. Reaffirms also the inalienable right of the Palestinians to return to their homes and property from which they have been displaced and uprooted, and calls for their return;

3. Emphasizes that full respect for and realization of these inalienable rights of the Palestinian people are indispensable for the solution of the question of Palestine;

4. Recognizes that the Palestinian people is a principal party in the establishment of a just and durable peace in the Middle East...."

In its 15-point Political Declaration, the Palestine National Council in March 1977, accepted this General Assembly resolution as a basis for negotiations rather than Security Council resolution #242 of November 1967 which the PLO rejects because it refers to the Palestinians only as "refugees."[*]

In November 1976, the General Assembly further endorsed the right of the Palestinian Arabs to establish their own state and to reclaim their former homes and properties in what has become Israel. This step came as the General Assembly endorsed a report of the Committee on the Exercise of the Inalienable Rights of the Palestinian People by a vote of 90 to 16 with 30 abstentions. A Security Council resolution along the same lines was vetoed by the United States the previous June. In this November 1976 General Assembly resolution, the world body called on the Security Council to arrange for Israeli withdrawal from all occupied territories by June 1, 1977, without any reference to previous Security Council resolutions stipulating the need for Arab-Israeli negotiations to bring about such a withdrawal. In a separate resolution, the General Assembly called for early resumption of the Geneva Conference "with the participation of all the parties concerned, including the Palestine Liberation Organization."

*The 15-point Political Declaration of the March 12-20, 1977, Palestine National Council meeting which took place in Cairo can be found in *Journal of Palestine Studies*, Vol. VI, No. 3, Spring 1977, pp. 188-190.

---

30 For a unique presentation of Israeli views on the Palestinian issue see Larry L. Fabian and Ze'ev Schiff (eds.), *Israelis Speak* (Carnegie Endowment for International Peace: 1977). Also see Rael Jean Isaac, *Israel Divided* (Johns Hopkins University Press: 1976) and Samuel Katz, *Battleground: Fact and Fantasy in Palestine* (Bantam: 1973).

31 See Vivian A. Bull, *The West Bank—Is It Viable* (Lexington: 1975).

32 See Mark A. Bruzonsky, "Carter and Begin: Getting Down to Business," *Sunday Boston Globe*, 17 July 1977, p. A1.

33 See Edward R. F. Sheehan, "A Proposal for a Palestinian State," the *New York Times Magazine*, 30 January 1977, p. 8. Also see *Time Magazine*, 27 December 1976, pp. 20-21.

Plan," unveiled during his July 20th visit to Washington, offered no hope at all that Israel would ever be willing to allow a third state in any form in the area of historic Palestine.

But within Israel there were political groups still promoting such an idea. A small party led by a former

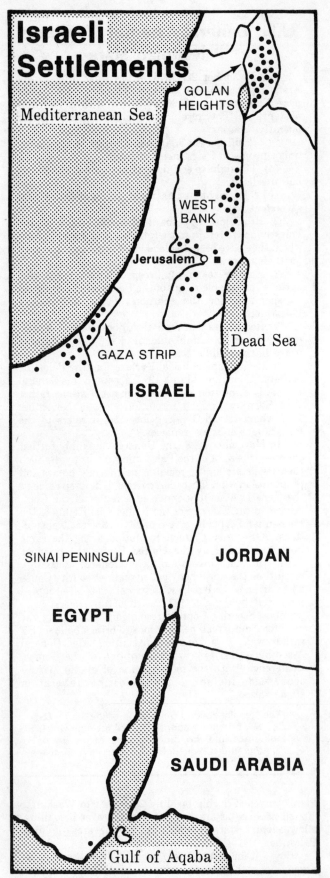

Israeli Settlements

Mediterranean Sea

GOLAN HEIGHTS

WEST BANK

Jerusalem

GAZA STRIP

Dead Sea

ISRAEL

SINAI PENINSULA

JORDAN

EGYPT

SAUDI ARABIA

Gulf of Aqaba

▮Indicates settlements "legalized" by new Israeli government July 1977.

Secretary-General of the Labor Party, Arie Eliav, has since 1976 been promoting the idea of negotiations with the PLO which could lead to a West Bank-Gaza Strip Palestinian State. These dovish Zionist leaders had begun discussions with PLO persons in Paris during 1976; but these discussions had not produced any concrete results by mid-1977.

Israeli Views on the Palestine Problem

Ever since the early 1970s, solving the Palestinian problem has become a topic of much concern in Israel. In 1970, for instance, Shlomo Avineri, then chairman of the political science department at Hebrew University and who in 1975 became Director-General of the Foreign Ministry in Yitzhak Rabin's government, stated that the replacement of King Hussein by a Palestinian state "could be more conducive to eventual peace than his continuing presence has proved itself to be." Even if the more radical Palestinians were to control the new government, Avineri continued, "it is a historical truism that an underground movement, once come into formal political power, tends to become responsible and respectable, in the exercise of authority. In all, I think it eminently worthwhile for Israelis, who shudder at the thought of Al Fatah taking over Jordan, to have second thoughts on the matter."[34]

By 1975 Avineri was thinking more about a separate West-Bank and Gaza Strip Palestinian state than about replacing the Hashemite Kingdom of Jordan with a large Palestinian state. In 1975, just before becoming Director-General of the Foreign Ministry, Avineri stated on Israeli radio that "there is is no reason to rule out in advance coming to an arrangement that might include a West Bank-Gaza Palestinian state.... There is no reason to rule out in advance, in any event, negotiations with the PLO."[35]

Other establishment Israelis also began showing a much greater willingness to contemplate dealing with the PLO in a way that could eventually lead to a Palestinian state—but nearly all insisted that the PLO must first renounce its intention of destroying Israel and revise the Palestine National Covenant accordingly. Former Chief of Military Intelligence and expert on the Palestinian issue, Yehoshafat Harkabi, wrote the following in the important Israeli daily *Ma'ariv* on Jan. 7, 1977: "While we continue to attack PLO extremism, it is incumbent upon us to show sympathy for the Palestinian problems and suffering, to the making of which we have contributed. Apparently, the official Israeli position today is to acknowledge not only the existence of a Palestinian problem but also the existence of the Palestinians as an ethnic group. If so, then the logical conclusion is that they are entitled to political expression in the form of a nation state. We may express our opinion that it be better for this state to 'have ties with' Jordan as there are no national or ethnic differences between the Jordanians and the Palestinians, but it is unthinkable that we should dictate the form this state should take. We may demand certain conditions necessary to our security, as the United States demanded the removal of the missiles from Cuba; the United States, however, had no pretensions to determine the character of the Cuban government." Harkabi went on to predict that "Israel is compelled to eventually recognize the PLO,...because its representative status is

---

34 "The Palestinians and Israel," *Commentary*, June 1970, pp. 38-39.

35 Quoted in an interview by Mark Bruzonsky with Israeli Ambassador Simcha Dinitz in *Worldview*, July-August 1977, p. 39. Avineri changed his mind by 1977 and no longer favored an independent Palestinian state for a variety of reasons stated in a *New York Times* editorial on 15 May 1977.

already a matter of fact." "It would be better if we recognized that the few voices which distinguish themselves from the PLO are as a drop in the ocean," he insisted, "compared to the chorus of Palestinians who chant the PLO praise and identify themselves with it."

Whether or not the West Bank and Gaza Strip would remain under Israeli control, would eventually be incorporated into Israel, would be returned to Jordan and Egypt respectively, or would eventually form a Palestinian region, entity, or state, remained highly problematical in mid-1977. But it was widely expected that the next few years would result in historical decisions that could determine the outcome of the Jewish-Arab struggle for control of historic Palestine.

United Arab Kingdom Plan

Discussion during 1977 of a Palestinian state linked to Jordan brought about recollection of King Hussein's United Arab Kingdom plan of 1972.

In an address to the Jordanian parliament on March 15, 1972, Hussein proposed a restructuring of the country into a federal state made up of two autonomous regions, Transjordan and the Israeli-occupied West Bank, with equal representation in the national parliament in Amman. The central government would be responsible for defense, foreign affairs and other matters of purely national interests. Hussein implied that residents of the Gaza Strip could freely join the new state, which would be called the United Arab Kingdom.[36]

His plan received a cool reception in Israel and was denounced by the PLO and its supporters as a sellout. Nevertheless, some analysts believe Jordan still hopes to have a definite relationship with the West Bank, possibly along the lines of a federal state. Some observers also believe that the West Bank Arabs would opt for a separate non-PLO type of state with close ties to both Jordan and Israel if they were allowed a plebiscite. In recent years, however, the expressed desire for a separate, sovereign state has appeared to grow. The Israeli position appeared—before Begin became prime minister—somewhat more flexible in the mid-1970s than before the Yom Kippur War as to returning part of the West Bank to Jordanian "administrative control"—while maintaining Israeli security outposts along the Jordan River. The plan was known as the "Allon plan" because it was promoted by Israeli Foreign Minister Yigal Allon.[37] On Aug. 15, 1975, the Israeli parliament rejected a motion by right-wing members to require a national referendum in advance of yielding any part of the West Bank. The Rabin government did find it necessary, however, to promise to hold new elections before agreeing to turn over any of the West Bank to the Arabs. But the new Likud government of Menahem Begin came into office completely opposed to return of the area to any kind of Arab sovereignty.

Probably the most difficult and emotional issue to resolve in any Israeli-Jordanian or Israeli-Palestinian negotiations is the fate of Jerusalem, a city holy to Jews, Moslems and Christians. Israeli leaders of all political parties have repeatedly and insistently declared that Jerusalem is "not negotiable" and can never again be divided. Still, it is unlikely that the Arab-Israeli dispute can ever fully be settled without some arrangement for Arab control of East Jerusalem.

Possibility of Resettlement

The issue of Palestinian refugee repatriation or resettlement has been debated endlessly for a quarter century or more. In countless resolutions on the subject the United Nations vacillated, calling at one time for a return, at another for assimilation, then for a "just settlement of the refugee problem," and most recently for both "Palestinian self-determination" and the right to reclaim former homes and property. The U.N. had little choice. The Arab states, except for Jordan, refused to assimilate the refugees, Israel refused to repatriate them, and many of the refugees clung to an insistence that they be allowed to return to their former homes and country. In addition to the immense security problem that resettlement in Israel would create, such an influx might give the country an Arab majority within a few years, thus destroying the Zionist dream.

Due to high Arab and relatively low Jewish birth rates in recent years, Israeli officials have become worried about the growth of the Arab minority even inside the 1967 boundary lines unless Israel receives large number of Jews from abroad, particularly from the Soviet Union. But Terence Smith reported in *The New York Times* on Aug. 15, 1974, that Jewish immigration to Israel dropped by 33 per cent in the first half of the year because of deteriorating economic conditions and the "political uncertainty caused by the October [1973] war." By 1977 the immigration and emigration rates into and from Israel were approximately equal. Should Israel at some time incorporate the West Bank and Gaza Strip, as Menahem Begin's Likud government has hinted, the Arab minority problem would suddenly become much more difficult.

Some American leaders have at times advocated that Israel, the United States, the United Nations and the rich Arab countries allot as much money as needed to assist the host governments in assimilating the refugees. Israel contends that the less populated Arab countries, notably Syria, Iraq and Saudi Arabia, could easily accommodate thousands of Palestinians without endangering national stability. But this does not seem a realistic solution in view of Palestinian aspirations. Furthermore, the situation in Lebanon, where about 400,000 Palestinians live, seriously complicates any plans for assimilating the Palestinians into their host countries. To do so with any significant number of these refugees in Lebanon is out of the question in view of that country's fragile political and religious balance. Also, assimilation as a policy begs the question of what happens to the one million Palestinians living in the West Bank and Gaza Strip areas under Israeli rule.

Commenting on a recommendation by British historian Arnold Toynbee that the Palestinians be given the option of returning to their homes as "loyal citizens of Israel" or generous monetary compensation for resettlement elsewhere, Fawaz Turki wrote that Toynbee "was right in asserting that most refugees would not return to Israel on the condition that they live as 'loyal Israeli citizens.' In toto, his solution, not unlike others before it, denied the Palestinians the opportunity for a decent existence in their own homeland and guaranteed their ultimate disappearance as an entity and a people—a fate they had tenaciously fought against."

Similarly, Cherif M. Bassiouni, an Arab-American and professor of law at DePaul University in Chicago, told the

---

36 For a discussion of King Hussein's motivations and plan see John K. Cooley, *Green March Black September* (Frank Cass, London: 1973).

37 See Yigal Allon, "Israel: The Case for Defensible Borders," *Foreign Affairs*, October 1976, p. 38.

U.S. House Subcommittee on International Organizations and Movements on April 4, 1974: "I must emphasize the importance to the Palestinians of their Palestinian entity.... They don't want to be assimilated anywhere regardless of the economic benefits they can derive from it; they want to be once again a people in their own country." While Turki and many of his fellow refugees believe that the struggle for Palestine will continue and perhaps even intensify, others think that with the passage of time, the irredentist fervor will lessen. Yehoshafat Harkabi feels that "as recognition grows that there is no future for the idea of Palestinian sovereign independence, it will lose much of its hold on the Palestinians as a political ideal."[38] Interestingly, in the years since Harkabi stated this view, he himself has become much more sensitive to Palestinian Arab aspirations. But the Likud leaders of Israel who took power in June 1977, continue to adhere to this view.

Secular Democracy

Dreamers on both sides of the Israeli-Palestinian struggle often view the ideal solution as the establishment of a secular, democratic state where Jews, Christians and Moslems could coexist peacefully. In recent years, Palestinian spokesmen, including most PLO leaders, have tried to draw a distinction between Jews and Zionists. Palestinians and Jews have lived and can again live in friendship, they insist, once the "exclusivist, discriminatory, imperialist Zionist state" is abolished. Jews sympathetic to a secular state have urged PLO leaders to put their words into practice by amending the 1968 Palestinian National Covenant to permit all Jews, regardless of their date of immigration, to remain once the new state is established.

One of the primary ideological goals achieved by the U.N. resolution declaring Zionism to be a form of racism was to delegitimize the idea of an exclusively Jewish state in order to further promote the idea of a multi-ethnic democratic-secular state.

There are some anti-Zionist Israeli Jews. One of the most prominent is Israel Shahak, a professor at Hebrew University in Jerusalem. Shahak, in a meeting with American journalists in Jerusalem on June 6, 1974, said he favored a federation of Arab and Jewish states, similar to Swiss cantons, with Jerusalem as the capital and with a bill of rights protecting all groups within the federation. Uri Avnery, editor of the popular weekly magazine *Ha'olem Hazeh*, made a far-reaching proposal in 1968. Avnery submitted a resolution in the Knesset which called for the creation of a Palestinian state and a federation between Israel and the new state. After the establishment of the federation and the resettlement of the refugees, a Semitic Union, "a great confederacy of all the states in the region," would be set up. "A solution in Palestine is almost a prerequisite to a general Semitic peace settlement, and at the same time, a Semitic peace is necessary to make the Palestinian solution meaningful and enduring," Avnery wrote in *Israel Without Zionists* (1968).

Avnery's proposed Semitic Union would put an end to "mutual fear and suspicion" and would permit a peaceful pooling of political power and economic resources. "Joining a great Semitic confederacy would mean, for Israel, putting an end to the Zionist chapter in its history and starting a new one—the chapter of Israel as a state integrated in its region, playing a part in the region's struggle for progress and unity. For the Arabs it would mean recognition of a

38 "The Problem of the Palestinians," *op. cit.*, p. 10.

How Many Palestinians?

The round figure of 3 million is usually used as the number of Palestinians in the world.

In 1977, estimates were as follows for the occupied territories:

West Bank (including eastern Jerusalem)	762,000
Gaza Strip and Northern Sinai	400,000

In addition, the non-Jewish population of Israel proper (530,000) is overwhelmingly Palestinian. Thus about half of all Palestinians live under Israeli rule.

As for the others, the biggest group, about 900,000, are living in Jordan and many of them are citizens there. It is believed that about another half million live in Lebanon, Egypt and Syria, with a small number dispersed throughout the Arab world and in non-Arab countries.

About 1.6 million Palestinians were registered as refugees in 1976. But only about 640,000 were living in the 63 refugee camps in Lebanon, Syria, Jordan and the occupied territories.

---

post-Zionist Israel as part of the region, a part which could and should not be abolished because, in its new form, it is a factor in the struggle for the common good."

But such visionary proposals were seldom discussed as talk of resumption of the Geneva conference increased in 1977. Instead, the Palestinian state proposal was in the forefront of thought. In a sense, this brought the issue back to the original partition of Palestine concept, first advocated by the U.N. in 1947. There were some Israelis and some Palestinians advocating such a settlement during 1977.

In Israel, the Israeli Council for Israeli-Palestinian Peace released a 12-point manifesto early in 1976 declaring "That this land is the homeland of its two peoples—the people of Israel and the Palestinian Arab people." "The only path to peace," the manifesto continued, "is through co-existence between two sovereign states, each with its distinct national identity." The council, a Zionist body whose top leaders are former members of the Labor party, went on to advocate mutual recognition between Israel and the PLO, Israeli willingness to withdraw to approximately the 1967 lines, and establishment of a Palestinian state with Jerusalem the joint capital of both states.

Beginning in July 1976, members of the Israeli Council began meeting with representatives of the PLO in Paris. Reserve General Mattiyahu Peled, one of the leaders of the Council, was reported to be directly informing Prime Minister Rabin of these developments. Writing in the Israeli Daily *Davar*, Hebrew University professor Daniel Amit outlined the evolution of the PLO to a body now willing to discuss possible coexistence with Israel. "The PLO leadership, as a whole, has changed radically," Professor Amit noted.[39]

Late in 1976 and into 1977 it was often reported, though usually ambiguously, that the PLO leadership was finally prepared for a role at Geneva and a compromise, two-state solution to the long-standing conflict. The Associated Press quoted the head of the PLO's political department, Farouk Kaddoumi, on Nov. 17, 1976, saying that "We accept es-

39 16 September 1976.

European Economic Community Middle East Statement

June 29, 1977

For the first time, the European Economic Community in June 1977 declared itself in favor of a Palestinian homeland. Excerpts from the statement follow:

"The Nine have affirmed their belief that a solution to the conflict in the Middle East will be possible only if the legitimate right of the Palestinian people to give effective expression to its national identity is translated into fact, which would take into account the need for a homeland for the Palestinian people. They consider that the representatives of the parties to the conflict including the Palestinian people, must participate in the negotiations in an appropriate manner.... In the context of an overall settlement, Israel must be ready to recognize the legitimate rights of the Palestinian people: equally, the Arab side must be ready to recognize the right of Israel to live in peace within secure and recognized boundaries. It is not through the acquisition of territory by force that the security of the states of the region can be assured; but it must be based on commitments to peace exchange between all the parties concerned with a view to establishing truly peaceful relations.

"The Nine believe that the peace negotiations must be resumed urgently, with the aim of agreeing and implementing a comprehensive, just and lasting settlement of the conflict. They remain ready to contribute to the extent the parties wish in finding a settlement and in putting it into effect. They are also ready to consider participating in guarantees in the framework of the United Nations."

tablishment of a state in the West Bank and Gaza Strip." And *The New York Times,* on Dec. 4, reported that the most important groups in the PLO had accepted this idea of a Palestinian state to coexist with Israel. On March 15, 1977, *The Washington Post* reported that "The prevailing mood at the Palestine National Council meeting appears to be resigned, grudging acceptance of the decision by the major Arab powers to seek peace with Israel."

Throughout the first half of 1977, doubts remained about just what the position of the PLO really was. Did the PLO now accept the idea of coexistence or was it only a tactical maneuver designed to press Israel and achieve gains with world public opinion? There was also considerable speculation about whether the Israelis might be prepared, under any conditions, to take the risks for peace involved in dealing with the PLO and agreeing to withdraw from the occupied territories. Finally, would the United States take the lead by beginning to deal with the PLO in line with the November 1976 General Assembly resolution which stated that the PLO should become a part of the Geneva Conference when it reconvened?

Manahem Begin's election in Israel seemed to set the clock back somewhat when he emphatically ruled out ever returning the West Bank and Gaza Strip to Arab sovereignty or ever dealing with the PLO. Still, by mid-1977, it continued to appear that the PLO was moving in the direction of eventually recognizing Israel and that the United States might at some point advocate bringing the PLO into the negotiating process, even over Israel's strong opposition. President Carter at a press conference on July 28, 1977, presented the Palestinians with a challenge: "I think the major stumbling block at this point is the participation in the negotiations [at Geneva] by the Palestinian representative. Our position has been that they ought to be represented and that we will discuss with them these elements that involve the Palestinians and other refugees at the time they forego their commitment presently publicly espoused that Israel should be destroyed. But until the Palestinian leaders adopt the proposition that Israel is a nation, it will be a nation permanently, that is has a right to live in peace.... I see no way that we would advocate participation by them at peace negotiations."

This approach seemed to be an invitation to the PLO to take an historic step toward recognition of Israel which the United States would reward by breaking with Israel, if necessary, and advocating a role for the Palestinians at Geneva. In this way the door could be opened to a Palestinian homeland in the now-occupied territories. But Secretary of State Vance indicated the day after the President's press conference that unless the Palestinians took the step of recognizing Israel, the United States would remain constrained by its September 1975 agreement with Israel. In this agreement, the United States pledged "it will not recognize or negotiate with the Palestine Liberation Organization so long as the Palestine Liberation Organization does not recognize Israel's right to exist and does not accept Security Council Resolutions 242 and 338.[40]

PLO Diplomacy

The PLO, some analysts indicated, had offered the United States two specific signs in the first half of 1977—in addition to various ambiguous statements—of willingness to consider entering a process that could lead to coexistence with Israel. First, in clause 15 of the political declaration passed at the Palestine National Council meeting in March, '77 the PLO indicated a willingness "to participate in all international conferences...dealing with the problem of Palestine and the Arab-Zionist conflict." In short, the PLO was willing to go to Geneva, where it would find itself negotiating with Israel, and consider a peaceful solution. Second, the new executive committee which emerged from the March PNC meeting excluded members of the so-called "rejection front" who refuse to consider peaceful coexistence with Israel under any conditions. "This was the most significant political decision of the National Council meeting," the PLO's London representative, Said Hammami, noted, "That's the Palestinians' message, the signal."[41]

Throughout the first half of 1977, the PLO continued in its attempt to reach some kind of agreement with the United States that would bring the PLO into the diplomatic process. Beginning in November 1976, when two PLO officials visited the United States to talk with American Jewish representatives and indirectly with U.S. government officials, the PLO sought to open a Washington information office staffed by an official of the organization. Though opening the office presented no legal difficulties, getting the

---

40 Published in *The New York Times,* 18 September 1975.
41 See "PLO Awaits U.S. Reaction to New Stance on Peace," *The Washington Post,* 27 March 1977.

visa for a PLO official did. As of August 1977 it had not been accomplished.[42]

The PLO has had for some years important relations with the Soviet Union, where it has had considerable support. On April 7, 1977, Communist Party Secretary Leonid Brezhnev publicly met Yassir Arafat for the first time. Brezhnev confirmed at that time Soviet support for creation of an independent Palestinian state.

Relations with Jordan remained complex and uncertain through the first half of 1977. Arafat and King Hussein met, for the first time since 1970, in Cairo on March 8, but relations between the PLO and Jordan remained strained and untrusting. The PLO refused to subordinate itself to Jordan, as some Arab leaders urged, by agreeing to go to Geneva as part of the Jordanian delegation. It was thought that to do so would jeopardize the PLO claim to a separate, sovereign state. Furthermore, it would seriously exacerbate tensions within the PLO.

The PLO continued to gain support in the United Nations during 1977. On July 22 the U.N. Economic and Social Council voted 27 to 11, with 12 abstentions, to give the PLO full membership in the Economic Commission for Western Asia. This was the first time a non-state was ever granted full membership in any U.N. body.

Financially, the PLO, battered from Syrian intervention against it in the Lebanese civil war and increasingly under the political thumb of the major Arab states seeking negotiations with Israel, has been doing very well. Saudi Arabia annually gives the PLO member organizations about $25-million. All in all, the PLO pulls in about $90-million yearly according to *Time* magazine[43]—$70-million from Arab governments (including $29-million from the Arab League) and a head-tax from the 300,000 or so Palestinians working in the oil states who have 5 per cent of their pay withheld by host governments. The Palestinians also have taken in money for terrorist ransom demands and the PLO claims to operate a $5-million a year illegal drug business inside Israel using Oriental Jews as pushers. The majority of the PLO's funds are used for training and military—often terrorist—operations. Some goes to the PLO offices, quasi embassies actually, in about 100 nations. The PLO also runs businesses, refugee camps, welfare programs, etc. It has investment holdings totaling as much as $100-million, including Beirut hotels, a youth hostel under construction in 1977 in Cairo, shares in shipyards, oil tankers, television stations, and even blue-chip U.S. companies. *Time* reported that PLO money has even been used to quietly purchase land on the West Bank that might otherwise be sold to Israelis.

In August 1977, the PLO's fate was uncertain. Internal differences were a most serious problem. Whether to continue to try to form a working relationship with the United States rather than to rely on the Soviet Union was a major decision. What form of representation to accept at Geneva, if the conference should reconvene, was another. But the PLO had certainly come a great distance in the short 13 years since it was created. And it appeared that the eulogies gleefully given by many of the PLO's bitterest enemies that the organization had been destroyed or at least made impotent by the Lebanese civil war were premature.

---

42 For background see Mark Bruzonsky and Judith Kipper, "Washington and the PLO," *Middle East International* (London), February 1977; Mark Bruzonsky, "Lending An Ear to the PLO," *Newsday*, 14 February 1977; "U.S. Role As Palestinians Meet," *Christian Science Monitor*, 8 March 1977 and "Carter's Offers to Palestinians," *Christian Science Monitor*, 17 August 1977.

43 "The Well-Healed Guerrillas," *Time Magazine*, 18 July 1977.

PETROLEUM POLITICS: VOLATILE ARAB WEAPON

When the Arab governments in October 1973 embargoed oil to the United States and, together with other oil-producing states in the following months, quadrupled its price, a historic change began. Never in modern history had such an abrupt transfer of wealth and power taken place without war and in so short a time. The first shock is passing, but the future is full of uncertainties.

The most obvious consequence of high-priced oil is a redistribution of wealth. While industrial countries sank into recession, oil-producing states were suddenly gorged with money. Developing nations not blessed with oil staggered under the new cost of fuels and fertilizers needed to lift them from poverty...and yet the success of fellow Third World nations that used oil to gain wealth and power aroused new hopes that they, too, might somehow wrest a bigger share of the world's goods from the long-dominant West.

With the transfer of wealth went a shift in power. Not only do the major oil-producing states, especially Saudi Arabia, control a vital resource without which all Western economies would face collapse; they also had accumulated, by 1977, at least $150-billion in financial reserves and liquid assets. These petrodollars could, at least theoretically, be used to destabilize various currencies or weaken the economies of most Western countries. Assistant Secretary of State for Economic Affairs, Julius L. Katz, testified before Congress in January 1977 that the financial assets of the OPEC countries could total $300-billion by 1980, an unprecedented accumulation of financial muscle. In 1977, the current account surplus of the OPEC countries—the excess between income and expenditures—was about $40-billion and was expected to stay about the same in the foreseeable future. The figure would have been considerably higher, in fact, were it not for the massive spending, development and arms sales programs being carried out by many of the OPEC countries.

States such as Iran and Saudi Arabi, to say nothing of petty sheikdoms along the Persian Gulf, are, of course, no match for the armed might of a United States or a Soviet Union. But by 1976 the discussion of using military might against the oil producers had subsided. Some Western journals, in the immediate aftermath of the oil embargo and price rises had featured scenarios in which American armed forces seized Middle East oil fields in the event of another embargo, or even as a means to prevent ruinous prices. Secretary of State Henry A. Kissinger warned that this would be a "very dangerous course," but did not entirely rule it out if "there's some actual strangulation of the industrialized world." (Box, this page)

The Persian Gulf states eagerly buy hoards of modern weapons, which the United States as eagerly sells (a little less eagerly since President Carter took office), competing with its allies to recapture some of the funds the West is pouring into the coffers of the oil states. The new arsenals of the gulf could not block a seizure of oil fields, although they might raise its risks, especially if a country such as Iran developed nuclear weapons. But various Arab leaders have hinted that they would blow up their oil fields if any attempt were made to take them over by military force. And a Library of Congress study has indicated that the attempt itself would not be feasible.[1]

The greatest danger arising from military intervention, however, would be the uncertain response of the Soviet Union.

Shifting Power

It would be misleading, however, to formulate the shift in power in military terms. The "oil weapon" already has

---

Kissinger on Using Force

Neither Secretary of State Cyrus R. Vance nor President Jimmy Carter has suggested the use of American military force to take over Middle East oil fields under any foreseeable circumstances. But the element of military force always lurks behind such major international problems and so Secretary of State Henry A. Kissinger's warnings in 1975 deserve remembering.

In an interview with *Business Week* magazine made public Jan. 2, 1975, Kissinger discussed the possible use of military force to deal with the oil situation. Kissinger Jan. 2 and 3 reiterated these views to newsmen and said that they reflected the views of President Ford. Ford's press secretary, Ron Nessen, confirmed Jan. 4 that Kissinger's statement "did reflect the President's views."

The following text is from the Jan. 13 *Business Week.*

One of the things we also hear from businessmen is that in the long run the only answer to the oil cartel is some sort of military action. Have you considered military action on oil?
Military action on oil prices?
Yes.
A very dangerous course. We should have learned from Vietnam that it is easier to get into a war than to get out of it. I am not saying that there's no circumstance where we would not use force. But it is one thing to use it in the case of a dispute over price, it's another where there's some actual strangulation of the industrialized world.
Do you worry about what the Soviets would do in the Middle East if there were any military action against the cartel?
I don't think this is a good thing to speculate about. Any President who would resort to military action in the Middle East without worrying what the Soviets would do would have to be reckless. The question is to what extent he would let himself be deterred by it. But you cannot say you would not consider what the Soviets would do. I want to make clear, however, that the use of force would be considered only in the gravest emergency.

---

1 "Oil Fields As Military Objectives: A Feasibility Study," Prepared by the Congressional Research Service of the Library of Congress for the Special Subcommittee on Investigations of the International Relations Committee of the House of Representatives, 21 August 1975.

changed the balance of political and economic power between Israel and the Arab states. It has also wrought changes far beyond the Middle East. A mutual dependence has been built up between the Western industrialized countries and the Arab states. The U.S.-Saudi Arabian relationship is especially close.

During and after the 1973 Arab-Israeli war, disagreement over support for Israel created severe strains between the United States and its allies, who depend heavily on Arab oil; in another Middle East war these tensions might prove more damaging. The Western alliance was further strained by competition to secure scarce oil, a competition that could grow more intense if supply were restricted and demand rose.

The damage oil prices have done to Western economies handicaps them in the arms race with the Soviet bloc and increases pressures for reducing the West's overseas commitments, including those in the Middle East and the Indian Ocean. As their monetary reserves grow to several hundred billion dollars, Middle Eastern governments will have much greater financial power to threaten Western economies.

Within industrial societies, the discontent fostered by inflation and unemployment may feed authoritarian movements in countries such as Italy and, if it grows worse, in more stable democracies. In developing countries, governments struggling against worsening economic problems may grow more repressive. In both rich and poor countries if political unrest grows with economic decline, leaders will find assertive foreign policies more tempting. A rising level of international tensions thus may coincide with a proliferation of nuclear weapons, and those weapons will be accessible to more states as they hasten to develop nuclear energy as a substitute for high-priced oil.

If the world can respond creatively to this sea of troubles, the oil crisis may prove, at least partially, a blessing in disguise. The exploitation of natural resources by industrial societies was growing at a dangerous rate, and the huge gap between rich and poor nations showed no sign of abating. Both are built into ways of life that were taken for granted until the new price of oil hit home.

Impact on World Finances

In early 1974, the world's money managers were close to a state of panic at the enormous imbalance in the international monetary system. The oil-producing countries had decreed huge increases in the price of petroleum, from an average of about $2.40 a barrel to more than $10. There seemed no way that oil consumers—the poor countries in particular—could afford to pay for their energy supplies, and no way that oil producers could spend their expected surpluses. On every hand, there were predictions of chaos and impending collapse of existing world monetary arrangements.[2]

Somehow the international community limped through 1974, and the sense of panic subsided. Optimistic statements about the resiliency of the world monetary system began to be heard. Early in 1975, Secretary of the Treasury William E. Simon told Congress that he believed "the international financial aspects of the oil situation are manageable." The sense of relief may be premature, however. Among economists, there remains a deep concern

about the condition of the monetary system; it is still subject to severe strains. Experts warn that a protracted imbalance of payments[3] will inevitably further slow economic growth in the industrial countries, bankrupt the less-developed countries, cause the banking system to collapse and throw the world into a massive recession or depression.[4]

By 1977 it was clear that rising oil prices had contributed to slower economic growth in many countries, that many countries, especially developing ones, were no longer good credit risks for the loans they required to purchase oil imports and that a world recession was possible should some countries begin defaulting on outstanding loans thus affecting the stability of the entire financial system.[5]

Business Week reported in 1977 that "The oil cartel's first big price hike tossed the world economy into the worst recession in 40 years—one from which it has not yet recovered. The OPEC-induced loss in output over the past three years is estimated at $550-billion. The external debt of the less developed nations has mushroomed to $170-billion. Even without the latest boost, the world is transferring $100-billion a year to OPEC."[6]

The member countries of OPEC, the Organization of Petroleum Exporting Countries, rang up a spectacular $97-billion foreign-exchange surplus in 1974. But massive spending programs and arms purchases have since recycled some of these funds back to Western economies. Late in 1976 it was reported that "The OPEC countries have surprised the world—and often themselves—by the speed with which they have been able to increase their imports."[7] In that first year of escalating oil prices, the industrial countries registered a $67-billion trade deficit, three times larger than in 1973. The less-developed countries saw their deficit more than double to $26-billion.

This seriously imbalanced situation has continued. It is hardest on less developed countries (the LDCs) who must pay for importing oil but have little to sell in return for recycling petrodollars. By 1977, LDC debt had skyrocketed to more than $170-billion with no end in sight.[8]

So far, "Private banking institutions, acting singly or in consortium, have continued to absorb and relend large sums of investible OPEC funds," President Ford reported just before leaving office.[9] But Princeton economist Peter B. Kenen said, "We have reached the point where to get back what you loaned in the first place you have to throw good money after bad."[10] Increasingly, doubt is being expressed about how long this situation can continue; how long the poorer countries can keep borrowing to pay their bills for imported oil.

"The continuation of sizeable balance of payments deficits," which are mainly the result of the oil price increases, President Ford warned in January 1977, "poses serious financing and management problems for the countries concerned and is a major disruptive element in their development process."[11]

2 See "Arab Oil Money," *Editorial Research Reports,* 1974 Vol. I, pp. 365-381.

3 When income from abroad is less than a country's foreign spending on trade investments, tourism, military and foreign aid.
4 See, for example, Robert V. Roosa et al., "How Can the World Afford OPEC Oil?" *Foreign Affairs,* January 1975, pp. 201-202. Roosa was under secretary of the treasury for monetary affairs in the Kennedy and Johnson administrations.
5 During 1977 one of the best-selling fiction books had as its theme such an international banking collapse. It was written by a former international banker, Paul Erdman. *The Crash of '79* (Simon & Schuster; 1977). See review in *Business Week,* 7 March 1977, p. 10.
6 *Business Week,* 10 January 1977, p. 61.
7 Lawrence A. Veit, "Troubled World Economy," *Foreign Affairs,* January 1977, p. 265.
8 See David O. Beim, "Rescuing the LDCs," and Harold van B. Cleveland & W. H. Bruce Brittain, "Are the LDCs in over their heads?" in *Foreign Affairs,* July 1977.
9 *International Economic Report of the President,* January 1977, p. 10.
10 *Business Week,* 10 January 1977, p. 62.
11 *International Economic Report of the President,* January 1977, p. 19.

1985 *12-29 000*

World Crude Oil Production, 1960-1976

(bpd = barrels per day)

Major Areas and Selected Countries	1960 1,000 bpd	1960 Per cent	1970 1,000 bpd	1970 Per cent	1972 1,000 bpd	1972 Per cent	1973 1,000 bpd	1973 Per cent	1974 1,000 bpd	1974 Per cent	1975 1,000 bpd	1975 Per cent	1976[3] 1,000 bpd	1976[3] Per cent
North America	7,845	37.3	11,373	25.1	11,498	22.7	11,452	20.5	11,045	19.8	10,550	19.9	10,270	18.0
– United States	7,055	33.5	9,648	21.3	9,512	18.8	9,189	16.5	8,770	15.7	8,370	15.8	8,140	14.3
Canada	519	2.5	1,305	2.9	1,542	3.0	1,798	3.2	1,695	3.1	1,460	2.7	1,315	2.3
Mexico	271	1.3	420	.9	444	.9	465	.8	580	1.0	720	1.4	815	1.4
Central and South America	3,470	16.5	4,758	10.5	4,393	8.7	4,666	8.4	4,245	7.6	3,585	6.7	3,535	6.2
Venezuela[1]	2,854	13.6	3,703	8.2	3,219	6.4	3,364	6.0	2,975	5.3	2,345	4.4	2,270	4.0
Ecuador[1]	7	*	5	*	67	.1	204	.4	175	.3	160	.3	190	.3
Other	609	2.9	1,050	2.3	1,107	2.2	1,098	2.0	1,095	2.0	1,080	2.0	1,075	1.9
Western Europe	289	1.4	375	.8	371	.7	370	.7	380	.7	550	1.0	875	1.5
United Kingdom	2	*	2	*	2	*	2	*	*	*	20	*	250	.4
Norway	0	0	0	0	33	*	32	.1	35	.1	190	.4	295	.5
Other	287	1.4	373	.8	336	.7	336	.6	345	.6	340	.6	330	.6
Africa	289	1.4	5,982	13.2	5,616	11.1	5,902	10.6	5,370	9.6	4,990	9.4	5,795	10.2
Algeria[1]	185	.9	976	2.2	1,022	2.0	1,070	1.9	960	1.7	960	1.8	990	1.7
Libya[1]	0	0	3,321	7.3	2,215	4.4	2,187	3.9	1,520	2.7	1,480	2.8	1,970	3.5
Nigeria[1]	18	.1	1,090	2.4	1,818	3.6	2,053	3.7	2,255	4.0	1,795	3.4	2,055	3.6
Gabon[1]	—	—	110	.2	125	.2	150	.3	200	.4	225	.4	220	.4
Other	86	.4	485	1.1	436	.9	442	.8	435	.8	530	1.0	560	1.0
Asia-Pacific	554	2.6	1,340	3.0	1,832	3.6	2,272	4.1	2,250	4.0	2,215	4.2	2,485	4.4
Indonesia[1]	419	2.0	855	1.9	1,056	2.1	1,339	2.4	1,375	2.4	1,305	2.5	1,505	2.7
Other	135	.6	485	1.1	776	1.5	933	1.7	875	1.6	910	1.7	980	1.7
Middle East	5,269	25.1	13,937	30.7	18,106	35.9	21,158	38.0	21,855	39.1	19,590	36.9	21,710	38.3
Saudi Arabia[1]	1,319	6.3	3,798	8.4	6,013	11.9	7,607	13.7	8,480	15.2	7,075	13.3	8,585	15.1
Kuwait[1]	1,696	8.1	2,983	6.6	3,279	6.5	3,024	5.4	2,545	4.6	2,085	3.9	2,090	3.7
Iran[1]	1,057	5.0	3,831	8.4	5,021	9.9	5,861	10.5	6,020	10.8	5,350	10.1	5,850	10.3
Iraq[1]	969	4.6	1,563	3.4	1,446	2.9	1,964	3.5	1,970	3.5	2,260	4.3	2,075	3.7
Abu Dhabi[1]	0	0	691	1.5	1,050	2.1	1,298	2.3	1,410	2.5	1,370	2.6	1,590	2.8
Qatar[1]	173	.8	367	.8	484	1.0	570	1.0	520	0.9	440	.8	495	.9
Other	55	.3	704	1.6	813	1.6	834	1.5	910	1.6	1,010	1.9	1,025	1.8
Total Non-Communist	17,716	84.3	37,765	83.3	41,816	82.7	45,820	82.3	45,145	80.8	41,695	78.5	44,890	78.6
Communist World[2]	3,310	15.7	7,610	16.7	8,738	17.3	9,865	17.7	10,720	19.2	11,650	21.9	12,170	21.4
Soviet Union	2,960	14.1	7,049	15.5	7,850	15.5	8,420	15.1	9,020	16.1	9,630	18.1	10,170	17.9
Other	350	1.6	561	1.2	888	1.8	1,445	2.6	1,700	3.1	2,020	3.8	2,000	3.5
Total World	21,026	100.0	45,375	100.0	50,554	100.0	55,685	100.0	55,865	100.0	53,120	100.0	56,840	100.0

* Production or percentage of production is negligible.
1 Member of Organization of Petroleum Exporting Countries.
2 Includes Soviet Union and other Warsaw Pact Nations, China, Cuba and Yugoslavia.
3 Estimate

SOURCE: International Economic Report of the President, March 1975, January 1977.

WG, 2t
Fr, 9, B-d, Sp

J Revolu

Selected Consuming Countries' . . .

(Thousands of barrels per day and per cent of imports)

	Total Consumption	Total [3] Imports	Total Arab	Total Non-Arab	Arab Oil			
					Saudi Arabia	Kuwait	Libya	Iraq
United States	16,291	6,030	1,770	4,260	850	30	330	10
Per cent	—	100	29.4	70.6	14.1	.5	5.5	.2
Japan	3,712	5,010	2,540	2,470	1,460	480	60	90
Per cent	—	100	50.7	49.3	29.1	9.6	1.2	1.8
Canada	1,593	890	300	590	190	30	10	30
Per cent	—	100	33.7	66.3	21.3	3.4	1.1	3.4
Western Europe[2]	13,505	12,080	7,520	4,560	3,340	790	740	920
Per cent	—	100	62.3	37.7	27.6	6.5	6.1	7.6
United Kingdom	1,613	1,830	990	840	450	220	60	50
Per cent	—	100	54.1	45.9	24.6	12.0	3.3	2.7
West Germany	2,319	1,970	1,170	800	380	60	300	30
Per cent	—	100	59.4	40.6	19.3	3.0	15.2	1.5
Italy	1,468	1,990	1,420	570	520	90	190	350
Per cent	—	100	71.4	28.6	26.1	4.5	9.5	17.6
France	1,921	2,190	1,550	640	670	130	40	240
Per cent	—	100	70.8	29.2	30.6	5.9	1.8	11.0
Netherlands	710	1,200	580	620	260	130	10	40
Per cent	—	100	48.3	51.7	21.7	10.8	.8	3.3
Belgium-Luxembourg	535	590	300	290	240	60	—	—
Per cent	—	100	50.8	49.2	40.7	10.2	—	—
Spain	865	820	580	240	410	10	50	100
Per cent	—	100	70.7	29.3	50.0	1.2	6.1	12.2
Other	4,074	1,490	930	560	410	90	90	110
Per cent	—	100	62.4	37.6	27.5	6.0	6.0	7.4

1 Estimated imports of crude oil and refined products, traced to the original source.
2 The consumption figure for Western Europe includes Austria, Belgium-Luxembourg, Denmark, Ireland, Finland, France, Greece, Iceland, Italy, Netherlands, Norway, Por- tugal, Spain, Sweden, Switzerland, Turkey, United Kingdom, West Germany, Yugoslavia, Cyprus, Gibraltar and Malta.

In general, oil exporting countries can "recycle" their huge surpluses in three ways: (1) by spending them on imports of consumer goods, military hardware, industrial equipment, food and other commodities; (2) by investing them, either in development projects at home or in foreign countries, and (3) by relending the money through official and private channels to oil-importing nations.

To a surprising extent, the exporting countries have done all three things since 1974. By 1976, U.S. exports to OPEC countries had increased nearly 400 per cent from 1973, to $12.1-billion.[12] *(See chart, p.128)* OPEC nations are, in fact, greatly dependent upon products from the United States and other industrial countries. *(See graph, p. 139)*

At the same time, OPEC countries, and especially the Arab oil states, have stepped up their investments in, and their loans to, Western enterprises. Arabs now have substantial holdings in American real estate and industries and have extended loans to several U.S. corporations. These developments are not entirely to the liking of many Western political leaders, who fear the possibility of Arab takeovers of huge national and multinational enterprises. So far, though, the Arabs appear to be more interested in secure placement of their capital than in exercising management control. One reason is fear of possible freezing of assets should a major world crisis occur. Thus highly liquid investments are favored, even though they command lower interest rates. By 1977, though, there were indications of cautious movement into medium-term investments of one to two years.

Finally, a sizable amount of oil money has found its way into international lending channels, both official and private. By 1977 U.S. banks were so crammed with short-term funds from OPEC states that some began discouraging them. Some banks in Switzerland have begun charging a negative interest rate on certain types of deposits. In fact it was because the private banking system had acted as the intermediary for recycling petrodollar surpluses to debtor countries to finance their purchases of more oil from the OPEC countries—at the rate of about $10-billion annually

12 *Development Forum*, publication of the U.N. Centre for Economic and Social Information, March 1975, p. 3.

. . . Dependence on Arab Oil, 1975 [1]

| United Arab Emirates | Algeria | Other | Non-Arab Oil | | | | | | |
			Iran	Venezuela	Indonesia	Canada	Nigeria	Ecuador	Other
170	290	90	500	1,040	450	800	820	70	580
2.8	4.8	1.5	8.3	17.2	7.5	13.3	13.6	1.2	9.6
410	10	30	1,180	10	560	—	60	—	660
8.2	.2	.6	23.6	.2	11.2	—	1.2	—	13.2
40	—	—	200	280	—	—	20	—	90
4.5	—	—	22.5	31.5	—	—	2.2	—	10.1
760	500	470	1,950	250	—	—	740	—	1,620
6.3	4.1	3.9	16.1	2.1	—	—	6.1	—	13.4
80	30	100	360	70	—	—	120	—	290
4.4	1.6	5.5	19.7	3.8	—	—	6.6	—	15.8
160	210	30	290	50	—	—	200	—	260
8.1	10.7	1.5	14.7	2.5	—	—	10.2	—	13.2
70	60	140	270	40	—	—	10	—	250
3.5	3.0	7.0	13.6	2.0	—	—	.5	—	12.6
250	120	100	270	40	—	—	180	—	150
11.4	5.5	4.6	12.3	1.8	—	—	8.2	—	6.8
110	10	20	350	10	—	—	140	—	120
9.2	.8	1.7	29.2	.8	—	—	11.7	—	10.0
—	—	—	110	—	—	—	—	—	180
—	—	—	18.6	—	—	—	—	—	30.5
—	10	—	70	20	—	—	—	—	150
—	1.2	—	8.5	2.4	—	—	—	—	18.3
90	60	80	230	20	—	—	90	—	220
6.0	4.0	5.4	15.4	1.3	—	—	6.0	—	14.8

3 Imports exceed consumption in several countries because they export products; the Netherlands, for example, tranships some crude oil to other Western European countries.

SOURCE: White House Council on International Economic Policy; BP Statistical Review of the World Oil Industry, 1976.

by 1977—that the banks were increasingly worried about the quality of their loans.[13]

OPEC countries also have contributed increasing amounts to (1) a special oil "facility" fund set up by the International Monetary Fund to ease the balance-of-payments problems of consuming countries and (2) international development institutions, such as the World Bank and the regional development banks. New international institutions to deal with the situation were being considered in 1977.

All these measures evidently helped forestall the drastic results predicted in early 1974, where estimates of the peak accumulation of OPEC reserves reached $600-billion and even higher. By 1975, U.S. bankers were making more optimistic projections, with the time and size of the peak surplus depending on such factors as the impact on consumption of high oil prices, the growth of alternative oil supplies, the degree to which alternative sources of energy such as coal and nuclear power are developed, the policies of consuming countries and the cohesiveness of the OPEC cartel (to say nothing of the chances of another war in the Middle East).

The First National City Bank in its June 1975 newsletter projected four different scenarios. The most pessimistic estimated a peak surplus of nearly $300-million in 1981; the most optimistic placed the peak in 1977, at a level of about $130-billion, which rapidly fell to an actual deficit in OPEC balances by 1983. (Graph, p. 130)

In 1977 the more pessimistic forecasts appeared to be the more accurate ones. Prices were continuing to go up, though it appeared OPEC might not raise prices in 1978. President Carter was predicting oil shortages within a few years, and some analysts were writing of a "more-or-less permanent annual OPEC surplus amount[ing] to about $38-billion."[14]—the surplus after spending on all the projects and schemes the OPEC countries had been able to put together since 1973.

13 See Forbes, 1 July 1976, p. 76 and Business Week, 12 July 1976, p. 32; 17 January 1977, p. 65; and 28 March 1977, pp. 23-24.

14 Beim, op. cit., p. 718. Beim is executive vice president of the Export-Import Bank of the United States.

Oil economist Walter Levy's predictions were coming true, it seemed. Back in 1975 he had written that "To the extent that oil imports are financed by a continued recycling of surplus oil revenues via investments or loans or commercial terms, oil importing countries will face pyramiding interest or individual charges on top of mounting direct oil import costs."[15] Levy wondered then, and many are wondering now, how long this could go on.

At some point in the future, it is predicted, the private banking system will find itself unable to cope with the massive surpluses being acquired by OPEC members. And insofar as official efforts are directed at lending money to consuming countries to meet their rising energy costs, the problem is not solved but merely postponed.

Price Rise Impact on Industrialized Nations

The impact of the oil price rise has not been uniform on all the oil-importing countries. Among the industrial nations, the United States has fared better than most because of its lesser, though growing, dependence on imported oil and because of substantial export sales and arms sales. But even the United States by 1977 was running a $12-billion deficit to the OPEC nations in its trade account. Western Europe fared worse, however. The combined trade deficits of Western European countries rose to $23.8-billion in 1974, up from $7.6-billion the previous year. Japan has had to resort to a $1-billion loan from Saudi Arabia to finance its oil imports, and Italy has received a series of credits from the European Economic Community (Common Market), the IMF and West Germany.[16]

Relative to other industrial countries, however, the United States may have benefitted from the rising oil price—leading to speculation that possibly the U.S. was not so opposed to the developments that took place in 1973 and 1974. In a front-page *Washington Post* story on July 19, 1977, it was stated that "since the takeoff of oil prices in the fall of 1973, America's economic stature in the world has improved dramatically, reversing a long decline, while our competitors (and friends) in Western Europe and Japan have suffered. So much of American political rhetoric casts OPEC as an adversary intent on hurting us by embargo or price-gouging. Yet...the actual relationship more resembles a partnership, an intricate symbiosis in which the United States gains and gives, and so do the leading partners of OPEC.... Europeans have been talking for years about the economic advantages and the increased political leverage that America may have gained from OPEC."* In an article which partially supports this thesis, *Forbes* has concluded that...the Europeans are basically correct.... OPEC gained its power over prices through the maneuverings of the State Department."**

Effect on Less Developed Countries

Still, the most severe impact is being felt by the poorer countries of the world—those without oil to sell or significant exports of other raw materials. These include 75 to 80 countries, or about three-fourths of the poorer nations. India, as the largest of the non-Communist developing

U.S. Trade with OPEC Members (In billions of U.S. dollars)			
	Exports	**Imports**	**Balance**
1972	2.8	2.7	+ .1
1973	3.6	4.6	− 1.0
1974	6.7	15.6	− 8.9
1975	10.8	17.1	− 6.3
1976*	12.1	24.2	− 12.1

* Estimate

SOURCE: International Economic Report of the President, January 1977.

countries, is the foremost example of a nation being driven to the brink of bankruptcy by the energy price crisis. In the best of times, India has had a difficult job providing food and other basic necessities for its nearly 600 million people. The huge jump in oil prices has made the task virtually impossible. India's oil import bill in 1975 was about $1.5-billion, in contrast to only $300-million five years earlier.[17]

In recognition of India's plight, both Iran and Iraq have offered New Delhi concessions in the form of loans and deferred payments for the purchase of petroleum. Such assistance is not likely in the long run to be sufficient to pull India out of the hole. This is also true for other countries whose commodity export earnings drop while their oil bills rise.

Economist Richard N. Cooper at Yale sees the debt burden on the less developed countries as creating a major international economic problem. "These countries are now hanging by their financial fingernails. They are top-heavy with external debt, and their financial position is extremely fragile," he wrote.[18]

A financial problem for the LDCs also poses a problem for the banking system in the West. American banks in particular have granted huge loans to various LDCs who between 1973 and 1977 increased their foreign debts by more than 150 per cent. Citibank in New York, for example, had 70 per cent of its portfolio in foreign credits of all types in 1977. In 1976, for the first time, bad debts abroad were greater than defaults on domestic loans for Citibank. "The real problem for banks in 1977 is going to be their exposure to less developed countries trying to finance oil deficits. Banks are in deep," *Business Week* reported in March 1977.[19]

Agence France-Presse, the French news agency, reported as long ago as 1973 that a number of developing

15 Walter J. Levy, "World Oil Cooperation or International Chaos," *Foreign Affairs,* July 1974, p. 697.

16 See "Italy's Threatened Democracy," *Editorial Research Reports,* 1975 Vol. I, pp. 1-20. West Germany, in contrast to its neighbors, enjoyed a trade surplus in 1974—some $20-billion.

*William Greider and J. P. Smith, "A Proposition: High Oil Prices Benefit U.S."

**See "Don't Blame The Oil Companies: Blame the State Department," *Forbes,* 15 April 1976.

17 P. D. Henderson, *India: The Energy Sector* (1975), World Bank, pp. 107-115.

18 *Business Week,* 18 October 1976, p. 105.

19 *Business Week,* 28 March 1977, p. 34. One study found that "commercial banks increased their loans to non-OPEC developing countries by about $46-billion in the 1974-76 period, a sum that represented 42 per cent of total new financings. This compared with a 20 per cent share in 1971-73." See Robert A. Bennett, "Mountains of Debt Pile Up As Banks Rush Foreign Loans," *The New York Times,* 15 May 1977, p. 1 (financial section). Even French banks were becoming hostage to both Arab depositors and LDC borrowers. See *Business Week,* 17 January 1977, p. 65.

countries "are finding it hard to honor their commercial debts, as a result of the world economic crisis." It cited the example of the African republic of Zaire, formerly the Belgian Congo, which has experienced declines in the prices of copper, diamonds, oil seeds, zinc, coffee, cocoa and cotton. Zaire's case is significant, because it points up a growing rift in the Organization of African Unity between the member states that produce oil and those that have to buy it. The oil consumers complain that the producers have turned a deaf ear to their pleas for economic assistance.[20]

Among the oil-exporting countries of the world, there is a wide divergence in their ability to make productive use of surplus revenues. Hollis B. Chenery, vice president for development policy at the World Bank, puts OPEC members into three categories:

● Countries with large reserves, small populations and little economic absorptive capacity. This group includes the primary Arab oil producers—Saudi Arabia, Kuwait, Qatar and the United Arab Emirates. These countries account for most of the monetary imbalance that now plagues the world—83 per cent of the 1976 surplus. Saudi Arabia now ranks first, with West Germany and the United States in second and third place, for foreign-exchange holdings.

● Countries with substantial reserves but large populations and ambitious development programs which can absorb most of the new oil money. This group includes Iran, Venezuela, Iraq and Algeria, all of which have reached a relatively sophisticated stage of economic development and are using up their oil money to finance the development effort. Need for money explains why these countries have been pushing for higher and higher oil prices while those in the first category have urged restraint.

● Countries with smaller reserves and large populations, whose development needs will eat up virtually every penny of the oil money now flowing in. Nigeria and Indonesia are in this group.[21]

[20] For background see "African Nation Building," Editorial Research Reports, 1973 Vol. I, pp. 355-372.

[21] Chenery's analysis is contained in his article, "Restructuring the World Economy," Foreign Affairs, January 1975, pp. 242-263.

Coping with Petrodollars

The problem of recycling surplus oil revenues—"petrodollars"—and of cushioning the shock for consuming nations most seriously affected by high oil prices has been attacked in a number of ways. Two of the most important are the creation of the special oil facility within the IMF and the approval of a "safety net" fund by members of the Organization for Economic Cooperation and Development (OECD), a group of major non-Communist industrial nations.

The IMF's special oil facility was set up in mid-1974, over U.S. objections. The United States felt that such a fund, specifically designed to ease balance-of-payments problems caused by high oil prices, amounted to an endorsement of "exorbitant" price levels. The fund is financed by borrowings from oil producers and can be tapped by any of the IMF's 126 member countries. When the fund went into operation in August 1974, its authorized capital was $3.2-billion. Subsequently, the amount was doubled to $6.2-billion.

The OECD's "safety net" fund of $25-billion was approved in March 1975 at the urging of Secretary of State Kissinger. It is made up of security pledges by participating nations under a formula based on national wealth. It is designed as a fund of last resort, to be used only when a member country has exhausted all other credit possibilities. Loans are limited to seven years at commercial interest rates. Because of rigid requirements governing loans, few requests are anticipated. However, the fund will provide psychological security for industrial nations with tottering economies.

Also at Kissinger's urging, a 19-nation International Energy Agency (IEA) was created, consisting of most of the larger industrial oil-consuming nations. The agency was intended to present a united front of oil users in negotiations with OPEC but has never really achieved its goal. Preliminary talks in Paris between the oil-producing and oil-consuming nations collapsed April 15, 1975. The United States, the European Economic Community and Japan in-

Organization of Petroleum Exporting Countries' Revenues, 1970-1976[1]

(in millions of dollars)

Country	1970	1971	1972	1973	1974	1975	1976
Saudi Arabia	$1,200	$2,149	$3,107	$4,340	$22,600	$25,700	$33,500
Kuwait	895	1,400	1,657	1,900	7,000	7,500	8,500
Iran	1,136	1,944	2,380	4,100	17,500	18,500	22,000
Iraq	521	840	575	1,840	5,700	7,500	8,500
United Arab Emirates[2]	233	431	551	900	5,500	6,000	7,000
Qatar	122	198	255	410	1,600	1,700	2,000
Libya	1,295	1,766	1,598	2,300	6,000	5,100	7,500
Algeria	325	350	700	900	3,700	3,400	4,500
Nigeria	411	915	1,174	2,200	8,900	6,600	8,500
Venezuela	1,406	1,702	1,948	2,670	8,700	7,500	8,500
Indonesia	185	284	429	950	3,300	3,850	4,500
Total	$7,729	$11,979	$14,374	$22,510	$90,500	$93,350	$115,000

1. In November 1973, Ecuador became a member of the OPEC, and Gabon an associate member; they are not included in the above chart.
2. A federation of the Persian Gulf states was formed in 1971. Revenue figures for 1975 and 1976 include all UAE production; figures for 1973 and 1974 include only Abu Dhabi and Dubai; figures before 1973 are for Abu Dhabi alone, which is the largest oil producer among the members of the UAE.

SOURCES: Petroleum Information Foundation (1970); Petroleum Economist, Vol. XLIII No. 9, Sept. 1976, p. 338 (1971-72); Petroleum Economist, Vol. XLIV, No. 7, July 1977 (1973-76)

sisted that the international conference planned for later in the year be confined to energy and related problems; Algeria, Saudi Arabia, Iran and Venezuela demanded that the agenda be broadened to include consideration of other raw materials and development aid. They were joined in this demand by some of the non-oil-producing countries of the third world.

At a May 27 meeting of the International Energy Agency in Paris, Kissinger reversed the U.S. position, proposing that stability of raw material prices be included in future meetings with oil-producing countries. "It has become clear as a result of the April preparatory meeting," he said, "that the dialogue between the producers and the consumers will not progress unless it is broadened to include the general issue of the relationship between developing and developed countries." The OECD met the next day and issued a "declaration on relations with developing countries," pledging greater cooperation and a resumption of talks with Third World countries which would deal generally with the problems of raw materials.

OPEC members themselves have made significant efforts to recycle their petrodollars, through bilateral lending arrangements and through contributions to international finance agencies. Abdlatif Al Hamad, director general of the Kuwait Fund for Arab Economic Development, told the Joint Development Committee of the World Bank and the IMF that OPEC members contributed $14.3-billion in development assistance in 1974, including a $3.1-billion loan to the IMF oil facility and $1-billion to the World Bank.

In this hemisphere, Venezuela, the world's fifth largest exporter of petroleum, has set up a $500-million fund in the Inter-American Development Bank for lending to the least-developed countries of the region. In addition, Venezuela has allotted $540-million to the World Bank, $540-million to the IMF oil facility and $100-million to the U.N. Emergency Fund, and is underwriting one plan to defer oil payments by Central American countries and another to stockpile coffee in the six coffee-exporting nations of the hemisphere.

During 1975 and 1976, additional sums from the oil producing countries have been made available in various ways to countries which must borrow to pay their oil bills.

Private Channels

The glut of petrodollars is obviously more than official channels—governments and international agencies—can handle. A substantial portion of the oil wealth is finding its way back to the West through private means, in the form of loans, bank deposits, investments and purchases of consumer and industrial goods. Western businessmen are suddenly in fierce competition with one another for the potentially lucrative markets of the Middle East.

The markets are obviously limited by what the national economies can comfortably absorb. Only so many cars and television sets can be sold in the tiny sheikdoms of Qatar and United Arab Emirates. But in a large and rapidly developing economy, such as Iran's, the sky is the limit. In the largest business arrangement ever signed, Iran and the United States in March 1975 agreed to $15-billion, five-year pact for exchange of goods. Included in the agreement is a provision for the United States to build eight nuclear reactors in Iran. Britain, France and West Germany also have concluded large trade deals with the Iranian government. Iran is also entering into trade agreements and joint ventures with numerous other countries, including some in the Communist bloc.

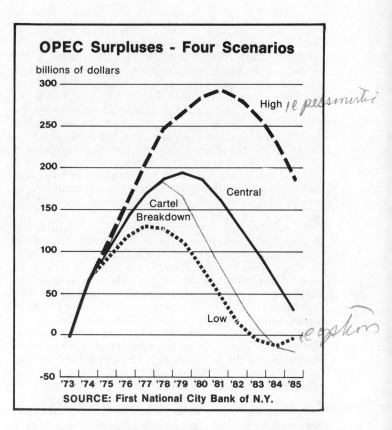

OPEC Surpluses - Four Scenarios

billions of dollars

SOURCE: First National City Bank of N.Y.

Other OPEC countries are active, too. Eight nations—Algeria, Bahrain, Iraq, Kuwait, Libya, Qatar, Saudi Arabia and United Arab Emirates—have contracted to acquire a fleet of supertankers that could cost upward of half a billion dollars in three years. Iraq has placed an order for 10,000 Mercedes-Benz trucks and has ordered $300-million worth of jet planes from Boeing. Many more billions in orders have been placed in more recent years. In 1977 it was estimated that Saudi Arabia had already placed orders for $27-billion with U.S. companies.

More unsettling is the rising level of armaments being purchased by Middle Eastern oil producers from the United States, Western Europe and the Soviet Union. U.S. arms sales to Iran alone in fiscal year 1974 amounted to $3.8-billion. Saudi Arabia has also become a multi-billion dollar arms purchaser in recent years. A report by the U.S. Arms Control and Disarmament Agency showed that 13 developing countries had spent more than 10 per cent of their annual budgets on weapons. Among the countries are Saudi Arabia, Iran and Iraq.[22]

Petrodollars are also being funneled to Western Europe and the United States for deposit in banks and for the purchase of government securities. Treasury Secretary Simon estimated that OPEC countries in 1974 bought $6-billion worth of government securities in the United States and put an additional $4-billion into private U.S. banks. Some observers think the actual total may be as much as 50 per cent higher. Tens of billions have come into the United States since 1975.

Finally, OPEC businessmen are investing increasing amounts in commercial and industrial firms and real estate

22 *World Military Expenditures and Arms Trade, 1963-1973,* Arms Control and Disarmament Agency, February 1975. See also Dale R. Tahtinen, *Arms in the Persian Gulf,* American Enterprise Institute, March 1974, "Resurgent Iran," *Editorial Research Report,* 1974 Vol. I pp. 305-322, and "World Arms Sales," *Editorial Research Reports,* 1976 Vol. I pp. 325-341.

interests in the West. In some cases, Arabs have offered loans to prop up sagging enterprises, as in the cases of Iranian aid to Grumman Aircraft Corporation and the negotiations between Iran and Pan American World Airways. More often, however, they have purchased shares in U.S. and European corporations, attempting in some isolated instances to acquire majority control.

Arab investment in Western corporations is a source of growing controversy among government officials in America and Western Europe. In West Germany, where Iran already owns 25 per cent of the Krupp steel works and has tried to buy nearly 30 per cent of Daimler-Benz, Chancellor Helmut Schmidt has expressed his misgivings. "In the long run," he said, "a sharper control by the state is unavoidable."[23] There has been the additional controversy on both sides of the Atlantic over attempts by the Arab Boycott Office to keep blacklisted "Jewish" banks and businesses in western countries from participating in international transactions involving the Arabs.

Response to Embargo

The immediate response of the U.S. government to the Arab oil embargo in the fall of 1973 was to rally the nation in support of a crash program aimed at removing dependence on foreign oil in a few years.

When President Nixon addressed the nation on Nov. 7, 1973, he couched his remarks in terms of that ambitious and lofty objective. "Let us set as our national goal...that by the end of this decade we will have developed the potential to meet our own energy needs without depending on any foreign sources," he said. He called it Project Independence, a name that fit snugly into the framework of the upcoming bicentennial anniversary. Nixon likened Project Independence to two other successful American ventures—the Manhattan Project, the race to develop an atomic bomb, and the Apollo Project, the race to put a man on the moon. By 1977, it was clear that Project Independence had failed.

Hardly had the ring of Nixon's words died out when leaders both inside and outside the administration began to question the wisdom and practicality of total self-sufficiency in energy. The year 1974 was to become one of study and planning while the country instituted initial energy conservation measures by lowering highway speed limits to 55 miles per hour and turning thermostats down to 68 degrees.

The Arabs ended their embargo on March 18, 1974. By summer, the lines at American service stations began to dwindle, and energy seemed to be downgraded from a crisis to a problem of national consciousness. The United States was distracted by the more immediate crisis of Watergate. When Nixon resigned on Aug. 9, 1974, and Gerald R. Ford became the 38th President, the nation was already hearing about "reasonable self-sufficiency" in energy matters. Not until President Carter took office were the American people told that they faced "the moral equivalent of war" in combating the somewhat hidden energy crisis.

Ford's Modifications

Ford immediately endorsed the aims of Project Independence, but in one of his first important speeches as President, at the World Energy Conference in Detroit Sept.

23, he said that "no nation has or can have within its borders everything necessary for a full and rich life for all its people. Independence is not to set the United States apart from the rest of the world; it is to enable the United States to do its part more effectively in the world's effort to provide more energy."

Ford met no outcry at home against that move away from the original Project Independence goal of total self-sufficiency by 1980. To the contrary, informed opinion had found Nixon's goal unrealistic.

The Federal Energy Administration (FEA) formally unveiled its 800-page "Blueprint for Project Independence" on Nov. 13, 1974. Five days earlier, FEA Administrator John C. Sawhill had said, "Project Independence implies zero imports. I do not feel that this is either necessary or desirable as a goal for U.S. energy policy."

Widening U.S. Oil Gap

Whatever the original goals of Project Independence—their implementation remained vague and confused through the Ford presidency. There was no doubt that the Arab states had used the oil weapon tellingly against the American superpower. American domestic oil production peaked in 1970, meaning that, before the Arab embargo of October 1973, the United States had expected to import larger quantities of foreign oil to meet ever-increasing demands.

In 1972, the United States received 850,000 barrels of Arab oil per day, which represented 17.9 per cent of total imports. In 1973, Arab oil imports increased to 1,590,000 barrels per day, despite the beginning of the embargo. But the percentage of Arab oil to total imports dropped to 9.2 per cent.

In 1974, with the embargo still in effect until March 18, Arab imports dropped to 1,250,000 barrels per day, but the percentage of Arab oil went back up to 20.5 per cent. The percentage has grown ever since, and by 1977 the United States was importing more than nine million barrels of crude oil daily (about five times the amount imported in 1973 and about 50 per cent of U.S. consumption) with about 80 per cent coming from OPEC countries and about 50 per cent of this coming from the Arab states.

World Oil Production and Consumption 1976*

Million barrels/day	Production	Consumption
United States	9.7	17.3
Canada	1.6	1.8
Western Europe	1.0	14.0
Japan	—	5.4
South America	4.6	3.8
Africa	5.9	1.0
Communist Countries	12.4	11.0
Far East & Asia	2.1	2.3
Australia	.5	.6
Middle East	22.0	1.8

*Estimate

SOURCE: International Economic Report of the President, January 1977.

23 "German Press Review," West German Embassy in Washington, D.C., Jan. 29, 1975. See "Foreign Investment in America," *Editorial Research Reports*, 1974 Vol. II, pp. 561-580.

"Without constraints," President Carter warned in April 1977, "U.S. oil demand probably would grow at the postwar rate of 4 per cent per year, and reach 25 million barrels per day by 1985."[24] The CIA projection, quoted in *U.S. News and World Report*, May 2, 1977, and relied upon by the President, estimated that it was unlikely that this amount of oil would be made available to the United States at a reasonable price.[25]

With American demand continuing to rise at 4 per cent yearly, it was estimated in 1977 that oil imports would have to go up from nearly 9 million barrels per day in 1977 to 16 million barrels per day by 1985, assuming anticipated increases in domestic oil production prove to be accurate. In short, the oil embargo and U.S. efforts such as Project Independence had not altered the basic and growing American dependence on imported oil, and increasingly on imported Arab oil.

What actually happened was that the Arabs employed their oil weapon at the very time that the United States was beginning to rely more and more on Arab sources. The deepening reliance came about as supplies began to dwindle from Venezuela and Canada, historically the two biggest suppliers of foreign oil to the United States. Canada announced plan to phase out all oil exports to the United States by 1983.

OPEC quadrupled its price between Jan. 1, 1973, and Jan. 1, 1974—from a posted price of $2.59 per barrel of Persian crude oil to $11.65. By 1977 the price was nearly $13 per barrel. The U.S. economy was caught in the two-edged vise of embargo and skyrocketing prices. An FEA report of August 1974 estimated that the five-month embargo cost 500,000 American jobs and a gross national product loss of between $10-billion and $20-billion.

The United States now faces, anytime that OPEC is emboldened to hike its prices, such results as a further drain on consumer purchasing power, another spur to inflation, deeper deficits in balance of payments and the federal budget, and an ever-increasing menace to the world economic system. This situation explains why President Carter, soon after taking office, announced a comprehensive "National Energy Plan" designed to lessen American dependence on imported oil.

FEA Blueprint (Nixon-Ford)

The designers of the FEA blueprint for Project Independence opted for a goal of independence from "insecure" foreign oil by 1985 and an emphasis on strict conservation measures to reduce the volume of energy imports. FEA administrator Sawhill said, "If we want to reduce our dependence on imports, we must reduce our energy demands. This is an inescapable conclusion." (He was to lose his job in a dispute about conservation measures with President Ford.)

The blueprint explored various alternatives, all of them presenting a multitude of problems and carrying enormous price tags. A Federal Energy Office (FEO)—forerunner of the FEA—report on Feb. 6, 1974, put the lowest estimate for energy self-sufficiency in terms of capital outlay at $253-billion. Other projections have ranged as high as $1-trillion. Highlights of the blueprint included:

● Conservation—An all-out conservation effort could reduce annual increase in energy demand from the existing 4 to 5 per cent to 2 to 3 per cent by 1980. The blueprint

U.S. Trade in Crude Oil

(*bpd* barrels per day)

Year	Exports Thousand bpd	Imports Thousand bpd	Net Imports Thousand bpd
1947	126	268	142
1948	110	353	244
1949	90	422	332
1950	96	488	392
1951	79	490	411
1952	74	575	501
1953	55	649	594
1954	38	658	619
1955	33	781	748
1956	79	937	858
1957	137	1,022	885
1958	11	953	942
1959	8	964	956
1960	8	1,019	1,011
1961	8	1,047	1,038
1962	5	1,126	1,121
1963	5	1,132	1,126
1964	3	1,203	1,200
1965	3	1,238	1,235
1966	5	1,225	1,219
1967	74	1,129	1,055
1968	5	1,293	1,288
1969	3	1,408	1,405
1970	14	1,323	1,310
1971	1	1,680	1,679
1972	*	2,222	2,222
1973	2	3,244	3,242
1974	3	3,422	3,419
1975	6	4,105	4,009
1976**	2	7,300	7,300
1977**	—	9,600	9,600

* Less than 500 barrels per day.
** Estimate figures from Carter's "National Energy Plan."

SOURCE: International Economic Report of the President, January 1977.

recommended a 15-cents-a-gallon tax on gasoline; mandatory fuel economy standards requiring that cars get at least 20 miles to the gallon; a 25 per cent tax credit for insulating existing housing; a 15 per cent tax credit for making commercial buildings more efficient users of energy; continued reduced speed limits, and more use of car pools and mass transit.

● Conservation alternatives—The study mentioned two possibilities. One was more intensive exploration and drilling for oil and natural gas, but the FEA said that environmental hazards and social and financial costs were a drawback to that approach. The other strategy was the establishment of a 1-billion-barrel oil stockpile in underground salt caverns, which would cost $1.4-billion in maintenance each year.

● Coal—"Coal and conservation will be most important in the short and intermediate terms," Sawhill's replacement as FEA administrator, Frank G. Zarb, has said. William E. Simon, who was the federal energy chief before becoming secretary of the treasury, said that coal could replace as much as 6 billion barrels of oil each day by 1985 if employed as a direct substitute for oil and gas in electricity production, space heating and industrial processes. The

24 See "U.S. Leans More on Arab Oil," *Christian Science Monitor*, 13 June 1977, p. 1; "The National Energy Plan," Executive Office of the Presidency, Energy Policy and Planning, 29 April 1977.
25 "The International Energy Situation: Outlook to 1985," Central Intelligence Agency, April 1977.

FEA study showed that coal production in 1974 amounted to 599 million tons—and that supply would last another 800 years at that rate. Most of the nation's coal is high in sulphur, and water shortages were possible consequences of some coal gasification plants.

• Oil—From the peak year of 1970, domestic oil production was expected to decline for several years, despite price increases awarded producers as incentives, because the development of new fields requires several years. By 1985, domestic output could rise to between 12 million and 20 million barrels per day, with production increasing as government-set prices were raised. The study said there are 200 billion to 400 billion barrels of undiscovered oil in the United States, one-third of it in offshore regions of the East Coast, California coast and Gulf of Alaska. The report minimized the danger of spillage from offshore drillings.

• Natural gas—"The outlook for increased gas supplies is not promising," the report stated. Reserves have been falling since 1967, and consumption is running as much as three times greater than annual discoveries of new supplies. Prices are federally regulated, and higher prices could stimulate recovery from offshore areas and Alaska.

• Oil shale—There are an estimated 1.8 trillion barrels of oil in shale deposits in Colorado, Wyoming and Utah. Production could reach 250,000 barrels per day by 1985, but a strong push to recover oil shale could cause severe water shortages. Heavy federal subsidies also would be required.

• Nuclear power—Nuclear power plants for the production of electricity have fallen behind schedule for becoming operational; nevertheless, such plants were expected to produce 30 per cent of all electricity by 1985.

• Exotic energy sources—Solar energy was not expected to have an impact until after 1985, and the FEA study said that geothermal energy development might prove to be a significant energy source by the year 2000.

Ford vs. Congress

The 94th Congress convened in January 1975, and the burden of putting an energy program into action fell to the legislators and the White House. The Democrats had made sweeps in the November elections, especially in the House, where they won a two-thirds majority. While the Republican President prepared his legislative recommendations on energy, Democratic task forces in the House and Senate drew up their own legislative proposals. As soon as all programs were made public, it became apparent that Ford and the Democratic congressional majorities were heading in different directions less than a year after the Arab oil boycott had ended.

Ford stated as his goals "surplus capacity in total energy," which the nation had had in the 1960s, and the reduction of oil imports to "end vulnerability to economic disruption by foreign suppliers by 1985." He called for swift legislative action to allow commercial production at the Elk Hills, Calif., Naval Petroleum Reserve and to enable more power plants to convert to coal. Recommending a 90-day timetable to the lawmakers, the President in January 1975 also asked for deregulation of natural gas to spur new production, a natural gas excise tax and a windfall profits tax by April 1, because he planned to act to remove controls from the price of domestic crude oil on that date.

Ford announced that he was using his presidential emergency powers to curb oil imports by making them costly. He would order crude oil tariff levels to be increased

Persian Gulf Crude Oil Prices [1]

(Dollars per barrel)

	Jan. 1, 1973	Jan. 1, 1974	Per cent Change, 1973-74	Jan. 1, 1975	Per cent Change, 1974-75	Jan. 1, 1976	Per cent Change, 1975-76	Jan. 1, 1977	Per cent Change, 1976-77
1. Posted price [2]	$2.591	$11.651	+ 350.0	$11.251	− 3.6	$12.376	+ 9.9	$12.995	+ 5.0
2. Royalty [3]	.324	1.456	+ 349.0	2.250	+ 54.5	2.475	+ 10.0	2.599	+ 5.0
3. Production cost	.100	.100	—	.120	+ 20.0	.120	—	.120	—
4. Profit for tax purposes (1-(2 + 3))	2.167	10.095	+ 366.0	8.881	− 12.0	9.781	+ 10.1	10.276	+ 5.1
5. Tax [4]	1.192	5.552	+ 366.0	7.549	+ 36.0	8.314	+ 10.1	8.735	+ 5.1
6. Government revenue (2 + 5)	1.516	7.008	+ 362.0	9.799	+ 39.8	10.789	+ 10.1	11.334	+ 5.1
7. Cost of equity oil (2 + 6) [5]	1.616	7.108	+ 340.0	9.919	+ 39.5	10.909	+ 10.0	11.454	+ 5.0
8. Cost of participation oil [5]	2.330	10.835	+ 365.0	10.460	− 3.5	11.510	+ 10.0	12.085	+ 5.0
9. Weighted average cost [5]	1.794	9.344	+ 421.0	10.240	+ 9.6	11.270	+ 10.1	11.836	+ 5.0
10. Weighted government revenue (9 - 3)	1.694	9.244	+ 446.0	10.120	+ 9.5	11.150	+ 10.2	11.716	+ 5.1

1 Prices shown are for Saudi Arabian light crude oil 34 degree API (American Petroleum Institute) gravity. Saudi light is used as a standard for Persian Gulf crude because it is the largest single type of crude produced there and represents a good average between higher-priced low-sulfur crude and lower-priced heavier oil.

2 The so-called "posted price" is a fictitious, artificially high price set by oil-producing countries for the purpose of producing the revenues—royalties and taxes—they receive from oil companies.

3 The Saudi royalty was fixed at 12.5 per cent of the posted price for the 1973 and 1974 dates, and at 20 per cent for the 1975 date.

4 The Saudi tax was fixed at 55 per cent of the profit for tax purposes (line 4) for the 1973 and 1974 dates, and at 85 per cent for the 1975 date.

5 The oil companies pay two different prices, and the weighted average cost per barrel falls between the cost for equity oil and the cost for participation oil. Participation oil is oil in which the oil-producing country has part ownership in the oil companies operating in the country. The oil companies—because of their exploration and development roles—have a right to a certain percentage of production at a cost something less than the market rate, which also is figured in the weighted average cost.

SOURCE: International Economic Report of the President, January 1977

by $1 a barrel on Feb. 1, by $1 a barrel on March 1 and by $1 per barrel on April 1. That action raised the first outcry from congressional Democrats and brought about the first confrontation between Congress and the administration on the energy package.

House Speaker Carl Albert (D Okla.) said the tax on oil imports would have an "astounding inflationary impact" on the economy. The Speaker explained that the Democrats favored a "more moderate approach" which combined gasoline rationing, a graduated excise tax on automobiles based on their fuel consumption, gasless days and other measures. But Ford said that, after considering rationing and high gasoline taxes as alternatives, he had rejected them as ineffective and inequitable.

Other details of the Democratic program, developed in task forces headed by Sen. John O. Pastore (R.I.) and Rep. Jim Wright (Texas), included a 5-cent federal tax increase per gallon of gasoline (raising it to 9 cents from the existing 4 cents). Proceeds from this tax would raise $5-billion a year to build an energy trust fund. The fund would be used for research and development of alternate energy sources—solar power, coal gasification, nuclear power and nuclear fusion projects.

Ford forged ahead with his $1 fee on foreign barrels Feb. 1, rebuffing New England governors who met with him to protest the action. The region depends heavily on foreign oil to meet its needs. Meanwhile, Democrats introduced bills in both chambers barring Ford's action for up to 90 days while the legislators considered alternative strategies.

The Democrats quickly moved legislation to delay the import-tariff action, and Ford vetoed the bill on March 4. However, in a conciliatory mood, the President held off on further fee increases, not adding the second $1 until June 1. Democrats made no move to override the veto.

On May 20, the President vetoed the strip-mining bill, claiming that it would result in a loss of coal production—up to 25 per cent of annual tonnage—and cost 36,000 jobs. Sen. Henry M. Jackson (D Wash.), a principal sponsor (and a presidential candidate), accused Ford of a "lack of sensitivity to the need to balance energy production with protection of our land." The House gave Ford a victory by sustaining his veto.

The President went on television May 27, assailing Congress for having "done nothing positive to end our energy dependence." Accusing the lawmakers of "drift, dawdle and debate," he added, "Our American economy runs on energy. No energy, no jobs. It's as simple as that."

Ford's Energy Policy Accomplishments

In his final International Economic Report as President, Gerald Ford outlined his accomplishments in the area of U.S. domestic energy policy. "During 1976," he reported, "the Congress enacted and the President signed a number of major energy bills. They were the Naval Petroleum Reserves Production Act, Energy Conservation and Production Act, Alaskan Natural Gas Transportation Act and Coastal Zone Management Act. The Federal Energy Administration estimated that when the programs called for in these acts and other programs were fully implemented U.S. imports would average approximately seven million barrels per day by 1985."

"Major administrative actions affecting energy undertaken during 1976," Ford further pointed out, included:

• Announcement of a liquefied natural gas policy, limiting the volume and sources of imports, as well as provisions for pricing and storage;

World Proved Oil Reserves
(Estimate at the end of 1976)

Major Areas and Selected Countries	Billion Barrels	Per Cent of Total
North America	69	10.3
United States	40	6.0
Canada	9	1.3
Mexico	20	3.0
Central and South America	22	3.3
Venezuela[1]	14	2.1
Ecuador[1]	2	.3
Other	6	.9
Western Europe	29	4.3
United Kingdom	19	2.8
Norway	7	1.0
Other	3	.4
Africa	65	9.7
Algeria[1]	7	1.0
Libya[1]	26	3.8
Nigeria[1]	20	3.0
Other	12	1.8
Asia-Pacific	22	3.3
Indonesia[1]	15	2.2
Other	7	1.0
Middle East	396	59.3
Saudi Arabia[1]	178	26.6
Kuwait[1]	79	11.8
Iran[1]	64	9.6
Iraq[1]	35	5.2
United Arab Emirates[1]	6	.9
Qatar[1]	31	4.6
Other	3	.4
Total Non-Communist	603	90.3
Communist World[2]	65	9.7
Soviet Union	40	4.0
Other	25	3.7
Total World	668	100.0

1 Member of Organization of Petroleum Exporting Countries (OPEC).
2 Includes Soviet Union and other Warsaw Pact Nations, Cuba, China and Yugoslavia.

SOURCE: International Economic Report of the President, January 1977

• Increase in the price of natural gas sold in the interstate market by the Federal Power Commission, though the decision is being challenged in court;

• Accelerated coal leasing;

• A new policy directive regarding nuclear reprocessing and proliferation.

• A study of a possible reorganization of the various energy functions of the federal bureaucracy;

• Funding of a Solar Energy Research Institute.[26]

President Ford also outlined the "impressive" progress made by the International Energy Administration. "The emergency program was tested and is operationally ready in the event of an embargo," he reported. "The oil information system is now functioning. A long-term cooperation program in conservation, energy production, and R&D was

26 International Economic Report of the President, January 1977, p. 94.

adopted, and a process for reducing dependence on imported oil initiated."[27]

Additionally, Ford referred to progress in the areas of nuclear energy and energy research and development. He also noted the adoption of a joint strategy for a dialogue between oil producers and consumers reached in the Energy Commission of the Conference on International Economic Cooperation (CIEC) and the continuing cooperation between the United States and Iran and Saudi Arabia through the Joint Economic Commissions established in 1974 and 1975, respectively.

Although these efforts did show some awareness of the magnitude of the energy problems facing the United States and the entire industrial world, while the talk and planning went on, U.S. oil imports were skyrocketing to about 50 per cent of U.S. consumption—more than 9 million barrels in the beginning of 1977. When the Carter administration took office, it did not find the efforts to date satisfactory; and the public appeared insufficiently sensitive to the seriousness of the situation the country could face in just a few years should a major energy program not be initiated immediately.

Exporting Countries' Unity

A central factor in the growing strength of Middle East states in the world oil market has been their increasing ability to act in concert. Increases in oil prices, new formulas for calculating royalty payments and taxes, agreements on participation in ownership and changes in prices to reflect devaluation of the dollar have been relatively recent breakthroughs for the Middle East oil countries.

These countries allowed foreign oil companies to exploit their oil reserves under concessions granted at various times between 1901 and 1935. These agreements required the companies to pay only a nominal royalty—an average of 21 cents a barrel in the Middle East, according to Charles Issawi in *The Washington Papers: Oil, the Middle East and the World.*[28] In return, the concessionaires were exempted from taxes and were free to determine production and pricing policy.

Issawi pointed out that these arrangements may have been fair enough in times when demand was slow, prices were fluctuating or dropping, prospects for discovering oil were uncertain and huge capital investments were needed. But during and after World War II, prices rose sharply, thus reducing the purchasing power of the fixed royalty and the government's share of the value of the oil.

Oil-producing countries pushed for a greater share, and by the early 1950s all the producing countries had negotiated agreements providing that revenues from oil production be divided between a country and a company or consortium on a 50-50 basis. These agreements—resulting in raising payments to 70-80 cents per barrel—multiplied the revenues of the Middle East governments almost tenfold between 1948 and 1960, from about $150-million to nearly $1.4-billion, according to Issawi.

OPEC and OAPEC

The Middle East countries had become acutely aware of the importance of maintaining a high price. Reductions in prices in the late 1950s by oil companies, without consulta-

tion with the producing countries, sparked a countermove by the world's five largest oil exporting nations.

Iran, Iraq, Kuwait, Saudi Arabia and Venezuela in September 1960 established OPEC to keep oil prices from dropping further. The five countries agreed to prorate their future production, based on 1960 production levels, and to pool and prorate any future increases in world market demands. By 1974 OPEC also included Algeria, Ecuador, Indonesia, Libya, Nigeria, Qatar and the United Arab Emirates.

Early in 1968, Saudi Arabia, Kuwait and Libya formed another group, called the Organization of Arab Petroleum Exporting Countries (OAPEC). The membership had been broadened by 1974 to include Algeria, Bahrain, Egypt, Iraq, Qatar, Syria and the United Arab Emirates. *(underlined of OPEC)*

OPEC was successful in preventing further cuts in posted prices for oil on which taxes were calculated but failed to restore prices to their earlier level or to agree on a formula for prorating to limit output. A 1965 attempt to agree on a uniform production plan also failed, but OPEC did manage to achieve some important financial gains, according to Issawi. (A "posted price" is a fictitious, artificially high price set by producing countries for the purpose of boosting the royalty payments and taxes they receive.) *(Chart, p. 133)*

In the meantime, the 50-50 split of the oil profits came under increasing fire. As Simon pointed out in Senate testimony in May 1973, there had been virtually no increase in per-barrel payments in the 1950s and only a 12-cent-per-barrel increase in the 1960s. In the late 1960s, OPEC began its move toward increasing revenues.

An OPEC conference in Vienna in June 1968 produced a resolution that (1) cited the United Nations principle that all countries had the right to exercise sovereignty over their natural resources, (2) declared the desirability of exploitation of resources directly by the producing countries, (3) set forth a doctrine of "changing circumstances" to allow the countries to alter existing contracts and (4) expressed the right of governments to share in ownership.

OPEC also agreed on a minimum taxation rate of 55 per cent, establishment of more uniform pricing practices, a general increase in the posted or tax-reference prices in all member countries and elimination of allowances granted to oil companies. *(85% in 75)*

Price Agreements

The 1968 conference marked the first significant step toward consolidation of OPEC power, and great strides have been taken since then. According to a January 1974 General Accounting Office (GAO) report on foreign sources of oil for the United States, OPEC members had negotiated eight agreements with oil companies that significantly increased the cost of crude oil.

Libya led the way toward exacting higher prices from the oil companies. In 1970, the companies held out for seven months while the Libyan government cut back production to force a price increase. The companies finally agreed to raise the posted price by 30 cents a barrel. Libya's success provided incentive to the Persian Gulf states.

A triple victory was scored in 1971 with the negotiation of the Tehran, Tripoli and East Mediterranean agreements. The Tehran agreement, reached in February 1971, involved Abu Dhabi, Iran, Iraq, Kuwait, Qatar and Saudi Arabia.

Following on the heels of a 9-cents-per-barrel price hike negotiated in 1970, the Tehran agreement raised the basic

---

27 *Ibid.*
28 Washington Paper #5, published for the Georgetown Center for Strategic and International Studies by Sage Publications, Beverly Hills/London.

posted price of oil 35-40 cents per barrel and contained a formula for a four-step increase in posted prices through 1975. Prices increased from about 86 cents per barrel in 1970 to about $1.24 per barrel in 1971—an increase of 43 per cent—with $1.50 as the goal by 1975.

The Tehran agreement also hiked the countries' taxes from 50 per cent to 55 per cent of the net taxable income and established a system for adjusting the posted price according to the oil's gravity.

The Tripoli agreement, negotiated by Libya in April 1971, and the East Mediterranean agreement, negotiated by Iraq in June 1971, used the Tehran agreement as their base. Additional elements included an increase in the basic posted price, a further increase for low-sulfur oil and temporary increases to reflect the geographical advantage resulting from the closing of the Suez Canal and the high freight rates prevailing for oil tankers.

Critics' Position

Some critics of Western negotiators have felt that these early agreements set the stage for spiraling Arab demands. In a controversial article in the winter 1973 issue of *Foreign Policy* magazine, oil economist M. A. Adelman of the Massachusetts Institute of Technology stated his belief that the steep price increases could be attributed to U.S. capitulation to OPEC during negotiations over the Tehran agreement.

"Without active support from the U.S., OPEC might never have achieved much," said Adelman. He called a January 1971 meeting of oil-importing nations the "turning point" and said that "there is no doubt that the American representatives and the oil companies assured the other governments that if they offered no resistance to higher oil prices they could at least count on five years' secure supply at stable or only slightly rising prices."

In testimony before the Senate Foreign Relations Subcommittee on Multinational Corporations Jan. 31, 1974, a vice president of Hunt International Petroleum Corporation, third largest independent oil-producing venture in Libya, offered a similar view. Henry M. Schuler called oil policy and negotiations since 1971 an "unmitigated disaster" which had created the "unstoppable momentum" within the Arab world for higher prices.

"If a political and economic monster has been loosed upon the world," Schuler stated, "it is the creation of Western governments and companies. Together we created it and gave it the necessary push, so only we, acting in harmony, can slow it down."

The GAO examination of Tehran agreement negotiations noted that the oil companies had agreed among themselves to negotiate only on the basis of reaching a five-year settlement with all the producing countries, in order to avoid the leapfrogging effect of a series of agreements and to promote stability for those five years. But Libya and Algeria rejected any such joint bargaining. The negotiations, which lasted about a month, began under the threat of an oil production cutoff if OPEC demands were not met by a certain deadline. Negotiations broke down at one point but, with the urging of U.S. government officials, the companies continued the negotiations and an agreement was signed Feb. 14, 1971.

In an April 1973 *Foreign Affairs* article, James E. Akins, U.S. ambassador to Saudi Arabia and former director of the State Department's Office of Fuels and Energy, dismissed charges that the United States had invited the threat of a cutoff of oil and thus had built up OPEC's bargaining position. Akins asserted that, upon the protest of the State Department, leaders from Iran, Saudi Arabia and Kuwait had assured the United States that the so-called "threats" had been misunderstood, that they had been directly solely at the oil companies and that the oil would be available to consumers even if the negotiations broke down. He said that the State Department had also requested an extension of the deadline set by OPEC and an assurance that agreements reached with the companies would be honored for their full terms.

Dollar Devaluation

Another major agreement was negotiated in January 1972 in Geneva. Abu Dhabi, Iran, Iraq, Kuwait, Qatar and Saudi Arabia reached agreement with the major oil companies to increase the posted price to restore the purchasing power lost by the oil-producing countries because of the December 1971 dollar devaluation. According to the formula agreed upon, the posted prices would be adjusted every time the U.S. exchange rates differed from an index of nine major currencies by more than 2 per cent. Posted prices rose by 8.55 per cent in February 1972 and 5.69 per cent in April 1972.

A second Geneva agreement on dollar devaluation and oil was reached in June 1973. Oil companies agreed to raise immediately the posted price of crude oil from Iran, Iraq, Abu Dhabi, Qatar, Kuwait, Saudi Arabia, Libya and Nigeria by 6.1 per cent, making a total increase of 11.9 per cent since the February 1973 devaluation. The agreement, effective through 1975, also set a new formula under which posted prices would reflect more fully and rapidly any future changes in the dollar's value.

Escalating Demands   *after only 2½ yrs*

The Tehran agreement was laid to rest in October 1973 when Iran, Saudi Arabia, Iraq, Abu Dhabi, Kuwait and Qatar announced a unilateral price hike after negotiations had broken down. The new level was a 17 per cent increase in the market price of crude—raising it to about $3.65—and a 70 per cent increase in the posted price—bringing that up to about $5.11.

Yet another 70 per cent increase in prices, effective Jan. 1, 1974, was announced by these six Persian Gulf producers during a December 1973 meeting in Tehran. This brought the posted price to $11.65. Other petroleum exporting countries have followed the initiatives of the Persian Gulf countries in raising their prices.

In 1974, a series of changes by oil-producing countries actually reduced the posted price slightly but increased the cost of oil by raising royalties and taxes. The actual prices charged the oil companies on terms that went into effect Jan. 1, 1975, were 9.6 per cent higher than a year earlier. OPEC froze those prices for nine months, until Oct. 1. (*See weighted average cost chart, p. 133*)

Further increases were expected in the last quarter of 1975. The Shah of Iran, at a May 17 news conference in Washington, called for an increase to compensate for the loss of "between 30 and 50 per cent of our purchasing power because of world inflation." Western economists felt there was no justification for such a further increase, and President Ford warned June 25 that another price hike would be "very disruptive and totally unacceptable."

Largely because of the recession, but also because of conservation measures in consuming countries, demand for oil was falling in 1975. As a result, OPEC production dropped from 30.6 million barrels a day in 1974 to about 25 million in early June 1975, leaving its member countries with a surplus productive capacity of about 10 million barrels per day.

The oil-producing states with smaller revenues and larger development programs were beginning to feel the pinch from reduced sales. Western economists say that reduced prices would enable these countries to expand their output. In fact, by discreetly discounting its prices below the OPEC level, Iraq in early 1975 was producing at a rate 42 per cent above the same period a year earlier, while OPEC production in the same period fell by 16 per cent. In spite of this downward pressure on oil prices, experts expected some increase by the end of 1975 for the sake of keeping the OPEC cartel united and demonstrating its continuing effectiveness.[29]

Participation Agreements

Participation—part ownership by oil-producing countries in the oil companies operating in their countries—has become a significant issue in the world oil market. The concept was originated by Saudi Arabia's shrewd oil minister, Sheikh Ahmed Zaki al-Yamani.

A 1972 House Foreign Affairs subcommittee report on the Persian Gulf pointed out that many nationalists have found the concessionary system—which involved contracts between companies and exporting countries that enabled the companies to exploit oil deposits for a fixed period of time—degrading. They were said to believe that as concessions approached their termination it was natural for the countries to begin to participate in all operations rather than remaining in the passive role of collecting revenues and royalties.

The attitude of many oil companies, according to the report, was that oil exporting countries held all the trump cards and participation was inevitable. They faced the dilemma of balancing the cost of yielding to demands too soon against the risk of holding out too long and perhaps precipitating nationalization.

Algeria had nationalized its natural gas fields in 1971 and taken over 51 per cent of oil concessions on behalf of its national company, thus breaking the favored position of French oil companies. Libya expropriated British Petroleum Company holdings in 1971. Iraq nationalized the Western-owned Iraq Petroleum Company in 1972.

In the OPEC participation agreement reached in December 1972, Saudi Arabia, Kuwait, United Arab Emirates and Qatar won an immediate 25 per cent interest in the companies' ownership which was to increase in steps up to 51 per cent by 1982. But these timetables were speeded up and by the mid-1970s, all the countries were heading for complete ownership, though actual management remained in the hands of Westerners.

Iran

In 1973, Iran took control of its oil industry. Shah Mohammed Reza Pahlavi had announced in June 1972 that Iran would extend the Western-owned consortium agreement of 1954 when it came up for renewal in 1979. However, in January 1973, the Shah declared that the agreement would not be extended unless the consortium agreed to in-

29 Wall Street Journal, June 27, 1975.

crease daily production from 4.5 million barrels to 8.3 million by 1977. But in March, the Shah nationalized the oil industry. A 20-year agreement was signed in May which provided for the National Iranian Oil Company to take control of all operations and facilities of the oil companies, with the consortium to have the role of technical adviser.

Libya

In June 1973 Libya nationalized a U.S. oil firm in retaliation for U.S. support of Israel. Libyan leader Col. Muammar el-Qaddafi told a rally that the United States "deserved a strong slap in the face." He warned: "The time might come where there will be a real confrontation with oil companies and the entire American imperialism."

The Libyan government acquired 51 per cent of the holdings of the Occidental Petroleum Company in mid-August 1973. Independent oil companies followed suit in sharing ownership, but the major companies held out. In September 1973 the revolutionary leader announced the unilateral takeover of 51 per cent of the operations of all the major companies. In February and March 1974, the government completely nationalized the operations of Texaco, Shell and subsidiaries of Standard Oil of California and Atlantic Richfield.

Oil Embargo

Oil production cutbacks and the embargo of exports to U.S. markets which began in late 1973 marked not only a peak in Arab unity but also the first formal use of oil as a political weapon to pressure the industrial nations into forcing Israel to return to its pre-1967 frontiers.

The potency of the weapon became highly visible within hoped would be a more evenhanded position in the Middle rupting the life style of every major industrial power, caused fissures in the Atlantic alliance, precipitated upsets in international money markets and prompted the United States to make an intensive search for a peace settlement in the Middle East.

(see next page)

1967 Boycott

An attempt at using oil as a political weapon had been made after the 1967 war, but it met with little success. Within two days of the outbreak of fighting in 1967, every major Arab oil producer except Algeria completely shut down production. But consuming countries turned to the United States, Venezuela and Indonesia, and the Arab countries soon broke ranks.

Three weeks after the embargo went into effect, Iraq began shipping oil to Turkey and France through a pipeline to Lebanon. By mid-July the major Arab oil producers had restored shipments to all countries except the United States, Britain and, in some instances, West Germany. The participating countries agreed in August to leave up to the individual countries the question of whether to continue the boycott, and by mid-September the only interruption to normal oil supplies was the continued closure of the Suez Canal.

Saudi Arabia's Role

Saudi Arabia—the top Middle East oil producer and holder of about one-fourth of the non-Communist world's oil reserves—held the key to the success of a boycott. In 1967, Saudi Arabia had gone along reluctantly with the embargo

Monetary Reserves, Middle East Oil-Exporting Countries*

(in millions of dollars)

Country	1969	1970	1971	1972	1973	1974	1975	1976
Algeria	$410	$ 339	$ 507	$ 493	$1,143	$1,689	$1,353	$1,987
Egypt	145	167	161	149	391	342	294	339
Iran	310	208	621	960	1,237	8,383	8,897	8,833
Iraq	476	462	600	782	1,553	3,273	2,727	4,601
Kuwait	182	203	288	363	501	1,397	1,655	1,929
Libya	918	1,590	2,665	2,925	2,127	3,616	2,195	3,206
Saudi Arabia	607	662	1,444	2,500	3,877	14,285	23,319	27,025
Syria	59	55	88	135	481	835	735	361

* A country's international reserves consist of its reserves in gold, SDRs (special drawing rights which are unconditional international reserve assets created by the International Monetary Fund), its reserve position in the Fund (unconditional assets that arise from a country's gold subscription to the Fund and from the Fund's use of a member's currency to finance the drawings of others) and its foreign exchange (holdings by monetary authorities—such as central banks, currency boards, exchange stabilization funds and Treasuries—of claims on foreigners in the form of bank deposits, Treasury bills, government securities and other claims usable in the event of a balance of payments deficit).

SOURCE: Data from the International Monetary Fund's International Financial Statistics, June 1975, Aug. 1977.

because of pressure from Egyptian President Gamal Abdel Nasser. But the country had not enforced the embargo strictly and lifted it shortly after it was made voluntary.

However, Saudi Arabia, under conservative, pro-Western King Faisal, led the way in 1973. The formal decision to use oil as a weapon came Oct. 17, 1973, at an OAPEC meeting in Kuwait. The members agreed to cut production monthly by 5 per cent over the previous month's sales until Israel had withdrawn from the Arab territories it had occupied since the 1967 war and had agreed to respect the rights of Palestinian refugees.

The following day, Saudi Arabia announced that it would cut back its oil production by 10 per cent and would cut off all shipments to the United States if the United States continued to supply Israel with arms and refused to "modify" its pro-Israeli policy. Libya imposed an embargo Oct. 19.

On Oct. 29—the day after President Nixon requested $2.2-billion in emergency security assistance for Israel—Saudi Arabia announced a total embargo against oil exports to the United States. When four Persian Gulf states—Kuwait, Qatar, Bahrain and Dubai—joined the embargo Oct. 21, the Arab boycott of U.S. markets was complete.

The reasons for Saudi Arabia's turnabout rested with inter-Arab politics and Faisal's anti-Zionism. Nadav Safran, writing in the January 1974 Foreign Affairs, attributed Faisal's previous reluctance to use the oil weapon partially to a fear that Egyptian President Nasser, who died Sept. 28, 1970, and other Arab radicals might usurp control over when and how the weapon would be used.

Another factor, according to Safran, was that Saudi Arabia needed all the revenues it was receiving at that time and suspected the weapon could possibly be turned around and used against the Saudi regime.

In fact, according to an article by Benjamin Shwadran in the February 1974 Current History, the original purpose for the formation of OAPEC by Saudi Arabia, Kuwait and Iraq early in 1968 was to block the other Arab countries—particularly Egypt—from interfering with their economic interests.

But, as Safran pointed out, the circumstances in 1973 were quite different than in 1967. Saudi Arabia was receiving vastly increased revenues, and Nasser was no longer around to threaten Faisal's autonomy. Saudi Arabia's decision to use the oil weapon, coupled with substantial financial assistance to Egypt, was seen as a move to minimize the principal appeal of radical Libyan leader Col. Qaddafi—self-proclaimed heir of Nasser—and prevent the proposed union of Egypt and Libya.

Another factor was the failure of Saudi Arabia's quiet diplomacy aimed at nudging the United States into what it hoped would be a more evenhanded position in the Middle East. Faisal's reluctance to employ the oil option had been lessened considerably after Sadat's expulsion of Soviet military advisers from Egypt in July 1972 failed to bring a change in White House policy.

Saudi Arabia's final move came after a series of pleas to Washington to tone down its support of Israel. *The Washington Post* reported Oct. 20, 1973, that Faisal had sent at least two letters to President Nixon with pleas for a more evenhanded appearance. Petroleum Minister Yamani had made the same appeal during a vist to Washington in April 1973.

In a July 4 interview with two American reporters, Faisal delivered what was said to be his first public warning, when he stated that Saudi Arabia would like to continue its friendly ties with the United States but would find this "difficult" if the United States support for Israel remained at its existing level. The warnings continued, and the final decision to impose the embargo came only after Nixon's Oct. 19 request for money to rearm Israel.

Embargo Categories

The Arabs were systematic in their embargo, with countries being divided into categories. On the boycott list were nations considered to be friends of Israel. The United States was at the top. The Netherlands followed, because the Arabs were angered by what they saw as a pro-Israel stance and reports that the Dutch had offered to aid in the transit of Soviet Jewish emigrants to Israel. In late

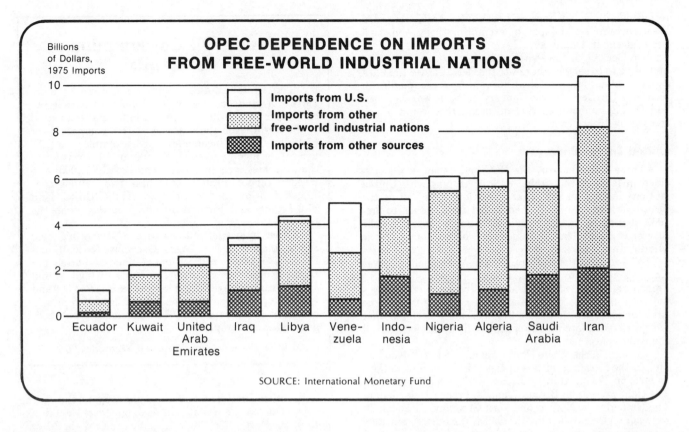

OPEC DEPENDENCE ON IMPORTS FROM FREE-WORLD INDUSTRIAL NATIONS

Billions of Dollars, 1975 Imports

- Imports from U.S.
- Imports from other free-world industrial nations
- Imports from other sources

Ecuador, Kuwait, United Arab Emirates, Iraq, Libya, Venezuela, Indonesia, Nigeria, Algeria, Saudi Arabia, Iran

SOURCE: International Monetary Fund

November, Portugal, Rhodesia and South Africa were officially placed on the embargo list. Shipments of oil to Canada were cut off because the Arabs feared the oil might be reshipped to the United States.

Exempted nations included France, Spain, Arab and Moslem states and—on a conditional basis—Britain. These nations were permitted to purchase the same volume of oil as they had purchased in the first nine months of 1973, but, since the fourth quarter of a year is normally a heavy buying period, these nations were also expected to feel the pinch.

All the remaining countries fell into the non-exempt category, which meant that they would divide what was left after the needs of the exempted nations had been met.

Embargo's Effect

In addition to the embargo, the Arab states continued their monthly reductions in production. The effects of the oil squeeze were soon felt in the consuming nations. Measures taken to cope with the oil shortage included gas rationing, bans on Sunday driving, reduced speed limits, increased prices, restrictions on energy usage, cutbacks in auto production and reductions in heating fuels.

Although estimates varied, the embargo was said to have resulted in the loss to the United States of about two million barrels of oil per day. However, Arab oil did leak through the embargo, reportedly from Iraq and Libya. In October, the United States began classifying data on its oil imports to prevent these leaks from being plugged up.

Hardest hit were Japan and Western Europe, areas most dependent on oil imports. Most of northern Europe suffered from the total embargo against The Netherlands, because the Dutch port of Rotterdam was Europe's largest oil-refining and transshipment center.

U.S. Peace Efforts

The embargo proved effective. Although Washington repeatedly denounced the Arab tactics and declared it would not submit to such coercion, the oil squeeze undoubtedly was the driving force behind U.S. peace-seeking efforts.

Secretary of State Kissinger shuttled relentlessly throughout the Middle East in attempts to mediate a settlement. A series of peace missions produced the Nov. 11, 1973, cease-fire agreement between Israel and Egypt, resumption of diplomatic relations between the United States and Egypt, the Dec. 21-22 first round of Geneva peace talks, the Jan. 18, 1974, Egyptian-Israeli disengagement accord and the May 31 disengagement agreement between Israel and Syria.

Atlantic Alliance

But in the meantime OAPEC's policies took a deep toll on an already troubled Atlantic alliance. The U.S. was unable to prevent its allies from bending to Arab pressure.

On Nov. 6, 1973, representatives of the European Economic Community (EEC), meeting in Brussels, adopted a statement calling on Israel and Egypt to return to the Oct. 22 cease-fire lines which had been drawn before Israeli troops completed the encirclement of Egypt's III Corps. They called on Israel to "end the territorial occupation which it has maintained since the conflict of 1967" and declared that peace in the Middle East was incompatible with "the acquisition of territory by force." Moreover, they declared that any settlement must take into account "the legitimate rights" of the Palestinian refugees.

Later in the month Japan followed suit. On Nov. 22, the Japanese cabinet announced that it might have to reconsider its policy toward Israel. The Arabs rewarded Western Europe and Japan by exempting them from the 5 per cent cut in December.

On Dec. 13, Japan switched from a neutral position to appealing to Israel to withdraw to the Oct. 22 cease-fire lines as a first step toward total withdrawal from occupied Arab territory. On Dec, 25, the January cutback was canceled and OAPEC announced that oil production would be increased by 10 per cent.

United Energy Policy

These statements by the EEC and Japan, while one-sided in favor of the Arabs, were in keeping with United Nations Resolution 242. That resolution, approved by the Security Council in 1967, had called for a return to pre-1967 boundaries in the Middle East and a respect for the sovereignty and territorial integrity of all states in the Middle East. *(Text, p. 76)*

What caused the real strain in the Atlantic alliance was Europe's refusal to forge a united energy policy. U.S.-French relations suffered the most because of France's moves to negotiate bilateral arms-oil agreements with the Arab nations. On Jan. 10, 1974, Kissinger declared that "unrestricted bilateral competition will be ruinous for all countries concerned."

On Jan. 9, the White House announced that Nixon had invited the foreign ministers of eight oil-consuming nations to meet in Washington to discuss world energy problems. The eight were Great Britain, Canada, France, Italy, Japan, The Netherlands, Norway and West Germany. Although all were not on the original invitation list, the foreign ministers of the nine EEC countries decided to attend. French Foreign Minister Michel Jobert, however, withheld his acceptance until less than a week before the conference.

The meeting convened Feb. 11, with representatives from Belgium, Britain, Canada, Denmark, France, West Germany, Ireland, Italy, Japan, Luxembourg, The Netherlands, Norway and the United States. It was scheduled to end the next day but was delayed by a dispute between France and other EEC countries over the response to a U.S. proposal for a joint effort to meet the energy crisis.

Their final communique contained several points opposed by France, including a proposal to formulate a comprehensive action program to deal with the energy situation, establishment of a coordinating group to prepare for a conference of oil-producing and oil-consuming nations, formulation of a group to coordinate development of the actions recommended by the conference and adoption of financial and monetary measures to deal with the balance-of-payments impact of oil prices.

Jobert charged that energy matters had been a "pretext" and that the conference's real purpose had been the "political" desire of the United States to dominate the relationships of Europe and Japan. Kissinger, however, called the meeting a success and said that it was possibly a major step toward "dealing with world problems cooperatively."

Embargo Lifted

With the progress made toward a peace settlement in the Middle East, the United States became increasingly insistent on an end to the embargo. Kissinger told a Jan. 22, 1974, press conference that he had been given assurances by Arab leaders that when an Israeli-Egyptian accord was reached, the embargo would be lifted.

Crude Oil Consumption Per Capita

The United States population consumed more barrels of crude oil per person (about 55 barrels per capita per year in 1974) than any industrialized nation in the world, even those nations with a comparable or higher per capita income level. For example, the U.S. per capita income (roughly $6,500 in 1974) was lower than income levels in Switzerland (nearly $7,000) and Sweden (about $6,800), but these two nations consumed considerably less oil per capita (Switzerland 17.5 barrels and Sweden about 27 barrels) than the United States.

Other industrial nations with relatively high per capita income levels, West Germany ($6,200) and France ($5,200), consumed only 20 and 25 barrels per capita respectively in 1974. The United Kingdom and Japan with lower per capita income levels of $3,500 and $4,200 respectively, each consumed about 17.5 and 25 barrels of oil per capita in 1974, less than one-half U.S. per capita consumption.

SOURCE: *United Nations Statistical Yearbook,* 1975.

Egyptian President Sadat led the way to ending the boycott. On Jan. 22, Sadat said Arab oil states should take note of the "evolution" in U.S. policy toward the Middle East. Although he did not mention the embargo, he said that "now that the Americans have made a gesture, the Arabs should make one too." And at a Feb. 24 press conference, Sadat said that the United States would now probably pursue a more evenhanded approach toward the Middle East.

OAPEC's formal announcement of an end to the embargo against the United States came at a Vienna meeting March 18. Libya and Syria, however, refused to end the boycott until later in the year.

OPEC's Strength

OPEC has remained a potent and effective cartel since 1974. There are differences among the members about pricing and supply policies, but until December 1976, OPEC managed to maintain a unified price for oil, having made periodic price rises since the quadrupling in the 1973-1974 period. Meeting in Vienna in October 1975 OPEC raised oil prices by 10 per cent to $11.28.

During 1976 OPEC did not raise the price on its benchmark crude. At a meeting in Bali it was determined that the world economy was too fragile to withstand additional oil price increases. But when OPEC met again in December to consider 1977 prices, disagreements between the members could not be fully resolved. A two-tier pricing system emerged with 11 members deciding to raise prices by 10 per cent in January and an additional 5 per cent on July 1, 1977. Saudi Arabia and the United Arab Emirates limited their price rise to 5 per cent in January, pricing oil at $12.70 per barrel.

When OPEC met again in July 1977 it was agreed that the majority would not implement the additional 5 per cent price rise planned since December and that Saudi Arabia and the United Arab Emirates would raise their prices an additional 5 per cent bringing a single price structure back to OPEC and ending hopes, held by some in the industrial countries, that OPEC could be encouraged to collapse.[30] The general conclusion in mid-1977 seemed that OPEC would "retain control over world oil prices into the next decade."[31]

Carter's "National Energy Plan"

Within months of taking office, President Carter went on national television, and the next day before the Congress, to declare the energy crisis to be "the moral equivalent of war." He urged a mobilized national effort to decrease energy use and to quickly develop alternative energy sources.

On April 29 the President's Energy Policy and Planning office released "The National Energy Plan." In the plan's introductory letter from the President, Mr. Carter noted that "Our energy crisis is an invisible crisis, which grows steadily worse—even when it is not in the news.... The changes the plan recommends will mean a new direction in American life."

Presidents Nixon and Ford had accomplished little with Project Independence. As the London *Sunday Times* reported in mid-1976, "America has again become the fastest growing customer for Arab oil and the one force in the world destined to ensure high oil prices for the rest of us. The U.S. is now bottom of the league in conservation."[32]

Whereas Project Independence had aimed at freeing the United States from OPEC domination by 1980, OPEC's grasp was growing ever stronger when Carter took office. The new President found America's appetite for foreign oil escalating at a frightening rate. He also found a projected 1977 U.S. trade deficit of more than $25-billion brought on by the continuing rise in oil prices, and a serious curtailment of America's freedom of action in the world due to the fact that both oil and petrodollars had become potential economic and political weapons. In early August 1977, former Secretary of State Henry A. Kissinger not only warned of the oil weapon but added, "In another Middle East crisis the vast accumulated petrodollars could become a weapon against the world monetary and financial system."[33]

The President's energy goals to be reached by 1985, as stated in "The National Energy Plan," included the following:

- Reduce the annual growth of total energy demand to below 2 per cent;
- Reduce gasoline consumption 10 per cent below its current level;
- Reduce oil imports from a potential level of 16 million barrels per day to 6 million, roughly one-eighth of total energy consumption;
- Establish a Strategic Petroleum Reserve of 1 billion barrels [about a 10-month supply];
- Increase coal production by two-thirds, to more than 1 billion tons per year,
- Bring 90 per cent of existing American homes and all new buildings up to minimum energy efficiency standards; and

---

30 See, for instance, "Can OPEC Be Broken up?" in *Forbes*, February 15, 1977, p. 48 and Christopher D. Stone & Jack McNamara, "How to take on OPEC," *New York Times Magazine*, 12 December 1976, p. 38.

31 *International Economic Report of the President*, January 1977, p. 36.

32 James Poole, "The Next Oil Crisis," *Atlas World Press Review*, July 1976, p. 11.

33 *The Washington Post*, 4 August 1977, p. 2.

U.S. Investment in Energy Production, Distribution and Conservation: 1975-1985

Energy Activity	Amount in Billions of 1975 Dollars*
Oil and gas	275
Electricity	315
Coal	25
Synthetic Fuels, Solar, Other Forms of Energy	10
Railroads	12
Nuclear-Fuel Waste Disposal	7
Natural Gas Distribution	12
Insulation, Conservation	165-240
Total	**821-896**

* Investment in the 10 years to 1985 by utilities, industry, government and people to meet Americans' energy needs if demand continues to climb at 4 per cent a year. Full enactment of President Carter's energy program is estimated to reduce growth of energy demand possibly by as much as 2 per cent, but huge capital-investment costs will be needed anyway.

SOURCE: Federal Energy Administration

- Use solar energy in more than 2½ million homes.

To accomplish these goals, the President recommended such measures as:

- Higher federal taxes on gasoline, with conservation targets determining the amount (the closer consumption to targets the lower the amount of additional tax);
- Penalties on "gas guzzling" cars;
- Government purchasing programs emphasizing efficient cars and other energy-saving programs;
- Tax incentives to encouraging switching from oil and natural gas to coal;
- Higher prices for domestic American oil and natural gas;
- Tax credits for home insulation and solar-heating equipment.

In May, the Federal Energy Administration estimated that electric utilities, gas and oil firms, coal producers and developers of new forms of energy would invest at least $600-billion during the coming 10 years to meet America's energy needs. Another $200-billion would be spent on insulating homes and redesigning cars, appliances, and other energy-saving equipment. The total was expected to be approximately $1-trillion from all of the programs and expansions the United States would require during the period 1975 through 1985.

Looking Ahead

President Carter signed the bill creating the new Cabinet-level Department of Energy on August 4, 1977. James R. Schlesinger became the new Cabinet officer—he had been the main architect behind the President's energy policies during the first half of 1977. The new department was designed to coordinate and centralize all government energy policies—something not possible before when numerous agencies were responsible for various aspects of the policy.

Nevertheless, the President was clearly anxious in August that the plan he had outlined only four months earlier would not be sufficient to accomplish the goals he had set. Even though Congress was in the process of approving most of the Carter plan, the President was hinting greater efforts might be required in the near future.

"For the first time in our history, a small group of nations controlling a scarce resource could over time be tempted to pressure us into foreign policy decisions not dictated by our national interest."

Henry A. Kissinger
August 1977

In an interview on July 30, the President stated, "I am concerned that the public has not responded well [to calls for conservation], and I think voluntary compliance is probably not adequate at all. The public is not paying attention" to the energy crisis and is wasting fuel at an increasing rate, the President stressed. He went on to speculate that it may take "a series of crises" for the American people to "quit wasting so much fuel." And he

further warned that the oil embargo of 1973 and the natural gas shortage of the winter of 1976 may just be "predictions of what is to come"[34]

Supporting the President's conclusions was a 291-page report by 35 industrial, government and academic experts from 15 countries—a study which took 2½ years to complete.[35] The report concluded that by the year 2000 oil production in the non-Communist world could fall short of demand by as much as 15 to 20 million barrels per day (about the amount being used by the U.S. alone in 1977). Moreover, the shortfall would occur, the report concluded, even if coal production doubled, nuclear generated power multiplied 15 times, conservation measures cut petroleum demand somewhat and the price of oil continued to rise cutting demand somewhat further.

The key to when the shortage would actually occur, the report concluded, was the single country of Saudi Arabia which alone has about one-fourth of the world's known oil reserves. If the Saudis only maintain production at about 9 million barrels per day, the shortage could begin to occur as early as 1981—if the Saudis doubled production, about 1989. But regardless of Saudi production decisions, the shortage would occur sometime in the 1990s. In its discouraging conclusion, this report was supported by the April 1977 CIA report[36] President Carter referred to when announcing his "National Energy Plan."

---

34 *The Washington Post*, 31 July 1977.
35 "Energy Global Prospects 1985-2000," Workshop on Alternative Energy Strategy (a group of international experts organized by MIT Professor Caroll Wilson). See *Time Magazine*, 23 May 1977, p. 63.
36 See footnote 25.

APPENDIX

CAST OF CHARACTERS IN THE MIDDLE EAST DRAMA

Following are biographical sketches of some of the leading 20th century political personages in the Arab world, Iran and Israel. The list also contains several persons from countries outside the Middle East, but who exerted influence on Middle East events.

Abboud, Ibrahim (born 1900). Sudanese army officer and politician. Advanced to lieutenant general. In 1958 seized power in military coup, dissolved Parliament, nullified constitution; became president of Sudan Supreme Council. Resigned in 1965 after popular uprising.

Abd-ul-Ilah (1913-1958). Regent of Iraq (1939-53); then crown prince. Educated at Victoria College, Alexandria. Became regent when his cousin and nephew, King Faisal II, succeeded to throne at age 3. Known for loyalty to boy-king, opposition to violent nationalism, support for cooperation with West. Relinquished powers to Faisal II when he reached majority in 1953. Assassinated with the king in Baghdad uprising July 14, 1958.

Abdullah ibn Hussein (1882-1951). Emir of Transjordan (1921-46); Hashemite king of Jordan (1946-51). Born in Mecca, second son of sharif Hussein (later king of Hejaz). Educated in Turkey. Played outstanding role preparing Arab revolt against Turkey in World War I. In 1920, boldly occupied Transjordan; recognized as emir by British, who held mandate. Established Transjordan as entity separate from Palestine, extracting pledge from British that Jews would not settle in his emirate. His unachieved ambition to unite Transjordan, Syria and Iraq into a Hashemite bloc embittered Saudi Arabian dynasty.

In World War II, sent his army, the Arab Legion, to assist British troops in Iraq and Syria. In 1946, rewarded with independence by British, renamed country Jordan and became king. After Palestine partition, in 1948 Arab-Israeli war, Abdullah's army captured Old Jerusalem and held central Palestine for Arabs. When territories annexed to Jordan, Abdullah drew wrath of Egypt, Saudi Arabia, Syria, which sought an Arab Palestine. Jordan, more impoverished by arrival of Palestinian refugees; Abdullah's enemies multiplied, and he was assassinated in Aqsa Mosque, Jerusalem, July 20, 1951, by young Palestinian Arab.

Abdullah al-Salem al-Sabah (1895-1965). Emir of Kuwait in 1961 at time British withdrew protection over emirate and recognized Kuwait's independence. When Iraq threatened to make Kuwait a province, Abdullah sought and got British aid, which deferred Iraq action. Using vast oil wealth, Abdullah modernized country, shared riches with people to give Kuwait one of world's highest living standards.

Allenby, Edmund Henry, 1st Viscount (1861-1936). British field marshal. His World War I campaigns in Palestine, Syria broke centuries-old Turkish rule over Arabs. While Allenby was high commissioner for Egypt (1919-25), Egypt was recognized as sovereign state (1922).

Amitay, Morris (born 1936). Executive Director of American Israel Public Affairs Committee (AIPAC) popularly known as the Israeli or the Jewish lobby in Washington. Contributing Editor to the weekly newsletter *Near East Report*. Born in New York City. Graduated from Columbia University and Harvard Law School. State Department foreign service officer (1962-1969). Congressional assistant before going to AIPAC in 1975. Home in Rockville, Maryland, bombed in July 1977.

Arafat, Yasir (born 1929). Native of Palestine, became head of Palestine Liberation Organization, alliance of Palestinian organizations. Trained as guerrilla fighter, he associated with Al Fatah, militantly nationalist Palestine organization, in 1950. As PLO chief, represents the more than three million Palestinians in Israel, occupied territories and in Arab states surrounding Israel. PLO has become widely recognized since 1974. Addressed United Nations General Assembly, 1974, calling for Palestinian state of Moslems, Christians, Jews. PLO became full member of Arab League in 1976 and in 1977 a member of the Economic Commission for West Asia of ECOSOC, the first non-state ever to attain membership in a U.N. body.

'Arif, Abdul Rahman (born 1916). Baath Party leader; president, premier of Iraq. Became president when younger brother, President Abdul Salem 'Arif, was killed in helicopter crash April 13, 1966. Abdul Rahman 'Arif made attempts to end Kurdish revolt in northeast Iraq. When 1967 Arab-Israeli war broke out, sent troops to Sinai, Jordan; cut off oil supplies to West; severed diplomatic relations with United States, Britain, West Germany. Arab defeat, Kurdish troubles, economic problems led to his overthrow in bloodless coup July 17, 1968.

'Arif, Abdul Salem (died 1966). Headed Baathist army coup overthrowing Iraq dictator Abdul Karim Kassim, who was executed, and making 'Arif Iraqi president. Improved relations with oil companies; dropped Iraqi claims to Kuwait. Killed in helicopter crash near Basra April 13, 1966.

al-Assad, Hafez, Lt. Gen. (born 1928). President of Syria since 1971. Became defense minister in 1965; headed "Nationalist" faction of Baath Party, which favored less strong ties to Soviet Union and more moderate Marxist economy. After unsuccessful February 1969 coup, led successful coup in November 1970, assumed presidency March 1971. Considered moderate by Syrian standards. Improved relations with Saudi Arabia and other conservative Arab states. A bitter foe of Israel, launched war against Jewish state in Golan Heights in October 1973, but entered into troop disengagement with Israel in 1974. Accepted massive Soviet military aid, Soviet advisers in Syria. Took Syria into the Lebanese conflict in mid-1976 and since then has exerted great influence over Lebanese developments and over the PLO.

al-Atasi, Hashim. Became president of Syria after August 1949 coup; ousted in another coup in December 1949. Further army revolt in February 1954 led to restoration of constitutional rule and al-Atasi to presidency; after elections, was succeeded by al-Kuwatly as president in 1955.

al-Atasi, Louai (born 1926). Syrian army officer, government official. Educated at Syrian Military Academy. In March 1963 led pro-Nasser military faction which seized Syrian government. As president of Revolutionary Council, helped establish Baath Party predominance, ending its 20-year clandestine existence.

al-Attassi, Nurredin (born 1929). Syrian medical doctor and government official. Led "Progressive" faction of Baath Party, favoring strong ties to Soviet Union and strong Marxist economy. Became president in 1966; deposed by Hafez al-Assad in November 1970 bloodless coup. *Avneri, Shlomo — negot w PLO p 18*

al-Azhari, Ismail (born 1900). Sudanese government official. Educated at Gordon College, Khartoum, and American University, Beirut. First prime minister of Sudan (1954-56). President of Supreme Council (1965-69) until overthrow by military coup led by Gen. Gafarr al-Nimeiry.

al-Bakr, Ahmed Hassan, Maj. Gen. (born 1912). President of Iraq since 1968. Seized power in bloodless coup July 17, 1968,

Avnery, Uri — Semitic union of A & J states p 120

assuming presidency and premiership. Under his regime, Iraq sought to end Kurd revolt by granting measure of autonomy; but after Iraq and Iran, which armed Kurds, settled differences, Bakr government militarily quashed revolt in March 1975. In June 1973, regime defeated an attempted military coup, resulting in execution of 36 officers. In April 1972, a 15-year friendship treaty with Soviet Union was signed; in June 1972 Iraq Petroleum Company was nationalized.

Balfour, Arthur James, First Earl (1848-1930). British government official. As foreign minister, decided in favor of Zionist aspirations for creating a Jewish national state in Palestine, embodying it in Balfour Declaration of November 1917.

Barzani, Mustapha Mulla. General and leader of Kurdish revolt against Iraq. Declared war on Baghdad government in 1974 after turning down offer of limited autonomy. Revolt crushed by Iraqi armed forces in March 1975; Barzani fled into exile in Iran.

Begin, Menahem (born 1913). Became Prime Minister of Israel in June 1977. Long-time opposition leader whose surprise election victory cast considerable doubt over what compromises Israel might be willing to make on the Palestinian problem and in regard to the West Bank and Gaza Strip territories. Former leader of the underground, terrorist organization *Irgun Zeva'i Le'umi.*

Ben-Bella, Ahmed (born 1916). Algeria's national hero and first president. Leader of nationalist movement against French. Exiled and imprisoned several times. After Algerian independence in 1962, was elected president Sept. 15, 1963. Pushed Socialist measures. Overthrown in bloodless coup June 19, 1965; in confinement ever since.

Ben-Gurion, David (1886-1973). Zionist leader and Israeli government official. Born in Plonsk, Poland; went to Palestine in 1906 as laborer. Started Labor Party. During World War I was expelled from Palestine by Turks; went to New York, where he formed Zionist Labor Party. Joined Jewish Legion, part of British forces in Palestine. From 1918, lived in Tel Aviv, where he headed Labor Party (Mapai).

Convinced Jews must have state, organized Haganah as fighting force. At Tel Aviv May 14, 1948, read publicly declaration of Israel's independence. In new state, became prime minister and defense minister, holding both posts until 1963 except for one interlude. His main aim was to bring in all Jews to Israel willing to come; nearly one million immigrants came in first decade.

Sent troops into Suez Canal conflict of 1956. Resigned as premier in June 1963 but remained member of Knesset until 1970.

Ben-Zvi, Isaac (1884-1963). Second president of Israel (1952-63). Born in Ukraine. Studied at Kiev University, Imperial Ottoman University, Istanbul. Went to Palestine in 1907, helped found Hashomer (Jewish self-defense organization). After exile by Turks in 1915, went to New York where, with Ben-Gurion, established Hehalutz (Pioneer) movement and Jewish Legion. Founder and chairman of Vaad Leumi (National Council of Palestine Jews). Signed Israeli declaration of independence. Elected to Knesset in 1949 and president Dec. 8, 1952, upon Chaim Weizmann's death.

Bernadotte, Folke, Count (1895-1948). Swedish humanitarian appointed U.N. mediator between Arabs and Jews at time of Palestine partition. Assassinated by Jewish extremists in Jerusalem Sept. 17, 1948.

el-Bital, Salah (born 1912). Syrian politician. Helped create Socialist Baath Party, which became Syria's ruling party, and short-lived United Arab Republic (Syria and Egypt). Led pro-Nasser coup in Syria which on March 8, 1963, resulted in Bital becoming premier, holding that post at various times until 1966.

Boumédienne, Houri (born 1925). Second president of Algeria, since 1965. Educated in Arabic studies at universities in Tunis and Cairo. Fought against French in war for Algerian independence; became chief of National Liberation Army in 1960. After independence, became defense minister under Ben Bella in 1962, vice premier in 1963 and president in July 1965, when Ben Bella was

ousted and jailed. Followed policy of democratic centralism; set up elections in which 700 communes elected councils of workers and militants.

Bourguiba, Habib Ben Ali (born 1903). First president of Tunisia, since 1957. Educated at Sadiqiya College, Tunis; law degree, University of Paris. Founded nationalist Neo-Destour Party in 1934; jailed and exiled several times by French. After France granted Tunisia independence, was elected president in 1957; re-elected in 1959, 1964, 1969. Favored social reform at home, association with West in foreign affairs. Incurred enmity of military Arab states when he urged moderate approach to Israel.

Brzezinski, Zbigniew (born 1928). Appointed by President Carter to head the National Security Council in 1977. Former professor at Columbia University and director of the Trilateral Commission.

Bunche, Ralph J. (1904-71). American official at United Nations. Negotiated 1949 armistice between Israel and Arab states, receiving Nobel Peace Prize for it. Directed U.N. peacekeeping efforts in Suez in 1956.

Carter, James Earl (born 1924). Became 39th President of the United States in January, 1977, and within months began discussion of a comprehensive Middle East settlement to include Israeli withdrawal to approximately the 1967 boundaries, creation of a "Palestinian homeland," and establishment of real peace and normal relations between Israel and her Arab neighbors. Former Governor of Georgia, nuclear submarine officer and peanut farmer.

Chamoun, Camille (born 1900). President of Lebanon (1952-58). Educated at College des Frères and Law School, Beirut. After election to presidency, his pro-Western policies led to open Moslem revolt, aggravated by radical Arab neighbors. At his request, United States sent Marines into Lebanon in July 1958 to help restore order. In July 1975, became defense minister in "rescue cabinet" formed to end bloody Moslem-Christian clashes over Palestinian refugee issue in Lebanon.

Chehab, Fuad (born 1903). President of Lebanon (1958-64). Educated at Damascus Military School, St. Cyr Military Academy, France. Served as commander-in-chief, Lebanese army; prime minister, interior minister, defense minister. As president, pursued neutralist policy acceptable to Arabs and West. Put down attempt to overthrow government by Syrian Popular Party, which sought Lebanese union with Syria.

Dayan, Moshe (born 1915). Israeli soldier, politician. Hero of Young Israel during building of Jewish state. Learned guerrilla warfare as member of "night squadrons" fighting Arab rebel bands in Palestine. A leader of Haganah, volunteer defense force, was jailed by British in 1939; released to serve in British army in World War II. In Syrian campaign against Vichy French, lost eye from sniper's bullet. Chief of staff for all Israeli forces (1953-58); prepared plans for invasion of 1956 Sinai war. Elected to Knesset; became defense minister and executed battle plans which defeated Egypt, Syria, Jordan in 1967 Six-day war. Quit cabinet in 1974 after criticism mounted over initial Arab victories in 1973 war. In 1977 became foreign minister to the new Begin government drawing much criticism from his Labor party from which he resigned while keeping the seat in Parliament he won on the Labor list.

De Gaulle, Charles (1890-1970). President of France (1958-69). Brought to power by Fourth Republic's inability to solve Algerian crisis; created Fifth Republic; became its first president; took steps which led to Algeria's independence in 1962; dismantled bloodlessly rest of French empire in Africa, where some former colonies became Islamic states.

Dulles, John Foster (1888-1959). U.S. secretary of state (1952-59). Architect of Eisenhower administration Middle East policy, which turned United States into a major Middle East power broker after departure of French and British left vacuum in area. Held that without U.S. economic and military assistance, Middle East states could not maintain independence and would come under Soviet hegemony.

Eban, Abba (born 1915). Educated at Cambridge University, England. Deputy prime minister of Israel (1963-66); foreign minister (1966-74). Worked to maintain strong U.S.-Israeli ties; architect of several Middle East peace plans. In August 1977 traveled U.S. discussing Israel's positions at behest of Prime Minister Begin. Earlier in year investigated by Israeli authorities for possible tax violations, but no violations found.

Eisenhower, Dwight D. (1890-1969). 34th President of the United States (1953-61). Under his leadership, United States entered Middle East affairs with economic and military aid on large scale. On request from Lebanon government, sent Marines into that country in 1958 to maintain order when Lebanon was threatened by internal revolt and radical Arab states, an application of the Eisenhower Doctrine aimed at maintaining independence of Middle East states and preventing Soviet dominance of area.

Eshkol, Levi (1895-1969). Finance minister of Israel when Ben-Gurion resigned premiership in 1963; succeeded him as prime minister (1963-69). Under his leadership, Israel crushed Arab states in Six-day war of 1967.

Fahd ibn Abdul Aziz (born 1922). Saudi Arabian prince and government official. Served as education minister, interior minister. Raised to crown prince by Saudi royal family after his half-brother, King Faisal, was assassinated in March 1975. Expected to be real power behind new King Khalid.

Faisal ibn Abdul Aziz al-Saud (1906-75). King of Saudi Arabia (1964-1975). Became crown prince when brother, King Saud, ascended throne in 1953. Served as prime minister, foreign minister, defense minister, finance minister. Became king in March 1964 when Saud was legally deposed. Supported pan-Arab and pan-Islam solidarity. Pressed for economic and educational advances domestically.

A Moslem ascetic, called for Israel to evacuate Islamic holy places in Jerusalem and all occupied Arab territory. An anti-Communist, fostered ties in United States and supported conservative Arab regimes. However, during 1973 Arab-Israeli war, enforced oil embargo and oil-price hike against United States, Western Europe and Japan. Assassinated by a nephew March 25, 1975, at palace in Riyadh.

Faisal I (Faisal ibn Hussein) (1885-1933). King of Iraq (1921-33). Horrified by Turkish anti-Arab actions, was a leader of Arab revolt, assisted by British and Lawrence of Arabia, against Ottomans in World War I. In effort to consolidate Arab state in Syria, was king of Syria briefly in 1920 until expelled by French who held Syrian mandate. With British help, elected to second throne in Baghdad, 1921. Identified himself with strengthening and liberating his kingdom.

Faisal II (1936-58). King of Iraq (1939-58). Inherited throne at age 3 upon accidental death of King Ghazi. Crowned May 2, 1953. During five-year reign, Iraq pursued anti-Communist course, culminating in Baghdad Pact (Britain, Iran, Iraq, Pakistan, Turkey) in 1955. Pact aimed at thwarting possible Soviet intrusion into Middle East. Assassinated with most members of royal family in Baghdad revolution July 14, 1958, which resulted in Iraq being declared a republic.

Faisal ibn Musaed. Saudi Arabian prince. At age 31, killed his uncle King Faisal, at royal palace, Riyadh, March 25, 1975. Tried by religious court and beheaded in Riyadh public square June 18, 1975.

Farouk I (1920-65). King of Egypt (1936-52). Educated in Egypt and England. Reign marked by quarrel with dominant Wafd Party and with British over fate of the Sudan. Standing was damaged by disastrous campaign against Israel in 1948 and corruption connected with arms purchases. Military coup of July 1952 forced abdication. Went into exile in Italy, where he died in Rome March 18, 1965.

Franjieh, Suleiman (born 1910). President of Lebanon since 1970. In crisis of 1975 between rightist Christians and leftist Moslems, had trouble finding formula for a government placating all sides. Refused to leave presidency, even after petition from Parliament and military assault on his residence, until his official term of office expired in September 1976. Followed as president by Elias Sarkis.

Fuad I (1868-1936). King of Egypt (1922-36). Proclaimed king when Britain relinquished protectorate over Egypt in 1922. Reign was marked by struggle between Wafd Party and palace parties centering around king. Interested in charitable and education matters, prime mover in establishing first Egyptian university of Western type, the Fuad I (now Cairo) University, in 1925.

Gemayel, Pierre (born 1905). Leader of Phalangist party in Lebanon. Pharmacist educated in Beirut and France. Member of Parliament since 1960 who held office in most Lebanese governments. Ran for presidency in 1970 but withdrew in favor of neutral candidate Suleiman Franjieh.

Ghazi (died 1939). King of Iraq (1933-39). Pleasure-loving monarch; reign considered inglorious. However, oil production began in 1934; revenues enabled progress in industry, irrigation, communications under series of able government ministers. Died in road accident.

Habash, George (born 1925). Palestinian leader of "rejection front" which refuses to consider the possibility of coexistence with Israel. In 1970 his Popular Front for the Liberation of Palestine (PFLP) became known for its daring and brutal hijacking of foreign planes. PFLP held responsible for triggering the Jordanian civil war in 1970 and for helping spark the Lebanese civil war in 1975.

Hafez, Amin (born 1911). Syrian army officer, politician. Prime minister (1963) and head of state (1963-66); ousted in coup by pro-Nasser and Baath Party followers.

Hassan II (born 1930). King of Morocco since 1961. Absolute monarch; dissolved National Assembly in 1965. Steered Morocco clear of radical Arab policies; cultivated friendship with United States and Western Europe. Rule limited by 1970 constitution. Survived coups in 1971, 1972.

Helou, Charles (born 1911). Lebanese journalist and government official. Educated at St. Joseph College and Ecole Francaise du Droit, Beirut. Was minister of justice and health. President of Lebanon (1964-70). Steered neutralist course between West and neighboring Arab militants.

Hitler, Adolf (1888-1945). German dictator (1933-45). Policy of persecution of Jews in Germany and central Europe led to immigration of more than 200,000 Jews to Palestine before World War II. After war, there was mass migration of Jews to Palestine (Israel after 1948)—many of them survivors of Hitler's death camps in Europe. Because of Hitler's murder of six million Jews, sympathy grew in many countries for creation of Jewish state.

Hoveida, Amir Abbas (born 1919). Iranian government official. Educated at Paris and Brussels universities. Served as general managing director of National Iranian Oil. Prime Minister 1965-1977. Replaced by Jamshid Amuzegar in Aug. 1977.

Hussein ibn Ali (1854-1931). Emir of Mecca (1908-16); king of Hejaz (now Saudi Arabia), (1916-24). During World War I, negotiated with British, leading to Arab revolt against Turkey. Opposed mandatory regimes imposed on Syria, Palestine, Iraq by Versailles Treaty. Kingdom attacked by Ibn Saud of Wahhabi sect; Hussein forced to abdicate in 1924; Ibn Saud later proclaimed king of Saudi Arabia. Hussein exiled to Cyprus. Hussein's son Ali was king of Hejaz briefly; another son, Abdullah, became king of Jordan; a third son became King Faisal I of Iraq.

Hussein Ibn Talal (born 1935). King of Jordan since 1953. Educated at Royal Military Academy, Sandhurst, England. Crowned May 2, 1953, after father, King Talal, was declared mentally ill. Opened kingdom to powerful Egyptian and republican influences. In 1956, abrogated Jordan's treaty with Britain. Accepted U.S. economic aid.

With more than half his subjects Palestinians, supported Nasser in war against Israel in 1967, losing half his kingdom (West

Bank of Jordan River) to Israel. Sought to treat indirectly with Israel to recover lost lands. When Palestinians sought to create state within a state, September 1970 civil war broke out, Hussein crushing Palestinian guerrilla enclaves. At Rabat summit conference of Arab League in 1974, right to negotiate for return of West Bank was taken from Hussein and given to Palestine Liberation Organization; however, it later appeared Arab states were returning to Hussein as a key figure in general peace settlement.

Jumblatt, Kamal (1917-1977). Leader of leftist forces in Lebanese civil war assassinated in March 1977. Strong supporter of Palestinian cause and of reforming Lebanon into a secular state.

Ibn Saud (1880-1953). First king of Saudi Arabia (1932-53). To create kingdom, overcame British policy maintaining status quo in Persian Gulf and made war on King Hussein of Hejaz, forcing him to abdicate and leading to merger of Hejaz and Nejd Kingdoms into Saudi Arabia. Worked to consolidate his realm and improve relations with old enemies in other Arab states. In 1933 granted 60-year oil concession to a U.S. oil company which became known as Aramco. Oil royalties greatly enriched his treasury. In 1940s, launched veritable social and economic revolution in Saudi Arabia. In 1945, helped form Arab League.

Idris I (born 1890). King of Libya (1951-69). As emir of Cyrenaica, fought Italian occupation of Libya; was declared constitutional monarch when Libya was made independent state in 1951. Deposed in coup Sept. 1, 1969, led by Col. Muammar el-Qaddafi, who declared Libya a Socialist republic.

Iryani, Sheikh Qadi Abdul Rahman (born 1911). Chairman of Republican Council (head of state) of Yemen Arab Republic. Came to power in 1967 after years of Yemen internal strife as Egyptian forces withdrew from Yemen; during 1960s, Yemen had been pawn in conflict between leftist Egypt and rightist Saudi Arabia for influence in Red Sea area. Moved to end hostilities with Southern Yemen (People's Democratic Republic of Yemen) in 1972 and merge two states; but new border clashes breaking out in 1973 seemed to make proposed union less likely.

Isa bin Sulman al-Khalifah (born 1933). Emir of Bahrain (head of state). Declared independence of Bahrain after Britain quit Persian Gulf in 1971. Considered, then rejected, union of Bahrain with Qatar and United Arab Emirates.

Jarring, Gunnar (born 1907). Swedish diplomat. Served as special mediator for U.N. Security Council during 1967 Middle East War.

Karami, Rashid (born 1921). Lebanese government official. Educated at Fuad al-Awal University, Cairo. Several times prime minister; last time, in May 1975, called upon to form "rescue cabinet" to restore order to Lebanon after three months of intermittent street fighting between leftist Moslems and rightist Christians in Beirut, April-June 1975.

Kassim, Abdul Karim (died 1963). Iraqi dictator (1958-63). As army general, led military revolution July 14, 1958, which, in one day, saw bloody killing of young King Faisal II and most of royal family. Republic was proclaimed; Kassim was named prime minister. Took Iraq out of Baghdad Pact; improved relations with Soviet Union, Communist China. Attempted assassination of Kassim in October 1959 failed; 13 former officials and army officers were executed.

Suppressed with great severity Kurdish revolt in 1961. Laid claim to Kuwait after its independence in 1961, but British sent troops to Kuwait's aid, deterring Iraqi action. Coup by "free officers" of Baath Party resulted in summary execution of Kassim in 1963.

Katzir, Ephraim (born 1916). Elected by Knesset as president of Israel in 1973. Term as president expires in 1978.

Khalid ibn Abdul Aziz (born 1913). King of Saudi Arabia since March 1975. Educated at religious schools. Represented Saudi Arabia at various international conferences. Became vice president of Council of Ministers, 1962; elevated to crown prince, 1965; succeeded to throne when King Faisal was assassinated.

Khalifa bin Hamad al-Thani (born 1937). Emir of Qatar (head of state). Assumed power after deposing cousin, Emir Ahmad bin Ali bin Abdullah al-Thani, in bloodless coup Feb. 22, 1972. Headed program of social and economic improvements made possible by oil revenues. Joined in oil embargo against West in 1973.

el-Khoury, Bichara (1860-1964). Lebanese government official. In 1943, while Labanon was under Free French, was elected president. Insurrection followed after French arrested el-Khoury and other government officials, leading to restoration of Lebanese government and transfer of powers to it by French.

In 1946, French relinquished Lebanese mandate and country became independent. During el-Khoury presidency, Beirut grew rapidly and became free-money market and trade center, Tripoli and Saida on Mediterranean became terminal points of oil pipelines from Iraq and Arabia. Abuse of power made regime unpopular; el-Khoury was deposed in popular movement in 1952.

Kissinger, Henry A. (born 1923). U.S. Secretary of State (1973-1977). During 1973-75, through on-the-scene diplomatic efforts, mediated troop disengagement agreements between Egypt and Israel in Sinai and between Syria and Israel on Golan Heights. In 1977 began writing his memoirs, lecturing, and teaching at Georgetown University.

Kuwatly, Shukri (1891-1967). Syrian government official. Opposed to French mandate in Syria, emerged as nationalist leader in 1920s and 1930s. While Syria was under Free French, was elected president in 1943; secured French withdrawal and Syrian independence in 1946. Regime discredited by failure of economic policy and of Arab policy in Palestine. Overthrown in 1949 coup, went into exile; returned to Syria and public life in 1954. Advocated broad Arab union led by Egypt; elected president again in 1955, serving until 1958, when United Arab Republic of Egypt and Syria was inaugurated.

Lawrence, Thomas Edward (1888-1935), known as Lawrence of Arabia. British scholar, soldier, author. Injected new life into Arab revolt against Turks in World War I; and, dressed in Arab garb, fought successful battles (1916-18) with Arab army led by Emir Faisal. In conjunction with British forces. Allies captured Jerusalem in December 1917, and Damascus in October 1918. As member of British delegation to Paris Peace Conference, opposed separation of Syria and Lebanon from rest of Arab states into French mandate. Wrote *Seven Pillars of Wisdom*, an account of his experiences in desert. Died of injuries received in motorcycle accident in England in May 1935.

Mansour, Hassan Ali (died 1965). Helped form New Iran Party from centrist groups in 1963; became prime minister 1963. Died from gunshot wounds inflicted by a student Jan. 21, 1965.

Meir, Golda (born 1898). Israeli government official. Native of Kiev, Russia, was brought to United States in 1906; immigrated to Palestine in 1921. Active in labor movement, World Zionist Organization and in Hagannah struggle to set up Jewish state. After Israeli independence, was ambassador to Soviet Union, minister of labor, minister of foreign affairs, prime minister (1969-75).

Maintained inflexible policy vis-a-vis Arab states; during 1973 war, led country as Arabs won initial success, later reversed by Israeli counter-offensives. However, Labor Party lost seats in parliamentary elections Dec. 31, 1974, and she was unable to form new government after several tries as criticism mounted over initial Israeli reverses in 1973 war. Relinquished premiership in March 1975.

Mohammed Reza Pahlavi, (born 1919). Shah of Iran. Educated in Iran, Switzerland. Became Shah in 1941 upon abdication of father, Reza Shah Pahlavi. Considered strong leader, has used money from oil resource for achievements in education, communications, social welfare, public health. Has pursued pro-Western, anti-Communist policy. Has sought to make Iran dominant power in Persian Gulf.

Mossadegh, Mohammed (1880-1967). Iranian government official. As prime minister, was largely responsible for act nationalizing Anglo-Iranian Oil Company in 1951. Efforts to obtain what amounted to dictatorial powers led to strained relations with shah. Overthrown by military coup in April 1953; later sentenced to three-year prison term for treason.

Naguib, Mohammed (born 1901). Egyptian soldier, government official. Became president of Egypt when it was declared republic in 1953. After power struggle with Nasser, was deprived of presidency in 1954.

Nasser, Gamal Abdel (1918-70). Egyptian soldier, government official. Educated at Military College, Cairo. As head of Revolution Command Council, made up of 11 army officers, led revolt which resulted in deposing King Farouk and establishing republic. After power struggle with President Mohammed Naguib, became president in 1956.

Negotiated withdrawal of British troops from Suez Canal Zone in 1954 and nationalized canal in 1956. Action resulted in Anglo-French-Israeli military intervention; after United States forced Allied withdrawal, Nasser emerged with unparalleled prestige in Arab world. Created 1958 United Arab Republic (Egypt, Syria) and was president of it until Syria seceded in 1961.

Avidly opposed Israel, which defeated Egypt in 1967 war. Defeat led to detente between Nasser and conservative Arab states, which had distrusted his revolutionary objectives. Received revenues from Saudi Arabia, Kuwait and arms from Soviet Union. At home, pursued course of social justice, redistribution of land, improved medical care, education. Construction of Aswan Dam was made a symbol of his achievements. After September 1970 conference in Cairo that ended Jordanian civil war, died of heart attack.

al-Nimeiry, Gaafer (born 1929). Sudanese army general, government official. Seized power in coup in May 1969; became first elected president in 1971. Has put down several coups, summarily executing at least 14 persons accused of plotting takeover.

Pasha, Nuri (Nuri as-Said) (1888-1958). Twelve times prime minister of Iraq. Known for suppression of communism. Formed defensive alliance with Turkey in 1955 which, when joined by Iran, Pakistan, Britain, became Baghdad Pact, aimed at possible Soviet aggression in Middle East. Killed along with King Faisal II in Baghdad uprising July 14, 1958.

Peres, Shimon (born 1923). Leading figure in Israeli Labor party associated with the small group known as "Rafi" built by Ben-Gurion, Peres, and Moshe Dayan. Became acting prime minister in April 1977 when Yitzhak Rabin resigned due to scandal. Considered a hard-liner within Labor party. Considered leader of opposition since the defeat of Labor in the May 1977 election in which Peres was Labor's choice for Prime Minister.

Qabus bin Said (born 1942). Sultan and absolute ruler of Oman. Overthrew father, Sultan Said bin Taimur, in 1970, known as one of century's most tyrannical despots. Qabus known for limited attempts to modernize country, which is isolated, inaccessible, feudal.

Qaddafi, Muammar (born 1942). Chairman of Revolutionary Command Council of Libya. Regarded as heir to radical mantle of Egypt's late President Nasser. Educated at University of Libya and Libyan Military Academy. In 1969, led coup which overthrew King Idris. Evicted United States, Britain from Libyan military bases. Nationalized oil industry. Unable to forge Arab unity through merger of Libya with Egypt, then with Tunisia, which backed out of union in fear of Qaddafi's radicalism. Known for implacable enmity to continued existence of Israel. Strong supporter of Palestinian "rejection front" and other radical Arab movements. Considered dangerous "madman" by many moderate and conservative Arab leaders. In July 1977, President Sadat of Egypt vowed to "teach Qaddafi a lesson he will never forget" when a military confrontation between the two countries began on their common border.

Rabin, Yitzhak (born 1922). Israeli soldier, government official. Was army chief of staff responsible for planning overwhelming victory in 1967 war. Served as ambassador to Washington.

Became prime minister in 1974 when Golda Meir retired. Held Israel to hard line in peace maneuvering of 1974, early 1975. Resigned as Prime Minister in April 1977 just one month before the Israeli election after a scandal involving an illegal bank account held by his wife in Washington

Rashid ben Saeed. Emir of Dubai, vice president of federation of United Arab Emirates. Used oil revenues to modernize Dubai, improve port, build international airport.

Reza Shah Pahlavi (1878-1944). Shah of Iran (1925-41). Gained throne after coup deposing Ahmad Shah. Autocratic ruler who ignored constitutional safeguards. Built Trans-Iranian Railway, developed road system. Sought machinery and technicians from Germany, which led to World War II occupation of Iran by British, Soviet troops. Abdicated in 1941 in favor of son, Mohammed Reza Pahlavi.

Rogers, William (born 1913). American Secretary of State (1969-1973) whose name is associated with the 1969-1970 "Rogers Plan" which aimed to bring about a Middle East settlement based on Israeli withdrawal from occupied territories in exchange for peace. Attorney General (1957-1961)noted for his strong stand on civil rights. After leaving government returned to the private practice of law.

Rubaya, Salem Ali (born 1934). Chairman of Presidential Council (head of state) of Southern Yemen (People's Democratic Republic of Yemen). Rules country, former British colony of Aden, according to principles of Koran and "scientific socialism."

Sabah al-Salem al-Sabah (born 1913). Emir of Kuwait since 1965. Served as police chief, public hygiene minister, foreign minister. As emir, continued modernization program, using Kuwait's vast oil wealth.

Sadat, Anwar (born 1918.) Educated at Military College, Cairo. Was deputy to Nasser in organizing secret revolutionary brotherhood which overthrew monarchy. Was speaker of National Assembly, twice vice president. Became president when Nasser died in 1970. Sadat surprised world by becoming strong leader. In 1972, ordered 20,000 Soviet military advisers out of Egypt. Became Arab world hero in 1973 war when Egypt won initial victories over Israelis. Agreed to troop disengagement accords in Sinai, first reliance on United States to mediate general Middle East peace agreement. Moved country away from radical socialism of Nasser to attract Western capital and alleviate grave economic problems. In June 1975, reopened Suez Canal after eight-year closure.

as-Said, Nuri *(see Pasha, Nuri)*.

Salam, Saeb (born 1905). Lebanese government official. Foreign affairs minister, defense minister. Prime minister (1952, 1953, 1960-61, 1973). Resigned last time during crisis over Israeli raids on Palestinian commando groups in Beirut, Saida.

Saud IV (1902-69). King of Saudi Arabia (1953-64), succeeding his father, King Ibn Saud. Expanded father's modernization program, with emphasis on educational and medical services. In foreign affairs, continued friendship with United States and all Arabs, suspicion of communism and firm opposition to Israel. Abdicated in 1965 after royal family transferred powers to brother, who became King Faisal.

Saud ibn Faisal (born 1940). Foreign Minister of Saudi Arabia said to be a possible Saudi king in the future.

Saunders, Harold (born 1930). In 1977 serving as Director of the State Department's Bureau of Intelligence and Research. Formerly member of the National Security Council and Deputy Assistant Secretary of State dealing with Middle Eastern affairs. While in this position he testified before Congress in November 1975, that "In many ways, the Palestinian dimension of the Arab-Israeli conflict is the heart of that conflict." This statement became known as "The Saunders Statement."

Sarkis, Elias (born 1924). Elected by Lebanese parliament on May 8, 1976, to be president. But assumed office only on September

23rd when President Franjieh stepped down at the legal end of his term. Supported by Syrians over the candidacy of Raymond Edde.

Shah of Iran *(See Mohammed Reza Pahlavi).*

Sharett, Moshe (1894-1965). Headed foreign policy department, Jewish Agency for Palestine; a founder of Israel. Prime minister (1953-55).

Shazar, Zalman (1889-1974). Born in Mir, Russia. Educated at Academy of Jewish Studies, St. Petersburg, Universities of Freiburg and Strasbourg. Long a leader in World Zionist movement. First education minister of Israel; president (1963-73).

al-Shishakli, Adib. Became president of Syria in 1949 after coup. Known as harsh leader; lost power in 1954 after Syria returned to democratic regime.

al-Solh, Rachid (born 1909). Lebanese government official. Educated American University and St. Joseph's College, Beirut. Became premier in 1973; resigned in 1975 after criticism of his handling of bloody Christian-Moslem riots put Lebanon on brink of civil war.

al-Tal, Wasfi. (1920-71). Jordanian government official. Educated at American University, Beirut. Thrice prime minister. Assassinated in Cairo in November 1974 by Black September, Palestinian terrorist organization, in reprisal for crushing Palestinian strongholds in Jordan.

Talal (1909-72). King of Jordan (1951-52). Jordanian Parliament declared him mentally ill, deposing him in 1952. Spent rest of life in mental institution in Turkey. Succeeded by his son, King Hussein.

Truman, Harry S (1884-1973). 33rd President of United States (1945-1953). After World War II, prevailed upon British to loosen restrictions on immigration of Jews to Palestine; worked in United Nations for creation of Israel; recognized Israel as state 11 minutes after its creation in May 1948.

Vance, Cyrus R. (born 1917). Attorney and international affairs specialist who was appointed by President Jimmy Carter Secretary of State. Vance's background in the Middle East was minimal but the month after taking office he made his first visit to the area. Then on July 31st he returned to the Middle East for a 12-day visit which took him to all of the confrontation countries plus Saudi Arabia and included indirect contacts with the PLO.

Weizmann, Chaim (1874-1952). Born in Russia, educated in Germany. Lecturer in chemistry at University of Manchester, England. Headed British Admiralty Laboratories that created synthetic acetene for explosives in World War I. Rewarded with Balfour Declaration, British white paper which stated national homeland for Jews should be created in Palestine. President of World Zionist Organization (1920-31). In 1947, headed Jewish Agency delegation to United Nations on Palestine question. Elected first president of Israel (1948-52).

Yamani, Sheikh Ahmed Zaki (born 1930). Educated at University of Cairo, New York University, Harvard. Minister of petroleum and natural resources for Saudi Arabia. Leader in formulating participation agreements with oil companies in Arab states, in devising oil embargo of 1973-74 against West and in raising oil prices in councils of Organization of Petroleum Exporting Countries (OPEC).

Zayed ibn Sultan al-Nahayan. Emir of Abu Dhabi, president of federation of United Arab Emirates. Used vast oil wealth to modernize his sheikhdom and provide social programs for its people.

MIDDLE EAST DEVELOPMENTS BETWEEN 1945 AND 1977

1945

March 22. Arab League founded in Cairo. Egypt, Iraq, Lebanon, Syria, Saudi Arabia and Transjordan are members.

Aug. 16. President Harry S Truman calls for free settlement of Palestine by Jews to the point consistent with the maintenance of civil peace.

Nov. 13. Truman in Washington and British Foreign Secretary Ernest Bevin in London announce agreement on creation of a commission to examine the problem of European Jews and Palestine; Bevin says Palestine will become a United Nations trusteeship eventually to have self-government.

Nov. 22. Palestine Arab leaders form a 12-man Higher Committee to present their views to the Arab League.

Dec. 10. Truman appoints six-man committee to represent the United States on a joint Anglo-American Committee of Inquiry to study the question of Jewish immigration into Palestine.

Dec. 12. Senate Foreign Relations Committee approves, 17-1, a resolution (S Con Res 49) urging U.S. aid in opening Palestine to Jews and in building a "democratic commonwealth." The resolution is adopted by the Senate Dec. 17 and the House Dec. 19.

1946

Jan. 7. Anglo-American Committee of Inquiry holds opening Washington session.

Jan. 30. Britain announces it will permit 1,500 Jews to enter Palestine each month during the inquiry by the Anglo-American committee. (This monthly quota was in addition to the 75,000 permitted by a 1939 British government White Paper.)

Feb. 2. Arab strike in Palestine; Arab Higher Committee protests against British decision to permit additional Jews to enter Palestine.

Feb. 25. Soviets tell Iran they will retain some troops in Iran after the March 2 deadline for foreign troop withdrawal set by the 1942 Anglo-Soviet-Iranian Treaty.

March 2. Arab League spokesman tells the Anglo-American committee the league will oppose creation of a Jewish state in Palestine; it calls for creation of an Arab state in Palestine.

March 5. United States protests Soviet retention of troops in Iran.

March 22. Britain and Transjordan sign a treaty ending British mandate.

April 5. Soviet Union and Iran reach agreement on Soviet troop withdrawal by May 6.

April 30. Anglo-American Committee of Inquiry report is released with recommendations for the immediate admission of 100,000 Jews into Palestine and continuation of the British mandate until establishment of a U.N. trusteeship.

May 2. Arab League protests the report by the Anglo-American Committee and the Arab Higher Committee warns Britain that the "national struggle" will be resumed if the committee's recommendations are adopted.

July 25. Anglo-American committee in London proposes tripartite partition of Palestine into Jewish, Arab and British-controlled districts.

Aug. 7. Britain announces plans for a blockade of ships carrying Jewish immigrants to Palestine and diversion to Cyprus of ships of immigrants in excess of the 1,500 monthly quota.

Sept. 10. London conference on Palestine opens with Arabs present but Jews boycott.

Sept. 20. Committee appointed to study Arab proposals for Palestine; Jews agree to end boycott of London conference.

Oct. 4. Truman releases appeal sent to Britain for "substantial immigration" into Palestine "at once" and expressing support for the Zionist plan for creation of a "viable Jewish state" in part of Palestine; British government expresses regret that Truman's statement was released because it might jeopardize a settlement.

1947

Jan. 27. London conference on Palestine reconvenes with Arab delegates and Zionist observers.

Feb. 14. London conference closes without agreement on a plan for Palestine; conference informed of Britain's decision to refer Palestine question to the United Nations.

April 28. A special session of the U.N. General Assembly to study the Palestine question opens. A Political and Security Committee of the U.N. General Assembly votes May 13 to establish an inquiry committee.

May 14. Andrei A. Gromyko, Soviet delegate to the United Nations, proposes that Palestine be divided into two independent states, if the Arabs and Jews cannot agree on one Arab-Jewish state.

June 16. U.N. Palestine inquiry committee opens hearings.

Aug. 31. U.N. committee issues majority report recommending Palestine be divided into two separate Arab and Jewish states by Sept. 1, 1949, with Jerusalem and vicinity maintained as an international zone under permanent U.N. trusteeship; minority report calls for federated Arab-Jewish state in three years.

Sept. 1. Zionist leaders approve majority plan of U.N. Palestine inquiry committee; Arab Higher Committee denounces plan and threatens military action.

Sept. 20. British cabinet announces acceptance of U.N. Palestine inquiry committee's majority report.

Oct. 9. Arab League Council recommends member nations station troops along Palestinian borders to prepare for action if British troops evacuate.

Oct. 11. U.S. representative to the United Nations committee on Palestine endorses proposal to partition Palestine. Soviet Union endorses proposal Oct. 13.

Nov. 29. U.N. General Assembly votes, 33-13, with 10 abstentions, to partition Palestine into separate, independent Jewish and Arab states, effective Oct. 1, 1948, with the enclave of Jerusalem to be administered by the U.N. Trusteeship Council; Arab members denounce decision and walk out.

Dec. 5. United States places embargo on arms shipments to the Middle East because of violent disorders which followed U.N. decision.

Dec. 8. Arab League pledges to help Palestine Arabs resist any move to partition Palestine.

1948

Jan. 15. Britain and Iraq sign a 20-year alliance.

March 19. United States proposes to the U.N. Security Council suspension of the plan to partition Palestine and urges special session of the General Assembly to restudy issue.

March 25. Truman calls for truce between Jews and Arabs.

April 1. U.N. Security Council adopts U.S. resolution calling for a truce and a special session of the General Assembly to reconsider Palestine question.

May 13. Arab League proclaims existence of a state of war between league members and Palestinian Jews.

May 14. State of Israel proclaimed at 4:06 p.m., effective at midnight.

May 14. Midnight. Israel comes into existence; British mandate for Palestine ends as British high commissioner sails from Haifa.

May 15. 12:11 a.m. (6:11 p.m. in Washington). President Truman recognizes Israel, eleven minutes after its independence. Simultaneously, five Arab League states—Transjordan, Egypt, Iraq, Syria and Lebanon—invade Israel. Egyptian planes bomb Tel Aviv.

May 17. Soviet Union recognizes Israel.

May 25. Israel's President Chaim Weizmann visits Truman and appeals for a $90-million to $100-million loan to arm Israel and assist immigration.

May 27. Truman says Israel's loan application should be sent to the World Bank and the U.S. Export-Import Bank.

June 11. Four-week Arab-Israeli truce goes into effect.

June 30. Last British troops pull out of Palestine.

July 8. U.N. mediator Count Folke Bernadotte reports Arabs had refused to extend truce.

July 18. Truce renewed.

Sept. 17. U.N. mediator Bernadotte assassinated, allegedly by Jewish terrorists; Dr. Ralph J. Bunche named to succeed him.

Sept. 20. Bernadotte's final proposals for Palestine —including recognition of Israel—are published; Arab League announces establishment of an Arab government for Palestine—a move denounced by Transjordan and Iraq as amounting to recognition of Palestine's partition.

Oct. 22. Israel and Egypt agree to halt renewed fighting and comply with a cease-fire ordered by the U.N. Security Council Oct. 19.

Nov. 16. U.N. Security Council approves resolution calling for armistice.

Dec. 11. U.N. General Assembly sets up new Palestine Conciliation Commission.

1949

Jan. 6. Israel and Egypt announce a cease-fire on all fronts to begin Jan. 7. Israel withdraws its troops from Egypt Jan. 10.

Jan. 13. U.N. mediator Bunche meets with representatives of Israel and Egypt to negotiate an armistice.

Jan. 24. Egypt and Israel extend the cease-fire indefinitely.

Jan. 25. In the first Israeli election to be held, Prime Minister David Ben-Gurion's party wins the largest number of seats in the Knesset (legislature).

Jan. 31. The United States extends full diplomatic recognition to Israel and Transjordan following a flurry of diplomatic activity during which a number of Western nations including France on Jan. 24 and Britain Jan. 29 recognized Israel.

Feb. 1. Ending its military governorship of Jerusalem, Israel formally incorporates the city as part of the new state.

March 7. Egypt signs an agreement on the Suez Canal with the British-owned Suez Canal Co. calling for 80-90 per cent of the company's jobs to be manned by Egyptians and for Egypt to receive 7 per cent of profits.

March 11. Transjordan and Israel sign a "complete and enduring" cease-fire agreement to be binding even in the event of a failure to reach agreement on other points.

March 21. The first meeting of the U.N. Palestine Council Commission to settle the question of Arab refugees opens in Beirut, Lebanon.

April 5. Armistice talks between Israel and Syria run aground because of Israel's refusal to negotiate with representatives of Syria's new military government, which took over the country in a coup March 30.

April 20. While consenting to internationalization of the Holy Places in Jerusalem, Israel again rejects internationalization of the entire city.

April 26. Transjordan announces that the correct name of the country is Jordan, or Hashemite Jordan Kingdom.

April 26. In response to Jordan's statement on April 7 that a "greater" Jordan may evolve, Syria closes its Jordanian border and warns against attempts to annex its territory.

April 28. Israel rejects a proposal to return Arab refugees to their homes in the new state.

May 11. Israel is admitted to the United Nations after a vote of 37-12 in the General Assembly. Great Britain abstained from voting. On announcement of the final vote, delegates of six Arab states walk out of the meeting.

July 20. Syria and Israel sign an armistice agreement setting up demilitarized zones in contested areas and calling for both countries to keep their forces behind the frontiers.

July 27. U.N. mediator for the Middle East, Dr. Ralph J. Bunche, reports that "the military phase of the Palestine conflict is ended."

Sept. 13. The U.N. Palestine Conciliation Commission issues a draft statute whereby the United Nations would control Jerusalem, neither Israel nor the Arab states could have government offices there, and neither would control the city except for local administration of areas where their citizens lived. Holy places would be under permanent international supervision. Israel rejects the statute Nov. 15.

Oct. 28. Israel threatens to quit U.N. Palestine Conciliation Commission negotiations unless Arab states agree to direct Arab-Israeli negotiations, rather than indirectly through the commission.

Dec. 9. The U.N. General Assembly adopts a resolution placing Jerusalem under administration of the Trusteeship Council. The resolution is opposed by the United States, Britain and Israel.

Dec. 16. Israeli Prime Minister Ben-Gurion announces that Jerusalem will become the country's capital Jan. 1, 1950. Transfer of government offices from Tel Aviv to Jerusalem's New City has been going on for some time.

1950

Jan. 15. Secretary of State Dean Acheson defends Britain's arms shipments to Iraq, Jordan and Egypt on the grounds that the West must have the friendship of the Arab states.

Jan. 30. Truman requests $27,450,000 for U.S. contribution to the U.N. relief and public works program for Palestinian Arab refugees.

March 9. Turkey becomes the first Moslem state to recognize Israel.

March 15. Iran recognizes Israel.

March 24. Jordan breaks off non-aggression pact talks with Israel.

April 1. Arab League Council votes to expel any member making a separate peace with Israel.

April 4. U.N. Trusteeship Council approves a statute for internationalization of Jerusalem.

April 13. Israel rejects Arab League terms for peace negotiations which included a return to 1947 U.N. partition boundaries.

April 20. Arab League secretary condemns Anglo-American policy in the Middle East and urges Arab states to turn to the Soviet Union.

April 24. Jordan formally annexes Jordan-occupied eastern Palestine, including the Old City of Jerusalem.

April 27. Britain recognizes Jordan-Palestine merger and changes its recognition of Israel from de facto to full recognition.

May 25. Truman reveals agreement with France and Britain, reached at the London foreign ministers' conference, to sell arms to Mideast states—Israel as well as the Arab states—on a basis of parity between Israel and Arab countries if purchasers promise there will be no renewal of the Palestine war.

June 2. Israel and Jordan formally notify the United Nations of their rejection of the Trusteeship Council's plan for internationalization of Jerusalem.

June 14. Trusteeship Council concedes failure of five-month attempt to internationalize Jerusalem and turns the problem back to the General Assembly.

June 17. Five members of the Arab League—Egypt, Saudi Arabia, Syria, Lebanon and Yemen—sign a collective security pact.

June 21. Arab League pledges not to use arms purchased from the United States, France, Britain or other countries for purposes of aggression.

Nov. 21. In response to Egyptian demands for Britain's immediate withdrawal from the Suez Canal Zone and the Anglo-Egyptian Sudan, British Foreign Secretary Bevin tells Parliament that British troops will remain until the 1936 Anglo-Egyptian treaty is altered "by mutual consent."

1951

Feb. 7. The U.N. Mixed Armistice Commission breaks down as small-scale guerrilla fighting breaks out on the Israeli-Jordanian border.

March 7. General Ali Razmara, Iranian Premier since June 26, 1950, is assassinated by a religious fanatic who belonged to a group favoring the nationalization of Iran's oil industry. Razmara had angered members of that group by his support of U.S.-suggested economic reforms. He is succeeded March 11 by Hussein Ala, a strongly pro-West official who has served as a former ambassador to the United States.

March 20. The Iranian Senate votes unanimously in favor of nationalizing the country's oil industry. Later in the day, martial law is proclaimed in Teheran to stem Communist political terrorism over the oil issue.

April 28. The Iranian Parliament votes unanimously to sanction government expropriation of the British-owned Anglo-Iranian Oil Co. With favorable Senate action April 30, the oil nationalization bill becomes law.

May 18. The U.N. Security Council adopts a resolution calling on Israel to halt the Hulek border zone drainage project that allegedly set off border clashes between Israel and Syria. Criticizing Israeli aerial attacks on Syria, the resolution denounces the use of force by both countries to settle their differences. Israel halts work on the drainage project June 6.

June 1. President Truman sends a personal message to Iran's Premier urging prompt negotiations between Iran and Britain on the oil question.

June 12. Iran's Premier, responding to President Truman's message of June 1, stands firm on the decision to nationalize the country's oil industry.

July 5. The International Court of Justice at The Hague hands down a 10-2 decision on Britain's dispute with Iran over nationalization of the Anglo-Iranian Oil Co. The decision calls for Iran to re-instate the Anglo-Iranian Oil Co. to full control of its assets and operations. Iran's ambassador to Britain says his country will "ignore" the recommendations and proceed with the take-over of the oil company.

July 9. President Truman sends a personal message to Iranian Premier Mossadegh urging him to examine the International Court proposal as a basis for settling the oil dispute. Truman offers to send W. Averell Harriman to Iran to mediate the British-Iranian dispute. Mossadegh accepts that offer July 11.

July 14. Disclaiming all previous obligations, Iran opens sales of its oil on an equal basis to old and new clients of the Anglo-Iranian Oil Co.

July 20. Jordan's King Abdullah is assassinated, reportedly by a member of a faction opposing his annexation of parts of Palestine. The King's son, Prince Talal, is crowned King Sept. 6 in Amman.

July 24. Secret plans for Iranian-British negotiations over the nationalization of the Anglo-Iranian Oil Co. are sent to London after formulation during eight days of talks between Iranian officials and U.S. envoy Harriman.

Aug. 2. Britain and Iran formally agree to begin negotiations over the nationalization dispute.

Sept. 1. The U.N. Security Council calls on Egypt to end its three-year-old blockade of the Suez Canal to ships carrying cargoes bound for Israel. Egypt refuses Sept. 2 to comply with the request until Israel obeys previous U.N. resolutions dealing with the partition of Palestine, repatriation and compensation of Arab refugees, and internationalization of Jerusalem.

Sept. 13. The U.N. **Palestine** Conciliation Conference with Israeli and Arab delegates opens in Paris.

Sept. 17. Averell Harriman refuses to relay Iran's ultimatum calling on Britain to resume negotiations within two weeks on Iran's terms or to have British nationals expelled from Abadan, Iran, site of an Anglo-Iranian Oil Co. refinery.

Sept. 21. Israel says it will sign non-aggression pacts with each of its four Arab neighbors, but warns that negotiations should not continue if Arabs will not meet in the same room with Israeli delegates. Israel offers to compensate Arab refugees and to make contributions to their resettlement in Arab countries but is unwilling to accept repatriation of the refugees in Israel.

Sept. 25. Iran orders the last 300 British oil technicians to leave the country by Oct. 4.

Oct. 3. Britain's 300 oil employees depart leaving Western Europe's largest oil supply source in the hands of untrained Iranians.

Oct. 8. Egypt announces plans to expel British troops from the Suez Canal and to assume full control of the jointly administered Anglo-Egyptian Sudan. The next day, Britain declares it will neither vacate the Suez Canal nor withdraw its administrators from the Sudan.

Oct. 16. Tension builds in the Suez Canal zone as eight persons are killed and 74 wounded during fighting between British troops and Egyptian rioters. A three-day state of emergency is proclaimed throughout Egypt. Meanwhile, the Parliament approves the change of King Farouk's title to "King of Egypt and the Sudan" and extends the Egyptian constitution to apply to the Sudan.

Oct. 23. Iranian Premier Mossadegh meets in Washington, D.C. with President Truman who stresses the importance of Iran's oil to the West. Truman appeals for new efforts to reach an amicable settlement with Britain.

Oct. 27. Egypt formally notifies Britain that the 1936 Treaty of Alliance and the 1889 agreement on joint administration of the Sudan are both considered broken. In the Sudan, the legislature rejects the decision to place the country under the Egyptian crown.

Oct. 31. Lebanon's Chamber of Deputies orders the government to undertake negotiations with oil companies to revise concession agreements. The American-owned Trans-Arabian Pipeline Co. comes under fire from the deputies for "smuggling" oil to Israel and Jordan and for not employing the promised quota of Lebanese citizens.

Nov. 2. Britain continues to amass additional troops in the Suez Canal to strengthen its garrison there.

Nov. 5. Egypt refuses to discuss the Suez Canal dispute until British troops leave the area.

Nov. 15. Britain proposes self-rule for the Sudan with the voters of that country deciding in a year or two whether to unite with Egypt.

Nov. 18. British troops and Egyptian police battle in Ismailia, killing 11.

Nov. 21. Citing the "rigid positions" on both sides, the U.N. Palestine Conciliation Commission ends mediation efforts between Israel and the Arab states.

Dec. 2. Following a bloodless military coup in Syria, Army Chief of Staff, Col. Adeeb Shishekly, becomes President, after the resignation of President al-Atassi. Col. Fawzi Silo is appointed Premier Dec. 3.

Dec. 10. Reportedly in response to President Truman's appeal, Iran ends a six-month boycott of the International Court of Justice's proceedings on nationalization of the Anglo-Iranian Oil Co. and agrees to contest the Court's ruling.

Dec. 13. Egypt recalls its ambassador to Britain in protest against Britain's "aggression" in the Suez Canal zone.

Dec. 24. The Federation of Libya, an Arab kingdom created by the United Nations, becomes independent. By agreement, Britain and the United States will retain their military bases in the country.

1952

Jan. 18. British troops and Egyptian guerrillas battle for four hours at Port Said.

Jan. 25. British troops disarm Egyptian police in Ismailia along the Suez Canal. Fighting breaks out, killing 42 persons.

Jan. 26. Martial law is imposed in Egypt following widespread rioting and burning in Cairo. More than 20 persons are dead. Extensive damage to American, British and French property is estimated to total over $10 million.

Feb. 7. Britain issues "strongest condemnation" of the violence in Egypt Jan. 26 that killed British nationals and destroyed British property in the Canal Zone.

Feb. 16. Jordan signs a collective security pact with other Arab League countries.

March 18. Iranian Premier Mossadegh accuses President Truman of delaying an American loan to his country pending acceptance of British proposals for settling the oil nationalization dispute.

April 2. Britain's submission of a new constitution for the Sudan, during renewed British-Egyptian talks on the Sudan, brings an immediate Egyptian protest. The Sudanese Parliament, welcomes the new constitution April 23, and asks for an amendment allowing the Sudanese to determine whether they wish union with Britain, Egypt or neither.

May 3. Britain, proposing a solution to the dispute with Egypt over the Suez Canal and the Sudan, offers to evacuate British troops from the base in the Suez Canal, but denies recognition of King Farouk as ruler of the Sudan until the Sudanese people are consulted.

July 22. Canceling its year-old decision calling for restoration of property to the British-owned Anglo-Iranian Oil Co., the International Court of Justice at The Hague rules it has no jurisdiction in the dispute over Iranian nationalization of the oil company.

July 23. Egyptian King Farouk flees the country following a military coup that empowers Maher Pasha as Premier. Farouk abdicates July 26 and goes into exile in Italy. The King's infant son, King Fuad II is proclaimed ruler of Egypt and the Sudan by the cabinet.

Aug. 11. Declaring that King Talal, suffering from mental disorders, is unfit to rule, the Jordanian Parliament proclaims Crown Prince Hussein the new King.

Aug. 11. Iranian Premier Mossadegh is granted full dictatorial powers by the Senate. The Chamber of Deputies had approved the dictatorial powers Aug. 3.

Sept. 7. General Mohammed Naguib assumes leadership of Egypt.

Sept. 18. Ending a nine-year rule, Lebanon's President el-Khoury resigns in the face of general strikes to protest political corruption. Parliament elects Foreign Minister Camille Chamoun President Sept. 23.

Oct. 7. Iranian dictator Mossadegh demands $1-billion from Britain before talks can resume on the question of nationalizing the Anglo-Iranian Oil Company.

Oct. 22. Iran severs diplomatic ties with Britain after Britain's rejection Oct. 14 of Iran's demand for $1-billion from the Anglo-Iranian Oil Co.

Oct. 29. Egyptian President Naguib and delegates of the Sudanese independent parties sign a final agreement to work for political development of the Sudan.

Nov. 7. Israel's first President, Dr. Chaim Weizmann, dies. The Israeli Parliament, (Knesset), Dec. 8 names Itzahk Ben-Zvi to succeed him as President.

1953

Jan. 19. The Iranian Parliament votes to extend Premier Mossadegh's dictatorship powers for one year, rejecting his bid for a 2-year extension.

Feb. 12. British and Egyptian officials sign an agreement in Cairo establishing immediate self-government in the Anglo-Egyptian Sudan and calling for self-determination by the Sudanese people within three years. The Sudanese must choose to unite with Egypt, become independent or follow another path.

May 2. King Faisal II of Iraq is crowned on his 18th birthday, thus ending the 18-year regency of his uncle Emir Abd-ul-Ilah.

May 10. Egyptian President Naguib tells his nation to prepare for a "big battle" to achieve their goal in the British-Egyptian dispute over the Suez Canal.

June 29. President Eisenhower notifies Iran that the American public is opposed to additional U.S. aid to that country while the oil dispute between Iran and Britain continues.

Aug. 16. The Shah of Iran seeks sanctuary in Iraq after his unsuccessful attempt to dismiss dictator Premier Mossadegh.

Aug. 19. A revolt by Iranian Royalists and troops loyal to the Shah ousts Premier Mossadegh. Announcing plans to return to Iran, the Shah names Maj. Gen. Gazollah Zahedi as premier. The Shah returns Aug. 22.

Sept. 1. President Eisenhower sends personal assurances to newly appointed Iranian Premier Zahedi of U.S. readiness to send financial and economic aid to Iran. The message is followed by a grant of $45-million on Sept. 5.

Oct. 19. The United States submits a plan to the U.N. Security Council aimed at helping Arab refugees and easing tensions in Palestine. It calls for development of the irrigation and power resources of the Jordan River.

Oct. 28. The United States resumes economic aid to Israel after that country agrees to halt work on a power project on the Jordan River.

Nov. 7. The United States, in the face of growing anti-American Arab sentiment, denies that economic aid would be withheld if the Arab states did not agree to a special T.V.A.-type development of the Jordan River.

Nov. 14. Jordan rejects the U.S. proposal for development of the Jordan River resources.

Nov. 29. The Sudan's first general elections yield a decisive victory to forces seeking union with Egypt.

Dec. 21. An Iranian military court convicts former Premier Mossadegh on all counts of attempted rebellion. Instead of the death penalty, the court imposes a three-year solitary confinement sentence after the Shah's request for clemency.

1954

Feb. 25. Egyptian President Naguib resigns after the Egyptian legislature refuses his appeal for absolute power, but he returns to power Feb. 27.

Feb. 26. A Syrian army revolt ousts President Shishekly. Former President Hashim Atassi succeeds him Feb. 28.

April 17. Egyptian Col. Gamal Abdul Nasser becomes Premier in a new organization of the government, replacing Premier Naguib. Nasser had served as Premier for two days in February following Premier Naguib's brief ouster.

June 5. Following a visit to the United States by Turkish Premier Menderes, it is announced that Turkey will receive $200-million in military aid from the United States.

June 11. Egypt announces an agreement with Saudi Arabia to place both countries' military forces under a unified command.

July 27. Egypt and Britain sign an agreement ending the dispute over the Suez Canal. Britain will remove its forces from the area within 20 months, but will retain the right to use the Canal base in the event of aggression against an Arab state or Turkey.

Oct. 9. Israel attends a meeting of the Israel-Jordan Mixed Armistice Commission, thus ending a seven-month boycott.

Oct. 21. The Iranian Parliament votes to restore that country's oil production.

Nov. 2. Jordan summons the ambassadors of Britain, France, and the United States to ask their governments to halt Israel's unilateral diversion of the Jordan River.

Nov. 14. Egypt's ruling military junta quietly deposes General Naguib whose powers had been turned over to Gamal Abdul Nasser April 17.

1955

Feb. 24. Iraq signs a mutual defense treaty (Baghdad Pact) with Turkey despite Egyptian protests.

April 4. Britain joins Baghdad Pact.

April 21. French Premier Faure and Habib Bourguiba, leader of the militant Tunisian nationalists, reach agreement on autonomy for the French protectorate of Tunisia.

June 29. U.N. efforts to initiate negotiations between Egypt and Israel over the dispute in the Gaza Strip break down.

Aug. 16. The Sudanese Parliament unanimously calls for the evacuation of British and Egyptian troops from the Sudan within 90 days.

Aug. 26. As fighting in the Gaza Strip increases, Secretary of State Dulles suggests a plan under which internationally guaranteed borders would be established with U.S. participation.

Aug. 30. Secretary of State Dulles says the United States has offered to sell arms to Egypt.

Sept. 23. Pakistan joins Baghdad Pact.

Sept. 28. Assistant Secretary of State Allen arrives in Cairo to discuss Premier Nasser's announcement that Egypt would exchange cotton for Soviet arms.

Oct. 12. American and Israeli sources disclose that the Soviet Union may be willing to sell arms to Israel as well as to Egypt. The announcement followed a request Oct. 11 by Israel for the United States to match Soviet arms to Egypt.

Oct. 12. The Soviet Union warns Iran that its intention to join the Baghdad mutual security pact is "incompatible" with peace in the Middle East.

Oct. 20. Egypt and Syria sign a mutual defense treaty, triggering Israeli requests for an Israeli-American security pact. Israel cites recent Egyptian arms purchases from Czechoslovakia.

Oct. 31. Israel appeals to the Soviet Union not to foster a new war in the Middle East by arming the Arab states.

Nov. 3. Iran joins Baghdad Pact.

Nov. 22. The five Baghdad Pact countries announce the establishment of a permanent political military and economic organization, the Middle East Treaty Organization, to be based in Baghdad. Members are Britain, Iran, Iraq, Pakistan, and Turkey.

Nov. 26. The Soviet Union accuses Iran of violating their mutual treaty obligations by participating in the Baghdad Pact. Iran rejects the protest Dec. 6.

Dec. 12. Arab ambassadors from eight countries tell Secretary of State Dulles that U.S.-Arab relations are being strained by the behavior of Israel.

Dec. 14. The Jordanian government resigns to protest British pressure to join the Baghdad Pact. As rioting breaks out in additional protest to the British pressures, King Hussein dissolves Parliament, Dec. 20.

Dec. 15. In light of recent clashes between Israeli forces and the Syrian Army, Egyptian Premier Nasser warns Israel that further aggression may prompt a full Egyptian-Syrian military response.

Dec. 26. Egypt, Saudi Arabia and Syria unify their military power by placing troops under one commander, the Egyptian Minister of War.

1956

Jan. 13. Lebanon and Syria agree on a bilateral defense pact.

Jan. 19. The Sudan joins the Arab League. Egypt, Iraq, Jordan, Lebanon, Libya, Saudi Arabia, Syria and Yemen are other members.

Jan. 30. Israel urges the United States and Britain to allow it to buy arms.

Feb. 6. Secretary of State Dulles, not excluding "the possibility of arms sales to Israel," suggests that Israel look for security in the United Nations and the 1950 Anglo-American-French Three-Power agreement.

Feb. 14. Israel agrees to halt work temporarily on the Jordan Valley Project to allow for U.S. attempts to gain Arab cooperation on joint continuation of the project.

Feb. 16. It is disclosed that the United States is sending tanks to Saudi Arabia, an action Israel calls "regrettable" to the balance of power in the Mideast.

Feb. 28. French Premier Guy Mollet offers Algerian rebels a choice between all-out war or a cease-fire. Guerrilla warfare, staged by the nationalist rebels seeking Algerian independence has continued over the past 16 months.

March 3. Agreeing to honor the 1948 treaty of friendship with Britain, Jordan announces it will grant Britain bases in Jordan.

March 9. Jordan rejected a bid by Egypt, Syria and Saudia Arabia to replace British subsidizing of Jordan's defense forces.

March 10. Ending a two-year period of relative quiet on their border, Jordan stages raids on Israel.

March 12. Egypt, Saudia Arabia and Syria agree to unite their defenses against Israel.

March 15. France sends two Army divisions into Algeria.

March 17. France recognizes Tunisia's independence.

April 10. Yemen joins the military alliance of Egypt, Saudi Arabia and Syria.

April 18. The United States becomes a full member of the Economic Committee of the Baghdad Pact. The next day, the U.S. agrees to set up a military liaison office at the permanent headquarters of the Baghdad Pact.

April 19. Israel and Egypt agree to a U.N. cease-fire.

April 21. Egypt, Saudi Arabia and Yemen establish a five-year unified military command under an Egyptian leader.

May 6. Jordan and Egypt announce plans to unify their armies.

May 8. The French Resident Minister, Robert Lacoste, says French troops in Alegeria will total 330,000 by June and that he will request 40,000-50,000 more forces.

May 9. Secretary of State Dulles states the United States' reason for not wanting to sell arms to Israel. He explains that Washington seeks to avoid a Soviet-American confrontation in the Middle East and to avoid incidents that could bring about a global war.

May 12. As the United States confirms its sale of arms to Saudi Arabia, 12 more jets arrive in Israel in a second shipment from France which has tacit U.S. approval.

May 16. Britain announces that six of its former bases in the Suez Canal are now under Egyptian control.

May 21. Lebanon and Jordan agree to coordinate their defense plans.

May 31. Jordan and Syria sign a military agreement.

June 7. Iranian troops seize Soviet oil concessions in Khuryan, Iran.

June 13. Britain turns over full responsibility for the defense of the Suez Canal to Egypt. On June 18, Britain declares its 74-year occupation of the Canal ended.

June 24. After an uncontested election, Gamal Abdul Nasser becomes Egypt's first elected president, having received 99 per cent of the vote.

July 20. Following disputes over funding the Aswan Dam, the United States refuses to loan Egypt funds for the project and Britain withdraws its offer to supplement the American loan. Egyptian officials are surprised and angered by the move.

July 27. Egyptian President Nasser nationalizes the Suez Canal and imposes martial law there in retaliation for American and British withdrawal of support for the financing of the Aswan Dam. Income from the canal will be funnelled into building costs of the dam.

July 28. Britain freezes assets of Egypt and the Suez Canal held in Britain.

July 31. Secretary of State Dulles arrives in London to discuss British and French proposals for international supervision of the Suez Canal.

Aug. 9. President Nasser announces the creation of a National Liberation Army made up of the National Guard and youth organization volunteers.

Aug. 14. The United States creates a Middle East Emergency Committee to supply U.S. oil to Western Europe in the event that shipments from the Middle East are discontinued during the present crisis.

Aug. 16. In London, 22 nations open conference on the Suez canal crisis. Eighteen nations agree Aug. 23 to ask Egypt to negotiate for international operation of the Suez Canal. On Aug. 28, Nasser agrees to meet with a five-nation delegation.

Sept. 11. After initial efforts at negotiations fail to settle the Suez crisis, Britain and France agree to apply economic pressure to force Egypt to accept international control of the canal. The United States rejects the plan "under present circumstances."

Sept. 14. Egypt takes over complete control of the Suez Canal.

Sept. 21. The Suez Conference in London concludes with a draft plan for a Suez Canal Users' Association. The following day, Britain issues invitations to 18 nations for a third conference on the Suez Canal situation.

Sept. 30. Lebanon calls for talks between Syria, Jordan, Egypt and Lebanon to mobilize forces in retaliation against Israeli reprisal raids.

Oct. 1. Fifteen nations, including the United States, Britain and France, set up a Suez Canal Users' Association.

Oct. 8. Egypt and the Soviet Union again reject proposals for international supervision of the Suez Canal.

Oct. 12. In the face of increasing clashes between Israel and Jordan, Britain warns Israel that it will stand by its 1948 mutual defense treaty with Jordan.

Oct. 25. Egypt, Syria and Jordan sign an agreement to place their armed forces under the joint command of an Egyptian general.

Oct. 30. Aerial warfare breaks out between Israel and Egypt. Egypt warns Britain and France that it will also fight to keep the Suez Canal. This provokes a British-French ultimatum warning that troops will be sent to the Suez unless Egyptian and Israeli troops withdraw 10 miles from the canal and cease fighting by a designated time. Egyptian President Nasser rejects the ultimatum. The same day, Egypt reports that British and French planes are bombing Egyptian cities and other targets in an effort to force an Egyptian evacuation of the Suez.

Oct. 31. Israel accepts the British ultimatum on the condition that Egypt also agrees. Meanwhile, British and French troops are readied on Cyprus in the event that they are ordered to move against Egypt.

Nov. 1. Egypt breaks off diplomatic relations with Britain and France and seizes their property in Egypt as bombing of military targets continues. Jordan also severs ties with France and tells Britain that it will no longer be allowed to use ground or air bases in Jordan for further attacks on Egypt.

Nov. 2. The City of Gaza surrenders to Israeli forces.

Nov. 3. France and Britain reject the U.N.'s call Nov. 1 for a cease-fire in Egypt.

Nov. 5. The Soviet Union warns that it is prepared to use force "including rockets" to restore the Mideast peace. The Soviets call for joint Soviet-American action against "aggressors," a proposal the United States rejects as "unthinkable."

Nov. 6. As Israel ends its Sinai campaign, France and Britain agree to a cease-fire in Egypt. The same day, British and French parachute troops capture Port Said.

Nov. 7. The U.N. General Assembly calls on Britain, France and Israel to withdraw their forces from Egypt.

President Eisenhower in a personal note to Israeli Prime Minister Ben-Gurion expresses U.S. "concern" that "Israel does not intend to withdraw." Israeli rejection of the U.N. appeal would, says the President, "impair friendly cooperation between our two countries."

Nov. 9. Iraq breaks off ties with France and announces it will boycott any future meeting of the Baghdad Pact attended by Britain.

Nov. 10. The Soviet Union calls for the withdrawal of British, French and Israeli troops from Egypt and warns that Russian volunteers will be allowed to join Egyptian forces unless the withdrawal takes place. On Nov. 14, the United States says it would oppose any such Soviet intervention.

Nov. 21. Token British, French and Israeli troop withdrawals from Egypt begin. Eisenhower, according to sources, is reported to have sent private messages to the British and French governments urging complete troop withdrawal.

Nov. 30. The United States puts into operation its emergency oil plan to supply Europe with 500,000 additional barrels of oil a day.

Dec. 17. Britain implements gas rationing due to the cut-off of Middle East oil.

Dec. 22. The last British and French troops withdraw from Egypt.

Dec. 24. Egypt demands reparations from Britain, France and Israel for war damages.

Dec. 26. Clearance of sunken vessels and mines in the Suez Canal begins.

Dec. 31. Secretary of State Dulles says the United States has a "major responsibility" to prevent Soviet expansion in the Middle East. The same day, the Syrian Ambassador to the United States says that Arab states will not welcome American protection against the Russians.

1957

Jan. 4. The Suez Canal opens halfway, for medium-sized shipping.

Jan. 7. Britain announces that its air force troops are protecting the Aden protectorate against recent incursions from Yemen.

Jan. 15. Egyptian President Nasser undertakes an "Egyptianization" process. Only natural citizens will be allowed to hold shares of Egyptian-based companies. British and French banks and insurance companies are nationalized.

Jan. 21. Turkey, Pakistan, Iran and Iraq, the four Moslem nations of Baghdad Pact, endorse the Eisenhower Doctrine which calls for American action to counter Communist actions in the Middle East.

Jan. 23. Israel sets a prior condition for complete withdrawal of its troops from Egypt: Egyptian assurances that the Gaza Strip and Sharm el Sheikh will not be used for hostile actions.

Jan. 28. Yemen says peace will not be restored in Aden until its claims to the Western and Eastern Aden protectorates are recognized. Britain is accused of not yielding Aden because of hopes to find oil there.

Feb. 6. Saudi Arabian King Saud, meeting in Washington with President Eisenhower during a 10-day state visit, says the Eisenhower Doctrine "is a good one which is entitled to consideration and appreciation."

Feb. 17. A small Egyptian cargo vessel completes the first full course of the Suez Canal since ships were halted in November 1956.

March 1. Israel agrees to withdraw its troops from the Gaza Strip and the Gulf of Aqaba on "assumptions" that the U.N. Emergency Force will administer Gaza until a peace settlement is reached and that free navigation of the Gulf will continue.

March 5. Saudi Arabia demands more than $84.7-million in back taxes from the American-owned Aramco oil company.

March 7. The last Israeli forces withdraw from Egyptian territory.

March 13. Jordan and Britain cancel their 1948 treaty of alliance. British troops are to withdraw within six months.

March 15. As the situation in the Middle East appears to be worsening, Israeli Foreign Minister Golda Meir flies to Washington to confer with U.S. officials. In contravention of U.N. resolutions, Egypt March 14 sent civil administrators into Gaza and on March 15 announced that Israel would not be permitted to use the Suez Canal. Saudi Arabia halted Israeli use of the Gulf of Aqaba March 15.

March 19. Vice President Richard M. Nixon visits Tunisia as part of a 22-day, 19,000-mile African tour.

March 22. The United States announces it will join the Military Committee of the Baghdad Pact.

April 8. The United States and Saudi Arabia sign an agreement extending the U.S. lease on the Dhahran air base by 5 years.

April 13. Syria attacks Jordan.

April 24. As internal political turmoil continues in Jordan, a U.S. statement, authorized by President Eisenhower and Secretary of State Dulles, warns that the United States regards "the independence and integrity of Jordan as vital." On April 25, the United States orders the 6th Fleet into the eastern Mediterranean.

April 24. Egypt reaffirms its intention to unilaterally operate the Suez Canal and issues a declaration on the conditions for management and operation of the canal.

May 5. King Hussein announces that the battle against leftist elements in Jordan has succeeded.

May 13. Britain accepts Egypt's conditions for British passage through the Suez Canal.

June 10. A rift arises in Jordanian-Egyptian relations as Jordan charges that an Egyptian military attaché is plotting against Jordanian officials. His recall is requested. Egypt complies.

June 29. The United States signs an agreement with Jordan, extending $10-million in military supplies and services.

July 9. In the most violent encounter since November, Syria and Israel battle for almost 10 hours before a fourth U.N. intervention halts the fighting.

July 19. British-led forces step in to suppress a tribal revolt in Muscat and Oman on the Arabian Peninsula. On July 24, British planes attack military targets controlled by rebel tribesmen in Oman after the rebels refused to heed a British warning.

July 26. Egyptian President Nasser accuses the United States of intriguing to overthrow his government.

July 26. Tunisia's National Constituent Assembly deposes the Bey of Tunis and declares Tunisia a republic, the third along with Egypt and Nigeria in Africa. The Assembly elects Premier Habib Bourguiba President.

Aug. 13. Following Syrian accusations of U.S. efforts to overthrow that government, Syria ousts three American embassy officials. On Aug. 14, the United States expels Syrian diplomats.

Aug. 20. British troops complete their withdrawal from Oman, following recognition Aug. 11 of the Sultan of Oman's authority by the rebels.

Aug. 29. Lebanon affirms its support of the Eisenhower Doctrine.

Sept. 5. The United States announces plans to send arms to Jordan, Lebanon, Turkey, and Iraq.

Sept. 7. Affirming his doctrine on the Middle East, President Eisenhower says the United States will take action to protect pro-West Middle East countries if they are threatened by Syria.

Sept. 12. Saudi Arabian King Saud, in Washington, calls on U.S. officials to modify the U.S. stance toward Syria.

Oct. 3. Egypt says it will help defend Syria if necessary. Lebanon echoes that sentiment Oct. 11.

Oct. 10. Syria accuses Turkey of aggression on the **Syrian-Turkish border.**

Oct. 21. Turkey accepts Saudi Arabia's offer to mediate the dispute between Turkey and Syria. Syria Oct. 23 refuses to accept that proposal.

Nov. 12. The Shah of Iran instructs his cabinet to present a bill to parliament to bring Bahrain, a British oil protectorate, under Iranian jurisdiction.

Dec. 28. Tunisian President Bourguiba declares that British troops must leave the country no later than March 20, 1958.

1958

Jan. 18. Israel and Jordan agree that the 1948 agreement calling for demilitarization of Mt. Scopus should be followed.

Jan. 30. Secretary of State Dulles, addressing the Baghdad Pact countries meeting in Ankara, Turkey, tells the delegates that the Eisenhower Doctrine commits the **United States to the Mideast as effectively as would membership in the Baghdad Pact.**

Feb. 1. Egypt and Syria merge into the United Arab Republic. Citizens of the two countries approve the merger, nearly unanimously, in plebiscites Feb. 21.

Feb. 11. Yemen agrees to federation with the U.A.R.

Feb. 12. Tunisia demands the removal of all French forces from the country. The demand followed France's attack Feb. 8 on a Tunisian border town as reprisal for Tunisia's "cobelligerence" in the French-Algerian conflict.

Feb. 14. Iraq and Jordan form the Arab Federation with Iraqi King Faisal II serving as head of the two-state federal union. King Hussein retains sovereignty in Jordan. The federation is approved Feb. 17 by the Iraqi parliament and by the Jordanian parliament Feb. 18.

Feb. 17. The Sudanese government discloses that Egypt has demanded that the Sudan relinquish common border lands north of the 22nd parallel, in exchange for a small strip of land south of that parallel. The Sudan requests U.N. consideration of the Egyptian actions.

Feb. 19. France announces plans to establish a 200-mile long, 15-mile wide "no man's land" along the Tunisian-Algeria border. Tunisian President Bourguiba strongly protests the plan Feb. 27.

March 3. Morocco proposes the union of Tunisia, Algeria and Morocco as a means of resolving the crisis in Algeria.

March 20. Tunisian President Bourguiba announces his intention to cooperate with France and thanks the United States and Britain for their efforts to resolve the tension between his country and France.

March 25. Israeli Premier Ben-Gurion voices optimism for an Arab-Israeli peace, provided other powers will guarantee Israel's borders.

April 3. Jordan seizes control of petroleum supplies within its boundaries.

April 21. At the first conference of independent African states, Egypt, the Sudan, Libya, Tunisia, Morocco, Ethiopia, Ghana and Liberia adopt a common foreign policy based on "nonentanglement" with the United States and the Soviet alliances.

May 13. Rightist French civilians and Army officers in Algeria rebel and form an Algiers Committee of Public Safety in order to maintain French rule there. Leaders demand the return to power in France of Gen. Charles de Gaulle.

May 24. The United States and Britain caution Tunisia and France against further clashes as fighting between French and Algerian forces spreads to Tunisia.

May 24. As a state of siege continues in Lebanon, the U.N. Security Council meets to discuss Lebanon's complaint that the U.A.R. has caused the continuing antigovernment rioting that started May 10.

June 1. Gen. Charles de Gaulle becomes premier of France after a paralysis of the French National Assembly **and after a rebellion by right-wing French living in Algeria force the resignation of Premier Pierre Pflimlin.**

June 11. The U.N. Security Council votes 10-0 to send U.N. observers to Lebanon to guard against the smuggling of arms or troops into that country.

June 11. The International Bank for Reconstruction and Development announces an agreement by which the Suez Canal Company shareholders will be compensated for their losses under the accords of the treaty signed in Rome April 29.

June 17. Secretary of State Dulles pledges American troops, if necessary, to quell Lebanese rebel warfare.

June 17. France agrees to evacuate all bases in Tunisia except the one at Bizerte.

June 29. Heavy fighting breaks out again in Lebanon.

July 4. Lebanon orders the departure of six U.A.R. diplomats accused of inciting the Lebanese revolt and financing the fighting.

July 14. The government of Iraq is overthrown by revolutionaries who kill King Faisal and Premier as-Said, seize Baghdad and proclaim a republic. Brig. Gen. Abdul Karim el-Kassim is named Premier. In reaction to the coup, King Hussein of Jordan announces his assumption of power as head of the Arab Federation of Iraq and Jordan. Hussein and Lebanese President Chamoun each appeal for U.S. military assistance because of the Iraqi coup.

July 15. President Eisenhower dispatches 5,000 Marines to Lebanon. He tells Congress, in a special message, that the troops will protect American lives and help defend Lebanon's sovereignty and independence. Meanwhile, martial law is proclaimed in Iraq.

July 15. At a meeting of the U.N. Security Council, the United States says its troops will remain in Lebanon only until U.N. forces can guarantee Lebanese "continued independence."

July 17. British paratroopers land in Jordan at the request of King Hussein.

July 19. The U.A.R. and the new Iraqi regime sign a mutual defense treaty.

July 19. British commandos land in Libya in support of the government against rumors of an Egyptian plan to overthrow it.

July 20. Jordan severs relations with the U.A.R. due to its recognition of the new Iraqi regime.

July 22. Iraq's new Premier el-Kassim declares his government's intention to increase cooperation with the West, especially in oil production.

July 24. Iraq and the U.A.R. set up committees to work out closer cooperation between those countries in political, economic, military and educational fields.

July 28. Secretary of State Dulles, committing the United States to partnership in the Baghdad Pact, assures Britain, Turkey, Iran and Pakistan that the United States will not fail to act in defense of their independence and integrity.

July 31. General Fuad Chehab, sympathetic to the Lebanese rebels, is elected President of Lebanon by Parliament over the strong objections of Premier Said.

Aug. 2. The United States recognizes the new government of Iraq. Britain recognized the government Aug. 1.

Aug. 2. Jordan announces the formal dismemberment of the Arab Union of Jordan and Iraq, in light of the new regime in Iraq.

Aug. 2. Iraq says it has not renounced the Baghdad Pact, nor will it buy arms from the Soviet Union at this time.

Aug. 5. Newly elected Lebanese President Chehab sends a message to President Eisenhower assuring efforts to "maintain the traditional friendship" between the countries.

Aug. 8. An emergency meeting of the U.N. General Assembly opens in New York to discuss the Middle East crisis, following a month-long conflict between the Soviet Union, Britain and the United States over the form of the special summit conference.

Aug. 13. President Eisenhower presents the U.N. General Assembly with a "framework of a plan of peace" in the Middle East. It includes provisions for a U.N. peace-keeping force in the region and for an "Arab development institution on a regional basis, governed by the Arab states themselves." At the same session, the Soviet delegate asks the Assembly to call for an immediate U.S. troop withdrawal from the area, as U.S. forces begin withdrawing from Lebanon.

Aug. 21. The General Assembly unanimously adopts an Arab resolution calling on Secretary General Hammarskjold to take the necessary steps to restore order in Jordan and Lebanon and thereby "facilitate the early withdrawal" of foreign troops. In the next few days, tension in the Mideast appears to lessen.

Sept. 19. An Arab nationalist government-in-exile, the "Republic of Algeria" government, is formed in Cairo. Ferhat Abbas, a leader of the Nationalist Liberation Front, is named Premier. Six Arab states recognize the government, while France denounces it.

Sept. 22. Lebanon's pro-West cabinet resigns. Rashid Karami, a rebel leader, becomes premier Sept. 24. The United States Sept. 27 assures Karami of continued U.S. support.

Oct. 1. Morocco and Tunisia are admitted to the Arab League composed of the U.A.R., Iraq, Jordan, Saudi Arabia, Libya, Lebanon, the Sudan, and Yemen.

Oct. 3. During his fourth visit to Algeria since assuming power, French Premier de Gaulle outlines a 5-year plan for creation of a peaceful and prosperous Algeria closely linked to France, but maintaining a distinct "personality."

Oct. 15. Tunisia severs diplomatic ties with the U.A.R. following a dispute over alleged U.A.R. interference in Tunisian affairs.

Oct. 25. The last U.S. forces leave Lebanon following a 102-day stay.

Nov. 17. A bloodless military coup, headed by Lt. Gen. Ibrahim Abboud, ousts the Sudanese government.

Nov. 18. Moroccan King Mohamed V orders the evacuation of all Spanish, French and U.S. troops and/or bases, thereby rejecting a U.S. bid to maintain military bases for another seven years.

Dec. 1. 19 months of martial law end in Jordan.

Dec. 6. The Soviet Union protests Iran's acceptance of a military agreement with the United States.

1959

Jan. 9. West Germany recalls its ambassador to the U.A.R. following the establishment of consular relations between the U.A.R. and East Germany.

Jan. 13. U.S. arms shipments arrive in Tunisia.

Jan. 17. Egypt and Britain resolve their two-year dispute generated by the Suez crisis in 1956. Terms of the agreement are not disclosed.

Feb. 10. It is reported in Washington that Soviet arms shipments to Iraq have doubled that country's military strength since last November.

Feb. 15. Iran announces its decision to stay within the Western alliance. It the past week, the Soviet Union had urged Iran not to sign a mutual defense treaty with the United States. The U.S., Britain, Turkey and Pakistan had countered, sending personal messages to the Shah urging him to sign further economic and defense agreements. Turkey, Iran and Pakistan sign mutual defense treaties with the United States March 5.

March 20. U.A.R. President Nasser denounces Soviet interference in Arab affairs and protests Soviet Premier Khrushchev's remarks the day before calling Nasser's hostility toward Iraq "hotheaded." Nasser on March 11 had accused Iraq and foreign Communist agents of attempting to divide the Arab world.

March 24. Iraq withdraws from the Baghdad Pact, which is left with four members—Britain, Turkey, Iran and Pakistan.

March 24. President Eisenhower praises Jordanian King Hussein for his resistance to outside pressures on his country.

April 14. Jordan's King Hussein arrives in Washington on an official visit for talks with President Eisenhower.

June 1. The U.S. embassy in Iraq announces the termination of U.S.-Iraqi military assistance agreements.

June 6. Jordan's King Hussein protests to the United Nations Syria's closing of their common border, which occurred after fighting sparked by Palestinian commandos.

July 13. Communist demonstrations in Kirkuk, an Iraqi oil center, nearly erupt into civil war. After the Iraqi army bombs the rebels, the government regains control.

July 21. Jordan and the U.A.R. agree to resume diplomatic relations, an action coupled with the reopening of the Syrian-Jordanian border.

Aug. 13. The Arab League, meeting in Bhamdun, Lebanon, supports reinforcement of the boycott of Israel.

Aug. 18. With the departure of Iraq from the Baghdad Pact, the organization is renamed the Central Treaty Organization (CENTO), with Britain, Iran, Pakistan and Turkey remaining as members. The United States supports the organization, holds memberships in certain committees, but is not an official member.

Aug. 26. Jordan offers citizenship to all Palestinian refugees.

Sept. 4. After four days of talks, U.A.R. President Nasser and Saudi Arabian King Saud agree to resume relations with Britain and to seek to end Communist penetration in Iraq.

Sept. 6. The U.A.R. announces an end to foreign control of banks in Syria, directing that 70 per cent of banks' stocks and board of directors seats must be held by Arabs.

Sept. 16. De Gaulle tells Algerians that within four years of the restoration of peace there, they will be allowed to determine their future by a free vote. They may choose between association with France and independence.

Oct. 5. U.A.R. Foreign Minister Fawzi tells the U.N. General Assembly that Israel will be permitted to use the Suez Canal after the Palestine refugee problem is resolved.

Oct. 9. President Eisenhower commends Iranian Premier Eghbal, in Washington for a meeting of the CENTO Council of Ministers, for his country's resistance to Communist propaganda.

Oct. 10. The United States and Turkey agree to establish an intermediate range ballistic missile base in Turkey.

Nov. 12. A spokesman for Saudi Arabian King Saud announces that the King disapproves of keeping U.N. troops in the Gulf of Aqaba.

Nov. 16. Iraq's Premier Kassim favors "Fertile Crescent" plan to unite with Syria and Jordan.

Dec. 1. Britain and U.A.R. reestablish diplomatic relations after 3-year break.

Dec. 14. President Eisenhower addresses the Iranian Parliament during a 6-hour stop-over visit.

1960

Jan. 18. The U.A.R. announces that the Soviet Union will finance the second stage of the Aswan Dam.

Jan. 24. Algerian rightists rebel against the liberal Algerian policies of French President de Gaulle. Tanks and troops seal off Algiers and the rebellion finally collapses.

Feb. 2—The United Nations asks the U.A.R. to withdraw forces from the Israeli-Syrian demilitarized border zone following air fights between Israeli and Syrian planes Feb. 1.

Feb. 11. Jordanian Foreign Minister Musa Nasir says the Arab states are "completely united" on a "declaration of war" against Israel if Israel attempts to divert the Jordan River to irrigate the Negev desert.

March 7. It is disclosed that at the Feb. 28 meeting of the Arab League, Jordan rejected a U.A.R. proposal to establish a "Palestine entity," claiming instead the allegiance of its Palestine refugees for Jordan's King Hussein.

March 9. As Premier Ben-Gurion leaves Israel for visit in the United States, Secretary of State Christian Herter assures Arab leaders that his visit will not undermine U.S.-Arab bloc relations.

March 14. Ben-Gurion meets with West German Chancellor Konrad Adenauer, the first meeting between leaders of those two countries.

March 27. Iraqi Premier Kassim announces plans to raise an "Army of the Palestine Republic" to train Palestinians for a war to restore their homeland.

March 27. It is disclosed that the U.A.R. has refused a Soviet offer to protect its borders against possible attack.

April 8. U.N. Secretary General Dag Hammarskjold protests U.A.R. seizure of ships carrying Israeli supplies and products through the Suez Canal. Hammarskjold says he will renew persuasive efforts to end such actions.

July 23. Iran recognizes Israel. The new diplomatic ties lead to a break-off July 27 in Iran-U.A.R. ties and Iran is put under an economic boycott by the U.A.R. July 28.

Oct. 22. Arab representatives at the second Arab Petroleum Congress protest price reductions by oil companies and ask for improved concessions.

Nov. 17. Jordan and Iraq agree to resume diplomatic ties in December. Relations had been cut off in July, 1958, following the Iraqi revolution.

Dec. 21. Saudi Arabian Prince Faisal resigns, returning complete control of the government to his brother, King Saud. Saud had relinquished executive power to Faisal early in 1958.

1961

March 16. It is disclosed that a U.S.-Saudi Arabian pact of 1957, which called for the setting up of a U.S. military base at Dhahran, will not be renewed. On April 11, Saudi Arabian King Saud explains that the decision was partially due to American aid to Israel.

April 22. Rightist Algerians, headed by four retired French generals, launch a second, right-wing revolt in Algeria. The attempted coup is aimed at preventing peace talks between France and Algeria's Moslem rebels and to keep Algeria part of France. The insurrection collapses April 26.

June 19. Great Britain grants independence to Kuwait, but signs a treaty with the new nation assuring British protection if requested.

June 26. Kuwaiti Sheik Abdullah al-Salem al-Sabah says he will fight to maintain Kuwait's independence after Iraq claims that Kuwait is an "integral part" of Iraq.

June 27. A state of emergency is declared in Kuwait. U.A.R. President Nasser supports Kuwait's independence.

July 20. The Arab League unanimously admits Kuwait to membership. Iraq walks out of the meeting, accusing the League of aiding "British imperialism."

Aug. 27. Ferhat Abbas resigns as Prime Minister of the Algerian Provisional Government and is succeeded by Ben Yusuf Ben Khedda.

Sept. 19. British protective forces in Kuwait are replaced by troops sent by the Arab League to assure Kuwait's sovereignty against Iraqi claims.

Sept. 29. Following a coup Sept. 28 by dissident Syrian army units, the revolutionary command sets up a civilian government for Syria and announces independence from the U.A.R. Jordan and Turkey recognize the new Syrian government.

Oct. 1. President Nasser announces in Cairo that the U.A.R. has broken ties with Jordan and Turkey for their recognition of new Syrian government.

Oct. 13. Syria is reseated at the United Nations, regaining the seat it gave up when it merged with the U.A.R. The New Syrian government was recognized by the Soviet Union Oct. 7 and by the United States Oct. 10.

Dec. 11. Adolf Eichmann, former Nazi SS lieutenant colonel and head of the Gestapo's Bureau of Jewish Affairs, is convicted on 15 counts of war crimes and genocide. He is sentenced Dec. 15 to death by hanging. He had gone on trial in Israel April 11.

Dec. 26. U.A.R. President Nasser dissolves his country's union with Yemen, formed in 1958, thus reducing the United Arab Republic to only the state of Egypt.

Dec. 27. British naval craft are ordered into the Persian Gulf to resist Iraq's possible seizure of Kuwait.

1962

Jan. 23. Morocco and Algeria establish a joint commission for a "United Arab Maghreb," a North African union.

Feb. 14. President Kennedy says U.S. use of the Saudi Arabian air base at Dhahran will terminate in April 1962.

March 9. Egyptian President Nasser issues a constitution for the Gaza Strip, under Egyptian administration since 1948.

March 16. Iraqi Premier Kassim and Syrian President Kodsi endorse close Iraqi-Syrian cooperation.

March 18. The Provisional Government of Algeria signs a cease-fire agreement with France to end fighting and violence which has continued since 1954. Ben Bella and four other Algerian rebel leaders are released in France.

March 28. Syrian army leaders oust the new Syrian government which was elected after the break with Egypt in the fall of 1961. The Syrian army leaders declare their intentions to work closely with Egypt and Iraq.

April 13. Syrian President Kodsi, ousted by an army coup March 28, is reported returned to office. Kodsi tells Syrians April 14 that he will seek a union of "liberated Arab states, beginning with Egypt."

May 22. Syria nationalizes all foreign banks.

May 31. Former Nazi Adolf Eichmann is hanged in Israel following conviction on charges that he committed crimes against humanity while heading Nazi efforts to annihilate Jews.

June 2. Iraq orders the departure of the American ambassador and recalls its ambassador to the United States following U.S. accreditation of an ambassador from Kuwait.

July 1. Algerians vote in a special referendum on the question of Algerian independence. Overwhelmingly, citizens vote for an independent Algerian cooperation with France. On July 3, de Gaulle recognizes Algeria's independence.

Aug. 7. Great Britain and the U.A.R. sign an agreement to transfer funds and compensate British nationals for property seized during the Suez crisis of 1956.

Aug. 27. Syria asks the Arab League to condemn the U.A.R. for alleged interference in Syrian internal affairs.

Aug. 29. Saudi Arabian King Saud and Jordanian King Hussein agree to merge military troops and economic policies.

Sept. 26. The U.S. State Department announces that the U.S. will sell short-range defensive missiles to Israel.

Sept. 30. Despite urging by the secret French nationalist Army in Algeria for the French population to remain in Algeria, all but 250,000 of nearly one million are reported to have left.

Nov. 6. Saudi Arabia breaks off diplomatic ties with the U.A.R. following charges that U.A.R. planes bombed Saudi Arabian villages near the Yemen border. The U.A.R. is reportedly cooperating with the revolutionary government in Yemen.

1963

Feb. 8. The Iraqi air force overthrows the government of Premier Kassim. Kassim is killed by a firing squad. A Nasserite and conspirator in the coup, Colonel Abdul Salem Arif, is appointed provisional president.

March 1. Iraq's Revolutionary Council guarantees "the rights" of Kurds, a dissident faction residing in northern Iraq that has been seeking autonomy within Iraq.

March 8. A coup by pro-Nasser and Baath Party followers ousts the Syrian government. The U.A.R. and Iraq governments threaten war if other nations interfere in the Syrian revolt.

March 12. Syria's new Premier Salah el-Bitar voices hopes of a federation of Syria, Iraq and Egypt under one president.

March 20. Israel's parliament (Knesset) endorses a government statement urging the West German government to forbid its scientists to aid the U.A.R. in the development of offensive missiles "and even armaments banned by international law."

April 1. Following pro-Nasser rioting, a state of emergency is declared by the Syrian Revolutionary Command Council.

April 10. U.A.R. Prime Minister Aly Sabry outlines the new federation between the U.A.R., Syria and Iraq, to be called the United Arab Republic. Later it is announced that the federation proposal will be submitted to national plebiscites to be held Sept. 27, 1963.

April 13. U.A.R. and Saudi Arabia agree to end their support of opposing factions in the Yemeni civil war.

April 20. Street demonstrations break out in Jordan in support of Jordan's joining the new U.A.R. federation.

April 23. Israeli President Itzhak Ben-Zvi dies at 78.

[handwritten annotations in top margin: "p 115", "May PLO est. by Arab heads of state at first National Palestinian Congress in Old City." "Problem never resolved given by the four arms"]

May 14. The U.N. General Assembly unanimously elects Kuwait as the body's 111th member.

May 21. In violation of an agreement reached in April, U.A.R. President Nasser declares that U.A.R. troops will not leave Yemen until royalist factions have been put down.

June 10. The Iraqi government announces war against troops seeking Kurdish autonomy within Iraq.

July 22. U.A.R. President Nasser renounces an agreement to unite Egypt, Syria and Iraq and denounces the Syrian Baath Party.

Aug. 9. Iraq and Syria, in agreement on ending disputes over Arab unity, urge U.A.R. President Nasser to join them in improving inter-Arab unity.

Aug. 21. The Arab League meets to consider a unified stance in support of Syria against Israel as fighting breaks out near the Sea of Galilee. Iraqi forces are placed "at the disposal of" Syria. U.A.R. troops are on alert for possible support of Syria.

Aug. 23. U.N. Secretary General U Thant announces Israeli and Syrian acceptance of a cease-fire.

Aug. 25. Israeli and Jordanian troops clash in Jerusalem before the U.N. truce observers persuade both sides to agree to a cease-fire.

Sept. 2. Iraq and Syria agree to seek "full economic unity" and to jointly work to strengthen their defenses.

Sept. 4. U.N. Secretary General U Thant reports that despite U.N. efforts to restore peace, the civil war in Yemen continues one year after the outbreak of hostilities.

Oct. 20. Iraq announces that Syrian troops are aiding government troops against the Kurdish rebels.

Nov. 12. Syrian Premier el-Bitar resigns and a new Syrian government is set up with Major General Amin el-Hafez as the president of the Revolutionary Council.

Nov. 18. Iraq's President Arif announces that his forces have overthrown Iraq's civilian Baathist government. Arif becomes president and chief of staff of the army. The new government announces Nov. 21 that it will seek to fulfill the April agreement between Iraq, Syria and Egypt on the formation of a union and offers Nov. 22 to settle differences with "the Kurds, our brothers."

1964

Feb. 10. Iraq's President Arif announces a cease-fire agreement, apparently concluding the Kurds' struggle for autonomy within Iraq.

March 24. The Soviet Union announces the signing of a five-year treaty of friendship with Yemen and discloses promises of increased economic aid.

March 28. Saudi Arabian King Saud signs over his full powers to his half-brother Crown Prince Faisal. King Saud took the action to reduce his own position to that of a figurehead after unsuccessful attempts to regain the powers he gave to Prince Faisal during his prolonged illness.

April 14. In talks with President Johnson in Washington, Jordanian King Hussein stands firm on Arab intentions to dam two tributaries of the Jordan River in order to block Israel's plans to divert the Jordan River for irrigation purposes.

April 28. Syria's Revolutionary Council cancels its military treaty with Iraq.

May 3. Iraqi President Arif introduces a new provisional constitution which has as its main goal the union of Iraq with the U.A.R.

May 26. Iraq and the U.A.R. sign an agreement providing for joint command of their troops in time of war.

June 2. After two days of talks in Washington, **Israeli Premier Levi Eshkol and President Johnson issue a communique calling for joint efforts to apply nuclear power capabilities to desalinization of sea water.** In the communique, the United States reaffirms its support of the "political integrity" of all Middle East nations.

July 13. Yemen and the U.A.R. sign an agreement to coordinate political, economic and cultural ties "as a step toward complete unity."

July 14. Iraq nationalizes all private and foreign banks, insurance companies and 30 industrial and commercial businesses.

Aug. 18. The Lebanese parliament elects Minister of Education Charles Helou as president to succeed President Fuad Chehab, who retired.

Aug. 22. Libyan Premier Mahmud Mutasser announces that both the United States and Great Britain have agreed to give up their military bases in Libya.

Sept. 11. After seven days of talks, chiefs of state of 13 Arab nations issue a final communique urging immediate Arab efforts on water projects to cut off the Jordan River from Israel in an effort to thwart Israeli plans to dam the Jordan for irrigation purposes.

Oct. 3. The economic ministers of Algeria, Morocco, Tunisia and Libya sign a protocol creating machinery for economic cooperation.

Nov. 2. The Saudi Arabian cabinet and consultative counsel proclaim Crown Prince Faisal the King of Saudi Arabia, thus dethroning King Saud.

Nov. 7. The Sudanese cabinet rescinds the state of emergency in effect since 1958.

Dec. 24. Syria announces that it will not grant oil and mineral exploitation concessions to foreigners.

1965

Jan. 3. Syria nationalizes, in whole or part, 107 principal industries, reportedly to stem the flow of capital from the country.

Jan. 12. At the conclusion of four days of talks in Cairo, premiers of 13 Arab nations issue a communique disclosing common policy toward nations henceforth recognizing Israel or aiding in her "aggressive military efforts." The policy is regarded as directed primarily at West Germany.

Jan. 26. Iranian Premier Mansour dies of gunshot wounds inflicted by a student Jan. 21.

Jan. 22. The Lebanese parliament approves construction of a pumping station on the Wazzani River aimed at diverting the Jordan River from Israel.

Feb. 10. West Germany temporarily halts military aid for Israel in the face of Arab threats to recognize the East German government if the military aid continues. Israel's parliament, the Knesset, Feb. 15 denounces West Germany's "surrender to blackmail" by the U.A.R.

Feb. 17. The State Department acknowledges that the United States was a secret partner in the West German-Israel military aid agreement.

Feb. 24. After East German leader Walter Ulbricht arrives in Cairo despite West German protests, West Germany suspends its economic assistance to the U.A.R. and cancels its program of guarantees for private investments there.

March 1. East Germany and the U.A.R. sign a $100-million economic aid agreement during Ulbricht's seven-day visit.

March 4. Syria nationalizes nine oil companies: six Syrian, two U.S. affiliates, and one joint British-Dutch company. Procedures for compensation are outlined.

March 16. U.A.R. President Nasser is elected to another six year term.

April 21. Tunisian President Habib Bourguiba criticizes Arab policy toward Israel and proposes broad terms to end the Arab-Israeli conflict. He calls for the opening of direct negotiations between Israel and the Palestinian Arabs on the basis of the 1947 United Nations plan for partition of Palestine into Jewish and Arab states; and for cession of one-third of Israel's territory for a Palestine Arab nation. Bourguiba says the land he would have Israel give up was earmarked for an Arab sovereignty by the 1947 U.N. plan but won by Israel during the Arab-Israeli war of 1948. Israel rejects Bourguiba's plan April 25. The U.A.R. rejects the proposals April 27 and "strongly denounces the issuance of such a proposal from the head of an Arab state." The Arab League rejects the plan April 29 during a meeting boycotted by Tunisia.

May 12. Israel and West Germany establish full diplomatic relations. The U.A.R. breaks diplomatic ties with West Germany. Nine other Arab states later follow suit in breaking off ties with Bonn.

June 1. Syria declares that the only Palestinian solution is the elimination of Israel. Syrian President Amin el-Hafez terms the Arab unified military command ineffectual.

June 4. Syrian President Hafez accuses other Arab nations of planning to leave Syria alone to face Israel in a showdown over diversion of the Jordan River.

June 22. The State Department announces U.S. plans to ship surplus farm products to the U.A.R. in fulfillment of an agreement that ends June 30. The aid had been suspended six months before because of anti-American incidents in the U.A.R.

July 15. The chief of the Israel General Staff says Israeli raids on Syria have deterred that country's plan to divert the Jordan River from Israel.

July 20. Lebanon's Prime Minister Hussein Oueini unexpectedly resigns. He is replaced by ex-Prime Minister Rashid Karami July 22.

Aug. 23. The Arab Socialist Baath Party in Syria establishes a National Council (legislature) to consolidate the party's political control. The new provisional legislature re-elects Hafez as chairman of the Presidency Council, Sept. 2.

Aug. 24. Saudi Arabian King Faisal and U.A.R. President Nasser sign an agreement ending the civil war in Yemen. The accord, reached during talks between the two leaders Aug. 22-24, calls for an immediate halt to hostilities, for Saudi Arabia to end military aid to the royalists, and for the U.A.R. to withdraw its troop support of the revolutionary republicans. Repre-
sentatives of the opposing factions had agreed Aug. 13 to end the three-year civil war.

Sept. 17. Heads of state of 12 Arab nations meeting in Casablanca call for "a cease-fire in the war of words" between the Arab states and for abatement of interference in each other's internal affairs.

Nov. 7. The Iraq government reestablishes free enterprise, announcing that it will reconsider socialism. Iraq will also adopt new approaches to the Kurdish rebel problem.

Nov. 10. In response to Syrian statements that the Khuzistan province in southwest Iran is part of the "Arab homeland," Iran closes its embassy in Syria and recalls its ambassador.

Nov. 24. Kuwait's Emir al-Sabah dies and his younger brother Sabah al-Salem al-Sabah is proclaimed the new ruler.

Dec. 29. The State Department confirms that the United States has been supplying Jordan with 50-ton Patton tanks.

1966

Jan. 3. The United States and the U.A.R. sign a $55-million aid agreement. The accord is the first major diplomatic overture between the two nations since relations became strained a year before.

Feb. 22. U.A.R. President Nasser warns he may station U.A.R. troops in Yemen for five years if necessary in order to establish a republican regime there.

Feb. 24. The United States acknowledges that Saudi Arabian King Faisal has asked for diplomatic and military support if the U.A.R. renews fighting in Yemen.

Feb. 25. Following a coup Feb. 23 by left wings of the military and of the Baath Party, Syria announces that the military junta in power has named Nureddin Attassi chief of state. Attassi had been ousted from office by a coup in December 1965.

March 19. For U.S. food shipments to continue to Egypt the United States says Egypt must reduce its cotton production. Egypt has been exchanging cotton for arms from the Soviet Union.

April 16. Iraqi Major General Abdul Rahman Arif is elected by a joint session of the cabinet and the national defense council to succeed his brother as president. Former president Abdul Salem Arif died in a helicopter crash April 13.

May 1. U.A.R. President Nasser warns that his country will attack Saudi Arabia if "any aggression or infiltration" into Yemen comes from Saudi territory.

May 18. Israeli Premier Eshkol urges the big powers to limit the arms build-up in the Middle East.

May 19. It is reported in Washington that the United States agreed in February to sell several tactical jet bombers to Israel. This is the first such weapons sale to be disclosed.

May 25. Jordanian King Hussein lays the first stone of the Mokheiba Dam. The dam is part of an Arab effort to divert the Jordan River from Israel.

June 22. At a news conference in Washington during a three-day visit, Saudi Arabian King Faisal says Israel is not his country's enemy, but Zionists have transplanted Arabs from Palestine.

June 29. Iraqi Premier al-Bazzaz announces a plan for local autonomy for the Kurdish tribes in northern Iraq, an attempt to end the five-year civil strife.

July 25. The U.N. Security Council discusses the Israeli-Syrian border conflict for the 214th time in 18 years.

Sept. 27. The United States announces that it will sell $100-million in vehicles to Saudi Arabia for modernization of the army.

Oct. 10. Jordanian King Hussein warns against excessive Egyptian interference in "Syrian affairs," saying Jordan will not allow the Syria-Jordan border to be closed. Hussein also says that his country would attack Israel if war developed between Israel and Syria.

Nov. 4. Syria and the U.A.R. sign a mutual defense treaty which provides for joint command of their armed forces.

Nov. 24. U.A.R. customs authorities seize Ford Motor Co. assets in the country allegedly due to a dispute over Ford's obligation to pay customs duties on cars assembled and then sold in the U.A.R.

Nov. 29. Jordan's King Hussein charges Soviet fomentation of tension in the Middle East following a week of anti-government demonstrations and riots.

Dec. 7. Syrian Chief of State Attassi calls on Jordanians and Palestine Arabs to overthrow Jordan's King Hussein and offers them arms.

Dec. 8. Syria impounds the assets of the Western-owned Iraq Petroleum Company over claims to back royalties and higher pipeline transit fees. When the company rejects Syrian demands of doubled transit fees and export fees, the company's oil pipeline is closed.

1967

Feb. 19. Iran and the Soviet Union sign a defense agreement whereby Iran will purchase $110-million in arms and supplies.

Feb. 27. Jordan and West Germany resume diplomatic ties that were broken in 1965 when West Germany recognized Israel.

March 2. The Western-owned Iraq Petroleum Company announces plans to pay an additional 50 per cent rental fee for the use of a pipeline through Syria.

April 20. Iraq and Kuwait call home their ambassadors as fighting erupts along their common border.

May 15. The U.A.R. alerts its military forces because of mounting tension with Israel. Syria also announces that its military forces are ready for action.

May 19. The U.N. Emergency Force in the Middle East pulls out, ending a 10-year commitment for peacekeeping in that area. The withdrawal had been requested by the U.A.R.

May 20. The U.A.R. declares that a state of emergency exists along the Gaza Strip.

May 22. The U.A.R. closes the Strait of Tiran to Israeli ships and to ships carrying strategic cargo bound for Israel.

May 23. The United States and Israel each issue strong warnings against the U.A.R.'s blockade of the Strait of Tiran, entrance to the Gulf of Aqaba.

May 23. The U.S. 6th Fleet is ordered toward the eastern Mediterranean. Meanwhile the Soviet Union warns that it will resist any aggression in the Middle East.

May 23. Following the explosion of a bomb on the Jordan-Syria border, Jordan orders the shutting down of the Syrian embassy and the departure of Syria's ambassador to Jordan.

May 24. The U.A.R. has reportedly mined the Strait of Tiran and the Gulf of Aqaba.

May 26. Israeli Foreign Minister Abba Eban meets in Washington with President Johnson on ways to restore order in the Middle East. On his way to Washington, Eban made stops in London and Paris to confer with officials there.

May 29. Egyptian and Israeli troops open fire on the Gaza Strip.

May 30. Jordan and the U.A.R. sign a mutual defense pact.

June 3. Israel appeals to the Soviet Union to help bring peace to the Middle East.

June 5. Arabs and Israelis fight in the Egyptian section of the Sinai Peninsula and in Jerusalem. Israel destroys Egyptian, Syrian and Jordanian air forces in surprise early morning attacks. After 36 hours of battle, Israelis capture the Jordanian sector of Jerusalem.

June 6-7. The U.A.R., Syria, Iraq, Sudan, Algeria, and Yemen sever diplomatic relations with the United States.

June 6. Kuwait and Iraq cut off oil supplies to the United States and Britain.

June 6. The U.A.R. closes the Suez Canal, charging that U.S. and British planes are aiding Israel. The U.S. strongly rejects the charges.

June 7. In a sweeping seizure of territory, Israel takes over the Old City of Jerusalem, Mt. Scopus, Bethlehem in Jordan, the Sinai Peninsula between the Negev Desert and the Suez Canal, Sharm el Sheik, the Gaza Strip, and announces the breaking of the blockade of the Gulf of Aqaba.

June 7. The U.N. Security Council adopts a resolution calling for an immediate cease-fire in the Middle East. Israel announces it will accept the cease-fire if the Arab states do. The cease-fire is accepted by Jordan June 7, the U.A.R. June 8 and Syria June 9-10.

June 8. Israeli planes erroneously attack an American vessel in the Mediterranean, killing 10 and wounding 100. Israel apologizes for the error.

June 9. Claiming the sole responsibility for Egypt's defeat by Israel, President Nasser resigns. Later the National Assembly rejects his resignation.

June 10. The Soviet Union severs diplomatic ties with Israel, pledging assistance to Arab states if Israel refuses to withdraw from conquered territory.

June 11. Israel and Syria sign a cease-fire agreement.

June 12. Israel announces that it will not withdraw to the 1949 armistice boundaries and calls for direct negotiations between Israel and Arab nations.

June 15. Libya asks the United States and Britain to remove their military bases and troops immediately.

June 19. President Johnson, in a nationally televised speech, sets forth five points for peace in the Middle East: right of each country's national existence, fair and just treatment of Arab refugees, freedom of innocent maritime passage, limitation of arms build-up, and guaranteed territorial integrity for each Middle East country. Meanwhile at the U.N., Soviet Premier Kosygin calls for the condemnation of Israel, the withdrawal of Israeli forces from occupied Arab lands, and Israeli reparations to Syria, Jordan and the U.A.R. for damages incurred during the war.

June 23. President Johnson and Soviet Premier Kosygin meet for 5 1/2 hours at Glassboro State College in New Jersey to discuss the Middle East, Vietnam and arms control. They meet again for over four hours June 25, but later President Johnson says that "no agreement is readily in sight on the Middle East crisis." Kosygin, at a televised news conference, says the first step to peace in the Middle East is Israel's withdrawal to positions behind the 1949 armistice lines.

June 27. President Johnson announces $5-million in U.S. emergency aid for victims of the Arab-Israeli war.

June 28. Israel proclaims the unification of all of Jerusalem under Israeli rule.

July 4. U.N. Secretary General U Thant asks Israel and the U.A.R. to accept U.N. supervision of the cease-fire in the Suez Canal zone. The U.A.R. agrees July 10 and Israel accepts July 11.

July 13. President Johnson defines an "urgent need" for a "maximum number" of Arab refugees to be permitted by Israel to return to their homes.

July 17. Israel tells the General Assembly that the Arab states must recognize Israel's "statehood, sovereignty and international rights" before peace talks can begin.

Aug. 3. Israel and the U.A.R. agree to a U.N. proposal to halt for one month navigation through the Suez Canal.

Aug. 9. It is reported that the Sudan will obtain arms from the Soviet Union, Yugoslavia and Czechoslovakia.

Aug. 23. The Shah of Iran, ending two days of talks with President Johnson in Washington, arranges the purchase of a second squadron of the latest U.S. fighter jets.

Sept. 1. Arab heads of state, meeting in Khartoum, agree to seek a nonmilitary solution to the tensions with Israel. Meanwhile, Israel rejects a Yugoslav peace proposal calling for a return of Israeli troops to pre-June 5 positions, free passage of Israeli ships through the Strait of Tiran and free cargo movement through the Suez Canal.

Sept. 24. Israel announces it will move settlers into occupied Syria and the captured Jordanian sector of Jerusalem. The United States expresses its "disappointment" with that decision Sept. 26.

Sept. 27. Israeli Foreign Minister Eban suggests the economic cooperation of Israel, Lebanon and Jordan, the demilitarization of the Sinai, and the establishment of a "universal status" for the "holy places" of Jerusalem.

Sept. 29. Rejecting Israel's suggestion of direct negotiations, Egypt's Foreign Minister Raid accuses the United States of violating its pledge of territorial integrity for each Middle East state.

Oct. 3. Israel repudiates the 1949 armistice treaty and declares that a new treaty must be negotiated.

Oct. 22. Israel announces plans to build an oil pipeline between Elath on the Gulf of Aqaba and Ashdod on the Mediterranean to circumvent the Suez Canal.

Oct. 24. The United States announces it will fill arms orders, placed before the June war, to Israel and five pro-West Arab states.

Oct. 26. Mohammed Reza Pahlavi, crowns himself Shah of Iran and his wife, the Empress Farah, as Iran's first crowned queen.

Nov. 5. Jordan's King Hussein tells an American television audience that his country is ready to recognize Israel's right to existence.

Nov. 22. The U.N. Security Council unanimously adopts a British proposal (Resolution 242) for bringing peace to the Middle East. Under the plan, Israel would withdraw from all conquered territory, each country would agree to recognize the territory of the other states, and free navigation through international waterways would be assured.

Nov. 23. U.A.R. President Nasser says he will continue to deny Israeli ships access through the Suez Canal and that Israeli withdrawal from occupied lands is not open to negotiation.

Nov. 28. Britain declares the independence of South Arabia, which is renamed the People's Republic of South Yemen. British troops complete their evacuation of Aden Nov. 29. The country had been a British colony since 1839.

Dec. 24. Iraq and the Soviet Union agree to cooperate in the development of oil deposits in southern Iraq.

1968

Jan. 7. President Johnson confers with Israeli Premier Levi Eshkol on his Texas Ranch. It is later disclosed that Johnson promised Eshkol more Skyhawk A-4 fighter-bombers.

Jan. 11. Israel expropriates a section of the former Jordanian sector of Jerusalem, promising compensation for the private land owners.

Feb. 14. The United States announces its intention to resume arms shipments to Jordan.

Feb. 15. Significant fighting between Jordan and Israel erupts along their common border. U.S. embassies in both states are successful in negotiating a cease-fire after eight hours of fighting. The next day, Jordan's King Hussein calls for an end to terrorist activities originating within Jordan against Israel because, he says, such raids prompt Israeli retaliation.

Feb. 27. South Yemen, in an effort to remove "vestiges of colonial rule," dismisses all British military and administrative officers.

March 20. Continuing to reject direct negotiations with Israel, the U.A.R. nonetheless indicates its willingness to implement the British proposal adopted by the United Nations on Nov. 22, 1967.

April 10. U.A.R. President Nasser says his country is "fully prepared to support and arm the Palestine resistance movement" in its terrorist activity against Israel. Iraq April 13 announces the formation of a committee to raise funds for the Arab guerrillas.

April 25. Iraq announces an agreement to resume diplomatic relations with Britain, broken during the June war of 1967.

May 28. Israel pays the United States $3-million for the families of Americans killed by an erroneous Israeli bombing of a U.S. vessel June 8, 1967.

June 1. Martial law is proclaimed in Beirut, Lebanon, following an unsuccessful assassination attempt on President Chamoun.

July 5. Israel rejects plans for a U.N. peacekeeping force in the Israeli-occupied Sinai Peninsula, endorsed the day before by the U.A.R., in place of direct negotiations for a peace settlement with Arab states.

July 6. Nine Persian Gulf sheikdoms agree to a loose form of federation.

July 15. Arab guerrilla leaders meeting in Cairo agree to coordinate their activities against Israel.

July 17. The fourth coup in 10 years deposes Iraq's government. Ahmed Hassan al-Bakr is named president and premier July 31.

Aug. 11. Reportedly easing its position on a Middle East settlement, the U.A.R. says it will agree to a demilitarization of the Sinai Peninsula, will lift demands for the return of Arab refugees to their homeland, will agree to internationalizing the Gaza Strip and will grant Israeli cargoes access throught the Suez Canal and Israeli vessels through the Strait of Tiran.

Oct. 8. Israeli Foreign Minister Abba Eban offers a nine-point peace plan at the United Nations. The proposal calls for Israeli withdrawal from occupied territory following the establishment of "permanent" boundaries between the Arab states and Israel. The U.A.R. rejects the plan Oct. 9, but agrees Oct. 10 to accept a timetable worked out by U.N. special representative Jarring for implementing the British peace proposal adopted by the Security Council Nov. 22, 1967.

Oct. 27. After night-long shelling between Egypt and Israel, Egyptian oil refineries at Port Suez are afire.

Nov. 6. Reports out of Washington indicate that the United States has agreed to sell 58 Phantom jets to Israel.

Nov. 19. Israel declares it will allow the return of 7,000 Arabs who fled during the June 1967 war.

Dec. 2. Jordanian and Israeli troops clash in heavy battle. The next day Israeli planes counter a pre-dawn artillery attack along the Jordan border. On Dec. 4, Israeli bombers strike back against Iraqi shelling along the Israeli-Jordan border.

Dec. 5. Jordan's King Hussein sends messages to the U.A.R., Saudi Arabia, Iran, Kuwait and Lebanon calling for unified action to liberate Arab lands.

Dec. 9. President-elect Nixon's special envoy to the Middle East, former Pennsylvania Governor William Scranton, tells newsmen that he believes American policy in the area should be more "even-handed."

Dec. 26. Two Arab terrorists attack an Israeli jet at the Athens airport, setting it afire. In retaliation, an Israeli task force Dec. 28 attacks the Beirut International Airport, destroying several airplanes.

1969

Feb. 2. *Newsweek* magazine publishes an exclusive interview with U.A.R. President Nasser who suggests a five-point peace plan for the Middle East: "a declaration of nonbelligerence; the recognition of the right of each country to live in peace; the territorial integrity of all countries in the Middle East, including Israel, within recognized and secure borders; freedom of navigation on international waterways; a just solution to the Palestinian refugee problem." On Feb. 4, Israel rejects

Nasser's plan as "a plan for liquidating Israel in two stages."

Feb. 18. An Israeli airliner is attacked by four Arab terrorists at the Zurich, Switzerland, airport.

Feb. 26. Israeli Premier Levi Eshkol dies following a heart attack.

March 7. Foreign Minister Golda Meir accepts election as leader of the Labor Party and thereby becomes premier of Israel, succeeding Eshkol.

March 14. Israeli Foreign Minister Abba Eban, in Washington for talks with President Nixon, addresses the National Press Club and denounces attempts by the four major powers to reach a settlement in the Mideast.

March 19. The Shah of Iran says he is opposed to continued use by the United States of the Bahrain Island naval base—used by the U.S. since 1949—after British troops withdraw from the Persian Gulf in 1971.

April 10. In Washington for talks with President Nixon, Jordan's King Hussein addresses the National Press Club and sets forth a six-point peace plan which, he says, has U.A.R. President Nasser's approval. Similar to the U.N. proposal of November, 1967, the plan is contingent upon Israeli withdrawal from occupied lands. Israel rejects the plan the following day as propaganda.

April 19. It is reported that Soviet missiles have been installed along the Suez Canal.

April 22. U.N. Secretary General U Thant says that Israel and Egypt are engaged in "a virtual state of active war," and declares that the U.N. cease-fire has become "totally ineffective in the Suez Canal sector." The U.A.R. April 23 repudiates the U.N. cease-fire, contending that Israeli forces had not advanced to the eastern bank of the Canal when the cease-fire was adopted in 1967.

May 5. Israel's Prime Minister Meir says "signed peace treaties" are the only acceptable follow-up to the U.N. cease-fire of 1967.

May 11. Palestinian guerrilla leaders meet with Jordanian leaders in an attempt to reverse Jordan's new policy of forbidding Jordan-based raids on Israel.

May 25. The government of the Sudan is overthrown by a military coup. Abubakr Awadallah becomes premier and pledges to work more closely with the U.A.R. and other "progressive" Arab states.

June 3. Israel asks the United States to provide protection against Arab sabotage for the American-owned pipeline from Saudi Arabia to Lebanon. Until protection is provided, the line will be closed.

June 18. Jordan reports that Palestinian guerrillas are withdrawing from the country.

July 3. Turkey and the United States sign a new mutual defense treaty under which Turkey controls U.S. military installations within its borders.

July 7. U.N. Secretary General U Thant announces the resumption of open warfare along the Suez Canal.

July 20. After two weeks of sporadic fighting along the Suez Canal, Israeli jets attack U.A.R. ground installations for the first time since the June 1967 war.

July 31. Staging an offensive for the first time since the 1967 war, Syria launches an attack, supported by jets and heavy artillery, against the main Israeli post on Mount Hermon and on the Israeli-Lebanon border.

Aug. 3. Israel announces it will retain the Golan Heights, the Gaza Strip and part of the Sinai Peninsula in order to protect its security.

Aug. 10. Iraq announces a major aid agreement with the Soviet Union. Iraq will pay for the aid with crude oil.

Aug. 23. U.A.R. President Nasser calls for an all-out war with Israel, charging Israel with responsibility for the fire at the Al Aksa Mosque in Jerusalem Aug. 22.

Aug. 31. Arab hijackers blow up a TWA airliner after diverting it to Damascus following take-off from Rome. The hijackers demand the imprisonment of all the Israeli passengers, but Syria released all but six Israelis Aug. 30. Those six are being held hostage for the release of Syrian prisoners of war in Israel.

Sept. 1. Libya's King Idris is overthrown by a revolutionary council. The council, on Sept. 2, says it will honor existing agreements with oil companies. Muammar el-Qaddafi, head of the Revolutionary Command Council, emerges as leader of the new regime.

Sept. 18. President Nixon addresses the U.N. General Assembly, suggesting an arms curb in the Middle East by the big powers. The Soviet Union rebuffs the suggestion Sept. 19.

Oct. 7. An Australian sheepherder pleads guilty at his trial in Israel to setting fire to the Al Aksa Mosque in Jerusalem Aug. 22.

Oct. 13. Israel proposes home rule for the West Bank of occupied Jordan.

Oct. 22. The nine-nation Federation of Persian Gulf Emirates is established with Sheikh al-Nahayan of Abu Dhabi as president.

Oct. 28. The new military regime in Libya Sept. 1 notifies the United States that Wheelus Air Base, near Tripoli, must be evacuated by Dec. 24, 1970.

Oct. 28. The Sudan's Premier Awadallah is deposed by the Revolutionary Command Council and the chairman of the council, Gaafar al-Nimeiry, assumes the premiership.

Dec. 9. Secretary of State Rogers discloses a previously private U.S. proposal for peace in the Mideast, including a provision for Israel's withdrawal from occupied lands in exchange for a binding peace treaty signed by the Arabs. Israeli Premier Golda Meir says Dec. 12 that the plan is an attempt by the United States to "moralize."

Dec. 13. Britain announces its agreement to withdraw its forces from Libya by March 31, 1970.

1970

Jan. 16. Muammar el-Qaddafi, leader of the Revolutionary Command Council, becomes premier and defense minister of Libya, succeeding Mahmoud Soliman al-Maghreby.

Jan. 21. Israel, launching what it terms its largest ground operation since June 1967, captures the Egyptian island of Shadwan at the entrance of the Gulf of Suez.

Feb. 2. The heaviest fighting since the June 1967 war breaks out between Israeli and Syrian troops in the Golan Heights.

Feb. 4. It is reported that President Nixon has sent a note to the Soviet Union urging cooperation in resolving the Middle East problem.

Feb. 21. A bomb enclosed in a letter addressed to Israel explodes in mid-air in a Swiss aircraft bound for Israel and kills all 47 passengers aboard.

March 19. A large number of Soviet troops and antiaircraft missiles are reported by diplomatic observers to have arrived in Cairo.

March 31. Libya stages nationwide celebrations as the last British troops withdraw after a 30-year presence there.

April 14. The civil war in Yemen between republican and royalist forces ends. Saudi Arabia signs an agreement with the republican regime, pledging to discontinue the arms and funds to royalist rebels that had been supplied since the outbreak of hostilities in 1962.

May 9. Israel warns that the installation of Soviet SAM-3 missiles along the Suez Canal will not be permitted and threatens to attack Soviet planes if they interfere with Israeli attacks on Egyptian bases.

June 11. Jordanian King Hussein takes control of the army and removes the commander in chief, in response to guerrilla demands.

June 11. The United States formally turns over Wheelus Air Force Base to Libya, ending 16 years of operation.

June 24. Foreign intelligence reports received in Washington indicate that Soviet pilots have taken over the air defense of Egypt against Israel and are flying combat missions south of the Suez Canal.

June 25. Secretary of State Rogers tells a news conference in Washington of a broad-based diplomatic effort to encourage Arab and Israeli representatives "to stop shooting and start talking" under U.N. supervision. The heart of the proposal is a 90-day cease-fire tied to withdrawal of Israeli forces from territory occupied during the June 1967 war. Israeli Prime Minister Meir rejects the plan June 29.

July 5. Libya's Revolutionary Command Council announces nationalization of that country's four oil distributing companies.

July 17. At the conclusion of U.A.R. President Nasser's visit to Moscow, a joint Soviet-U.A.R. communique is issued calling for a political settlement of the Middle East crisis.

July 23. The U.A.R. accepts a U.S. proposal of June 19 calling for a 90-day cease-fire in the Middle East.

July 26. Jordan accepts the U.S. cease-fire proposal. One of the terms of the cease-fire requires Jordan to control guerrilla activities organized within its borders.

July 31. Yasir Arafat, chairman of the Palestine Liberation Organization, rejects the U.S. cease-fire proposal and all other compromise solutions to the conflict with Israel.

Aug. 4. Israel formally accepts the Middle East cease-fire. Israeli Prime Minister Golda Meir tells the Israeli parliament that she overcame her reluctance to accept the proposal after assurances of military and political support from President Nixon.

Aug. 7. A Middle East cease-fire goes into effect.

Aug. 18. Iraq's 12,000 troops in Jordan are placed at the disposal of the Palestinian commandos.

Aug. 23. The U.A.R. and Iran agree to renew diplomatic relations.

Sept. 6. The 14 nations of the Arab League call for an end to fighting between the Jordanian army and Arab commandos which has continued since the Sept. 1 aborted attempt to assassinate King Hussein.

Sept. 6. Members of the Popular Front for the Liberation of Palestine, an Arab commando group, are successful in three of four attempts to hijack commercial jets. A Pan Am 747 is forced to land in Cairo, passengers are disembarked, and the plane is blown up. A Swissair and a TWA plane are forced to land at a desert air strip in Jordan controlled by the commandos. The next day, the commandos release nearly half of the passengers from the planes brought down in the desert, and demand the release of Arab guerrillas held in Israel and Western Europe in exchange for the remaining hostages.

Sept. 9. Members of the Popular Front seize a fifth plane, a British BOAC jet, and hijack it to Jordan to join the two other planes in the desert. Hostages total nearly 300.

Sept. 12. Arab commandos release all but 54 of their hostages and blow up the three empty airplanes. Britain announces it will release the Arab commando seized in an aborted hijack effort Sept. 6.

Sept. 16. Jordanian King Hussein proclaims martial law and installs a military government as the civil unrest continues in his country.

Sept. 18. Israeli Prime Minister Meir meets with President Nixon in Washington and says Israel will not participate in the U.N. peace talks until new Egyptian missile installations along the Suez Canal are removed.

Sept. 26. President Nixon says he has ordered $5-million in relief for Jordan, which has been battered by a civil war.

Sept. 27. A 14-point pact is signed in Cairo by Arab heads of state—including Jordan's King Hussein and the leader of the Arab commandos—to end hostilities in Jordan. The agreement calls for King Hussein to remain on the throne, but for a three-member committee, headed by Tunisian Premier Ladgham, to supervise the government until conditions are normalized.

Sept. 28. U.A.R. President Gamal Abdel Nasser, 52, dies suddenly of a heart attack.

Sept. 29. Arab commandos release the last hostages held since four planes were hijacked earlier in the month. Forty-eight others had been released Sept. 25. Switzerland announces that a total of 19 Arabs will be released by Britain, West Germany, Switzerland and Israel.

Oct. 17. Anwar Sadat is sworn in as president of the U.A.R. following an Oct. 15 election in which he received 90 per cent of the vote.

Oct. 23. It is disclosed that the United States will send 180 tanks to Israel as part of $500-million in aid.

Nov. 4. The U.N. General Assembly votes for a 90-day extension of the Middle East cease-fire and for unconditional resumption of the peace talks between Arab states and Israel.

Nov. 13. Syrian President and Premier al-Attassi is reported to have been placed under house arrest. "Provisional leadership" will guide the government until a national congress can elect permanent leaders.

Nov. 27. Syria joins an alliance of the U.A.R., Libya and the Sudan to pool resources "for the battle with Israel."

Dec. 13. Jordanian King Hussein, via American television, urges a joint effort by the United States and Soviet Union to assume a peacekeeping role in the Middle East.

Dec. 22. Libya nationalizes all foreign banks.

1971

Jan. 15. U.A.R. President Sadat and Soviet President Podgorny dedicate the Aswan Dam.

Feb. 14. Representatives of six oil-producing countries (Iran, Iraq, Saudi Arabia, Kuwait, Abu Dhabi and Qater) reach agreement with Western oil companies. The accord has a five-year tenure and increases payments by more than $1.2-billion for 1971. Agreements have yet to be reached with four other oil-producing countries.

March 12. Golda Meir, stating Israel's position on a Middle East settlement, calls for Israeli retention of Sharm el Sheik, Israel's sole land link with East Africa and Asia, demilitarization of the Sinai, and Israeli possession of Jerusalem.

March 13. Premier Hafez al-Assad is proclaimed president of Syria.

March 19. Israeli Foreign Minister Abba Eban declares that Israel will not withdraw to boundaries that existed before the June 1967 war.

April 2. Libya signs an agreement with 25 Western oil companies raising the price of Libyan oil from $2.55 to $3.45 a barrel.

April 3. Amid continued Palestinian guerrilla activity, Jordanian King Hussein says the guerrillas must remove their weapons from Amman within two days. The commandos have said they will stay in Amman and continue their fight to overthrow Hussein and to use Jordan as a base for operations against Israel.

April 7. The United States agrees to increase military aid to Jordan.

April 17. Egypt, Syria and Libya sign an agreement to form the Federation of Arab Republics. Plebiscites will be held Sept. 1 in the three countries to gain popular approval for the union.

May 1. Secretary of State Rogers arrives in the Middle East to confer with Saudi Arabian King Faisal, King Hussein of Jordan, Egyptian President Sadat, Lebanese officials and Israeli Premier Golda Meir on proposals for reopening the Suez Canal.

May 27. Saudi Arabian King Faisal arrives in Washington on an official visit to discuss the Middle East with President Nixon.

May 27. U.A.R. President Sadat and Soviet President Podgorny sign a 15-year treaty of friendship and cooperation.

June 7. By agreement, Iraq raises the price of oil to foreign oil companies by 80 cents a barrel.

July 4. The U.A.R. and the Soviet Union issue a joint communique, declaring that the Suez Canal will be opened only after Israel withdraws all of its forces from Arab territory.

July 6. Syria announces an agreement with Western-owned Iraq Petroleum Company under which Syria will receive 50 per cent more in payments for permitting pipelines across the country.

July 22. Sudanese Premier Nimeiry is returned to power after his ouster two days before by a faction of the military.

July 26. U.A.R. President Sadat is granted "full powers" by the national congress of the Arab Socialist Union to take whatever action necessary to recover Arab lands from Israel.

Aug. 2. Soviet officials are ordered out of the Sudan following charges that they influenced the coup that temporarily ousted Premier Nimeiry.

Aug. 12. Syria breaks off diplomatic relations with Jordan following clashes on their common border.

Aug. 14. Jordan demands the end of economic pressures exerted by other Arab states as a condition for Jordan's efforts to resolve hostilities between the army and Palestinian commandos.

Aug. 28. It is reported that Lebanon will buy arms from the Soviet Union.

Sept. 2. Citizens of Egypt, Libya and Syria vote almost unanimously for the Federation of Arab Republics aimed at providing a solid front against Israel.

Oct. 4. Egyptian President Sadat is selected first president of the Federation of Arab Republics.

Oct. 4. Secretary of State Rogers presents a detailed account of the U.S. position on the Middle East. He calls on Israel and Egypt to agree to open the Suez Canal as a first step towards peace in the area. Egypt rejects the proposal Oct. 6, lacking assurances of Israeli withdrawal from Arab lands.

Nov. 1. Israel says it will not agree to the U.S. proposal for indirect talks with Egypt without U.S. guarantees of a continuing supply of Phantom jets.

Nov. 28. Jordanian Premier Wasfi Tal is assassinated while visiting Cairo. Ahmed al-Lawzi succeeds him as Premier Nov. 29. The Jordanian government Dec. 17 formally charges Al Fatah, the Palestinian guerrilla group, with responsibility for the slaying.

Nov. 30. Iranian troops occupy the Iraqi territory of Abu Musa, Greater Tumb and Lesser Tumb. Iraq severs relations with Iran and Britain as a result.

Dec. 2. Israeli Premier Meir confers with President Nixon in Washington.

Dec. 2. The United Arab Emirates proclaims its independence. Zaid ben Sultan al-Nahayan of Abu Dhabi is named president of the union, which consists of six Persian Gulf sheikdoms.

Dec. 7. Libya nationalizes British oil interests and withdraws all deposits from British banks in protest of what it regards as British collusion in the Iranian seizure of Iraqi lands.

1972

Jan. 5. The United States announces an agreement with Bahrain for the establishment of a U.S. naval base in the Persian Gulf.

Jan. 20. An agreement to adjust oil prices to account for devaluation of the dollar is signed in Geneva between six oil-producing Persian Gulf nations (Abu Dhabi, Iran, Iraq, Kuwait, Qatar and Saudi Arabia) and Western oil companies. It raises the posted price of oil by 8.49 per cent.

Feb. 11. Ras al Khaimah, a small sheikdom, becomes the seventh member of the United Arab Emirates.

Feb. 26. Sixteen years of civil war in the Sudan are ended by an agreement between the government and the South Sudan Liberation Front.

March 12. The Organization of Petroleum Exporting Countries votes to accept an Arabian-American Oil Company (Aramco) offer to give a 20 per cent share of the company to the six countries of Iraq, Iran, Kuwait, Saudi Arabia, Abu Dhabi and Qatar.

March 15. Jordanian King Hussein unveils his plan to make Jordan a federated state comprised of two autonomous regions on the East and West Bank of the Jordan River. Israel currently holds the West Bank region. Hussein proposes Jerusalem as the capital of the West Bank, or Palestine, region. Israel, the same day, denounces the plan. The Federation of Arab Republics denounces the plan March 18 and calls on all Arab governments to similarly reject it.

March 28. King Hussein of Jordan meets with President Nixon in Washington.

April 6. Egypt severs diplomatic relations with Jordan, criticizing King Hussein's proposal for a federation of Jordan and the West Bank.

May 30. President Nixon receives a warm welcome in Iran on a stopover visit en route to Poland and then Washington after the U.S.-Soviet summit in Moscow.

June 1. Iraq nationalizes the Iraq Petroleum Company, owned jointly by American, British, French and Dutch oil companies. The company produces 10 per cent of Middle East oil.

July 18. Egyptian President Sadat orders all Soviet military advisers and experts out of his country and places all Soviet bases and equipment under Egyptian control. Sadat says July 24 during a four-hour speech that the Soviet Union's "excessive caution" as an ally led him to his decision.

Aug. 2. Libyan leader Qaddafi and Egyptian President Sadat issue a joint declaration to establish "unified political leadership."

Sept. 5. Arab commandos of the Black September organization seize a building housing 26 Israeli athletes at the Olympic Games in Munich, West Germany. Two Israelis are killed; nine are held hostage. When day-long negotiations fail, the nine hostages are killed. Israel launches reprisal raids Sept. 8 on Arab guerrilla bases in Syria and Lebanon, the most extensive raids since the June 1967 war.

Sept. 13. Syria and the Soviet Union agree to security arrangements. The Soviet Union will improve naval facilities in two Syrian ports for Soviet use and Syria will receive jet fighters and air defense missiles.

Sept. 18. Egyptian President Sadat and Libyan leader Qaddafi agree to make Cairo the capital of their projected unified state, to popularly elect one president, and to allow one political party. This action marks the first step toward gradual unification since the announcement Aug. 2 of the intent to unite.

Sept. 19. A letter bomb, believed to have been sent by the Black September organization, explodes in the Israeli embassy in London, killing the Israeli agricultural counselor. Similar terrorist activities in the next two days are blamed on the Black September group.

Oct. 15. Israel, launching its first unprovoked attack on Palestinian guerrilla bases in Syria and Lebanon, says "we are no longer waiting for them to hit first."

Oct. 15. A New York City postal clerk's hands are maimed as a letter bomb, addressed to an undisclosed official of the Women's Zionist Organization, explodes. The incident follows a flurry of such letter bombs in New York.

Oct. 28. Ending several weeks of heavy fighting, Yemen and South Yemen sign an accord in Cairo calling for their merger.

Oct. 29. Two Arab guerrillas of the Black September group hijack a West German airliner, forcing the release of three commandos held in the Sept. 5 murder of 11 Israeli athletes at the Olympic Games. Israel protested the German action on Oct. 30, calling it "capitulation to terrorists."

Nov. 1. It is reported that the Soviet Union will restore missiles to Egypt's air defense system that were removed when the Soviets were ousted by the government in July.

Nov. 14. Western diplomatic sources report that the attack on Israeli athletes at the Olympic Games was aimed at obstructing secret Egyptian-Israeli negotiations.

Nov. 27. Jordanian King Hussein confirms reports of an aborted coup to overthrow him, planned by Libyan leader Qaddafi, Yasir Arafat, leader of the Palestinian guerrillas, and other Palestinians.

Dec. 16. Tunisian President Bourguiba rejects a proposal for the union of his country with Libya.

1973

Jan. 8. An agreement between Kuwait and two Western oil companies grants Kuwait a 25 per cent share of oil concessions in the country. Over 10 years, Kuwait will assume 51 per cent ownership.

Feb. 7. Jordanian King Hussein visits the United States.

Feb. 21. A Libyan passenger airliner, reportedly failing to heed Israeli instructions to land after straying over Israeli-occupied Sinai, is fired upon and crashes, killing 106 persons. Israel assumes no responsibility for the crash and instead blames the airline pilot for not landing and the Cairo air control for misguidance.

Feb. 26. Israeli Premier Meir arrives in the United States for 10-day state visit.

Feb. 28. Iraq announces that Western oil companies have accepted the nationalization of Iraq Petroleum Company in exchange for compensation.

March 1. The U.S. ambassador to the Sudan, Cleo A. Noel, Jr., and U.S. charge d'affaires, George C. Moore, are seized and later killed by Black September terrorists in Khartoum who demand the release of several hundred Arab prisoners, including the convicted slayer of Sen. Robert F. Kennedy. A Belgian diplomat also is killed.

March 7. Sudanese President Nimeiry bans all guerrilla activity, charging that Al Fatah, a Palestinian commando groups, planned the embassy siege that left two Americans dead.

March 14. Golda Meir offers Jordan guardianship of the Islamic shrines in the Old City of Jerusalem.

March 16. The Shah of Iran announces that Western oil companies have turned over "full control" of that country's oil industry.

March 28. Egyptian President Sadat proclaims martial law.

May 7. Martial law is proclaimed in Lebanon as renewed fighting between the army and Palestinian guerrillas breaks out.

May 7. Israel celebrates its 25th national anniversary.

May 15. In a symbolic protest against Israel, Iraq, Libya, Kuwait and Algeria temporarily suspend oil shipments to the West.

June 11. Libyan leader Qaddafi nationalizes an American oil company in protest of the U.S. policy on Israel.

July 1. An Israeli military attache in the Israeli embassy in Washington, D.C., is slain, apparently by Arab terrorists.

July 20. Arab and Japanese hijackers seize a Japan Air Lines jet and demand the release of a man serving a life sentence for taking part in the May 1972 massacre of 26 persons at the Tel Aviv airport. The hijackers blow up the plane July 24 at Libya's Benghazi airport after the passengers and crew have been evacuated. The hijackers are arrested by Libyan officials.

July 25. The Shah of Iran pays a state visit to the United States.

Aug. 5. Two Arab Black September terrorists kill three persons and wound another 55 after firing machine guns and hurling grenades in the Athens airport.

Aug. 11. Israel forces a Middle East Airlines jet, flying over Lebanon, to land in Israel. Israel announces it diverted the wrong plane in its search for the leader of the Popular Front for the Liberation of Palestine, the group held responsible by Israel for the slayings in the Athens airport.

Aug. 29. Libyan leader Qaddafi and Egyptian President Sadat announce the "birth of a new unified Arab state" and the gradual approach to unification of their countries. Egypt had insisted on gradual unification instead of completion of the union by Sept. 1, 1973, as originally agreed.

Sept. 1. Libya nationalizes 51 per cent of the assets of all foreign oil companies.

Sept. 29. The Austrian government closes the Schonau facility, a key transit point for Jews leaving the Soviet Union for Israel. The action is taken to meet the demands of Arabs who kidnaped and later released three Soviet Jews and an Austrian customs official.

Oct. 6. War breaks out in the Middle East on the Jewish holy day of Yom Kippur. Egyptian forces cross the Suez Canal and Syria attacks the Golan Heights. Israeli forces counter on Oct. 7, striking back in the Sinai and on the Golan Heights.

Oct. 7. Iraq nationalizes the American-owned Mobil Oil Corporation and Exxon Corporation.

Oct. 8. Tunisia, the Sudan, and Iraq pledge support of Egyptian and Syrian forces battling Israel.

Oct. 10. Israel announces it has abandoned the Bar-Lev line along the Suez Canal but has pushed back Syrian forces from the Golan Heights. Egyptian forces cross the Suez and advance nearly 10 miles onto the East bank. The Syrian army is pushed back to the 1967 cease-fire line.

Oct. 12. Israeli forces advance to within 18 miles of Damascus, the capital of Syria.

Oct. 13. Jordan announces it will join Egypt and Syria in the war against Israel. The same day, Israel claims to have nearly eliminated an Iraqi division in Syria.

Oct. 13. Saudi Arabian troops join the war against Israel after urging by Egyptian President Sadat.

Oct. 15. The United States announces it is resupplying Israel with military equipment to counterbalance a "massive airlift" to Egypt by the Soviet Union.

Oct. 17. Arab oil-producing states announce a 5 per cent reduction in the flow of oil to the United States and other countries supporting Israel.

Oct. 17. Egyptian President Sadat, in an open letter to President Nixon, proposes an immediate cease-fire on the condition that Israel withdraws to pre-1967 boundaries. The same day, foreign ministers of four Arab states meet in Washington with President Nixon and Secretary of State Henry Kissinger to present a similar peace proposal.

Oct. 18. Libya cuts off all shipments of crude oil and petroleum products to the United States.

Oct. 18. Saudi Arabia announces a 10 per cent cut in oil production and pledges to cut off all U.S. oil shipments if American support of Israel continues.

Oct. 19. President Nixon asks Congress to appropriate $2.2-billion for emergency military aid for Israel.

Oct. 19. Libya cuts off all exports to the United States and raises the price of oil from $4.90 to $8.92 per barrel.

Oct. 20. Saudi Arabia halts oil exports to the United States.

Oct. 20. Secretary of State Kissinger arrives in Moscow for talks with Soviet Communist Party chief Leonid I. Brezhnev on restoring peace to the Middle East.

Oct. 21. Iraq nationalizes the holdings of Royal Dutch Shell Corporation.

Oct. 21. Kuwait, Qatar, Bahrain and Dubai announce suspension of all oil exports to the United States, theoretically marking the total cut-off of all oil from Arab states to the United States.

Oct. 21. The United States and the Soviet Union present a joint resolution to the U.N. Security Council calling for a cease-fire in place in the Middle East and for implementation of a Security Council resolution calling for Israeli withdrawal from lands occupied since the 1967 war. The proposal, formulated during Kissinger's trip to Moscow, is adopted by the Security Council early Oct. 22.

Oct. 22. A cease-fire takes effect on the Egyptian-Israeli front, but fighting continues nonetheless.

Oct. 22. Kissinger confers with Israeli Premier Golda Meir in Israel on his way back to Washington from Moscow. Meanwhile, Jordan accepts the U.S.-U.S.S.R. cease-fire proposal. Iraq and the Palestinian Liberation Organization reject it.

Oct. 23. The U.N. Security Council votes to reaffirm the Middle East cease-fire, asks Egypt and Israel to return to the cease-fire line established the day before, and asks that U.N. observers be stationed along the Israeli-Egyptian cease-fire line. The U.N. secretary general announces Syria will accept the cease-fire if Israel withdraws from lands occupied during the 1967 war. Egypt and Israel accuse each other of cease-fire violations as heavy fighting resumes on the canal front. Israeli forces on the West Bank of the canal push south to cut off completely both the city of Suez and the 20,000-man Egyptian III Corps on the East Bank.

Oct. 24. Egypt's Sadat appeals for the United States and the Soviet Union to send troops to supervise the cease-fire. The White House announces it will not send forces. Israel claims it rebuffs the III Corps' attempt to break out of the trap, after which Israel says calm returns to the Suez front. Egyptians say Israelis reached Adabiya on the Red Sea coast.

Oct. 25. President Nixon orders a world-wide U.S. military alert as tension mounts over whether the Soviet Union may intervene in the Middle East crisis. Kissinger says there are "ambiguous" indications of that action. Speculation is that the Soviets might try to rescue the trapped Egyptian army.

Oct. 25. To avert a U.S.-U.S.S.R. confrontation in the Middle East, the U.N. Security Council votes to establish an emergency supervisory force to observe the cease-fire. The force would exclude troops from the permanent Security Council members, particularly the United States and the Soviet Union.

Oct. 27. The United States announces that Egypt and Israel have agreed to negotiate directly on implementing the cease-fire.

Oct. 28. The trapped Egyptian III Corps receives food, water and medical supplies after Israel agrees to allow a supply convoy to pass through Israeli lines. It is reported that Israel yielded following U.S. warnings that the Soviet Union threatened to rescue the troops. Israeli sources concede that on Oct. 23 their units drove to the port of Adabiya to isolate the III Corps.

Oct. 29. In a flurry of diplomatic activity, Egyptian Foreign Minister Fahmy meets with Kissinger in Washington.

Oct. 29. Syrian President Assad says Syria accepted the cease-fire after U.S.S.R. guarantees of Israeli withdrawal from all occupied territory and recognition of Palestinian rights.

Oct. 31. Israeli Prime Minister Golda Meir arrives in Washington for talks with President Nixon on her country's concern over U.S. pressure to make concessions. The same day, Egyptian President Sadat warns that his country will take up the fight again if Israel does not withdraw to the cease-fire lines of Oct. 22, 1973.

Nov. 1. Israeli Prime Minister Meir, meeting in Washington with President Nixon, says she has been assured of continued U.S. support.

Nov. 2. Secretary of State Kissinger meets separately in Washington with Meir and Egyptian Foreign Minister Fahmi.

Nov. 4. The Organization of Arab Petroleum Exporting Countries (OAPEC) reports that exports are running 25 per cent below the September export level, plans an additional 5 per cent production cut for later in the year and sends Saudi Arabian and Algerian oil ministers on a tour of western nations to explain the embargo.

Nov. 6. Israel lists 1,854 casualties from the war.

Nov. 7. After talks between Kissinger and Sadat, it is announced that Egypt and the United States will resume diplomatic relations. Ties are resumed Feb. 28, 1974.

Nov. 8. Kissinger flies to Jordan and Saudi Arabia to meet with leaders there.

Nov. 11. Israel and Egypt sign a cease-fire accord, drawn up by Kissinger and Sadat during recent talks. The six-point plan calls for (1) both sides to observe the cease-fire, (2) immediate discussions on the return to the Oct. 22 cease-fire lines, (3) immediate food and medical supplies for Suez City, (4) access for non-military supplies to the stranded Egyptian III Corps on the East Bank of the Suez Canal, (5) replacement of Israeli troops along the Suez by U.N. forces and (6) exchange of all prisoners of war. The signing is the first by the two parties on an important joint document since the 1949 armistice ending the first Arab-Israeli war. The ceremonies are in a tent at the kilometer 101 marker, signifying the distance to Cairo, on the Cairo-Suez road. High officers of the two countries immediately begin direct discussions to implement the cease-fire.

Nov. 14. Kilometer 101 negotiators agree to exchange POWs and to turn over check points on the Cairo-Suez road to a U.N. truce force.

Nov. 15. The first planeloads of Egyptian and Israeli POWs are exchanged.

Nov. 18. OAPEC cancels its 5 per cent output cut slated for December in a conciliatory gesture to most West European nations. The Netherlands and the United States are exempted because of their pro-Israeli stance.

Nov. 22. Saudi Arabia threatens to cut oil production by 80 per cent if the United States retaliates for Arab oil cuts or embargoes.

Nov. 26. The conference of Arab heads of state opens in Algeria to discuss the recent war. On Nov. 28, 15 Arab leaders declare Middle East peace is conditional on Israeli withdrawal from "all occupied Arab territories."

Dec. 9. Nine Arab oil ministers, meeting in Kuwait, order a new oil cutback of 750,000 barrels a day, a 5 per cent reduction, effective Jan. 1, 1974, to be lifted only after Israel begins withdrawal from lands occupied during the 1967 war.

Dec. 17. Arab guerrillas attack an American airliner at Rome airport, killing 31 persons. Then they hijack a West German aircraft to Athens with hostages aboard. At Athens, Arab guerrillas kill one Italian hostage but are unsuccessful in their demands for the release of two Palestinian commandos held by Greek authorities. The aircraft flies to Kuwait Dec. 18, where the 12 hostages aboard are freed and five hijackers surrender to police.

Dec. 21. The first Arab-Israeli peace conference opens in Geneva. Israel, Egypt, Jordan, the United States, the Soviet Union and the United Nations are represented. Syria boycotts the conference.

Dec. 23. Six Persian Gulf states raise the price of their oil from $5.11 to $11.65 a barrel, effective Jan. 1, 1974.

Dec. 25. The Saudi Arabian oil minister, speaking for the OAPEC countries, announces the cancellation of the 5 per cent oil production cut and instead discloses a 10 per cent increase. The U.S. oil embargo will continue, however.

Dec. 31. Israel holds general elections, resulting in a loss of parliamentary seats for Prime Minister Meir's Labor alignment and in a governmental crisis extending into spring of 1974 as Meir seeks to form a government. Results for the 120-member Knesset: Labor alignment, 51 seats, a loss of five; right-wing Likud coalition, 39, a gain of seven; National Religious Party (Labor's former coalition partner), 11, a loss of one; minor parties, 19 seats.

1974

Jan. 9. Egypt and Israel suspend the Geneva conference on troop disengagement along the Suez front to await new mediation by Kissinger, who arrives in Egypt two days later.

Jan. 17. Kissinger's "shuttle diplomacy" results in announcement of accords on Suez disengagement; agreements include the "U.S. proposal" for limiting the number of troops and deploying military equipment along the canal front.

Jan. 18. The accords are signed. The chief provisions are: Israel is to abandon its West Bank bridgehead and to withdraw on the East Bank about 20 miles from the canal; Egypt is to keep a limited force on the East Bank; a U.N. truce force is to patrol the buffer zone between the two; the pullback is to be completed in 40 days. Sadat says he will press Syria to open talks with Israel.

Jan. 25. Israeli forces begin the Suez withdrawal.

Jan. 28. The pullback lifts the siege of the city of Suez and ends the isolation of the Egyptian III Corps.

Jan. 30. In Jerusalem, Meir accepts a formal request to form a new cabinet.

Feb. 9. France and Iran sign a $5-billion energy agreement, which includes French construction of five nuclear power plants in Iran.

Feb. 11. Libya announces the complete nationalization of three American oil companies.

Feb. 13. At a Washington meeting dealing with the world energy shortage, 13 oil-consuming nations agree to seek a meeting with the oil-producing nations. France dissents.

Feb. 22. Israeli Defense Minister Dayan's refusal to serve in a new cabinet complicates the Meir bid to form a minority government.

Feb. 27. Kissinger, seeking a Syrian-Israeli disengagement on the Golan Heights front, arrives in Israel from Damascus and turns over to Meir a list of 65 Israeli POWs held in Syria.

Feb. 28. The United States and Egypt renew full diplomatic relations after a seven-year break. President Sadat announces he has invited President Nixon to visit Egypt.

March 4. The nine Common Market countries offer the Arab world a long-term plan for technical, economic and cultural cooperation. The move leads to a U.S. charge that it was not informed of the action by its West European partners. Paris claims Washington was consulted in advance.

March 4. Israel completes its Suez front pullback, restoring to Egypt control of both banks of the canal for the first time since the 1967 war.

March 10. A nine-week crisis ends in Israel with the formation of a new cabinet, including Moshe Dayan again as defense minister. He had wanted the right-wing Likud Party included in the coalition, a move repeatedly rejected by Meir.

March 11. Iraq offers self-rule to its Kurdish minority in the north, but the Kurds reject the proposal and sporadic fighting continues. Also during the month, an undetermined number of casualties results from Iraqi-Iranian border clashes.

March 18. After a joint meeting in Vienna, Saudi Arabia, Algeria, Egypt, Kuwait, Abu Dhabi, Bahrain and Qatar lift a five-month oil embargo against the United States, but Libya and Syria refuse to join in the decision.

March 18. The United States joins in international operation to clear obstructions in the Suez Canal, closed since the 1967 war.

April 6. Reports say Libya's Qaddafi has been relieved of "political, administrative and traditional" duties, but continues as armed forces chief. However, later it is said he has relinquished only some ceremonial duties.

April 10. Meir quits the Israeli premiership in an intraparty squabble over where to put the blame for military shortcomings in the October war.

April 11. Three Arab guerrillas storm an apartment building in an Israeli town near the Lebanese-Syrian border, killing 18 persons, mostly women and children. The terrorists die from their own explosives.

April 12. Israel retaliates by raiding Lebanese villages and blowing up houses said to belong to Palestinian sympathizers.

April 18. President Sadat says Egypt will no longer rely solely on the Soviet Union for arms.

April 19. Kurdish rebels declare all-out war on the Iraqi government as bloody clashes continue.

April 23. Yitzhak Rabin is asked to form a new Israeli government. Labor minister in the last cabinet, he is former ambassador to the United States and commanded the Israeli army in the 1967 war.

April 29. Meeting in Geneva, Kissinger and Gromyko pledge U.S.-Soviet cooperation in seeking a troop separation accord on the Syrian-Israeli front.

April 30. Kissinger in Cairo begins a month-long quest to end the Golan Heights confrontation.

May 7. The Egyptian State Council rules illegal the confiscation of private property that began under Nasser.

May 15. In a schoolhouse battle at the Israeli town of Maalot, 16 teen-agers are killed and 70 are wounded after three Arab terrorists seize the school and demand the release of 23 prisoners held by Israel. The Arabs are slain when Israeli soldiers attack the school.

May 16. Israel initiates a week-long series of raids in replying to the Maalot tragedy. Planes and gunboats hit Palestinian camps and hideouts in Lebanon, killing at least 61 persons; the Israelis kill six Arab guerrilla infiltrators, who, Israel says, were planning a massacre similar to the one at Maalot.

May 28. Premier-designate Rabin forms a new Israeli cabinet.

May 29. Syria and Israel agree on disengagement.

May 31. The accords, achieved by Kissinger in his latest round of "shuttle diplomacy," are signed in Geneva. Israel and Syria accept a separation of forces, a U.N.-policed buffer zone between them and a gradual thinning out of forces.

June 6. Israel returns 382 Arab prisoners to Syria, which hands over 56 Israeli POWs.

June 9. Saudi Arabia announces it will assume 60 per cent ownership of Arabian-American Oil Company (Aramco).

June 12. Nixon starts his Middle East tour in Egypt. Other stops are Saudi Arabia, Syria, Israel and Jordan.

June 14. Sadat and Nixon sign a friendship pledge. The United States promises Egypt nuclear technology for peaceful purposes.

June 16. Syrian President Assad and Nixon establish full U.S.-Syrian diplomatic relations, ruptured in the 1967 war.

June 17. At the end of his visit to Israel, Nixon says the United States and Israel will cooperate in nuclear energy, and the United States will supply nuclear fuel "under agreed safeguards."

June 23. Israel evacuates the last portion of Golan Heights territory occupied in the October 1973 war.

July 15. Rabin says there is "no possibility whatsoever" that Israel will hold discussions with Palestinian guerrilla organizations.

Aug. 7. Egyptian President Sadat blames Libyan President Qaddafi for plots against Sadat and for the recall of Libya's Mirage jets loaned to Egypt.

Aug. 18. Qaddafi and Sadat meet in an attempt to settle disputes.

Aug. 27. France ends its embargo on arms sales to Middle East "battlefield" countries.

Sept. 10. Rabin meets in Washington with President Ford (sworn in Aug. 9), who says the United States remains "committed to Israel's survival and security."

Sept. 15. Rabin, after a four-day U.S. visit, says Israel can count on continued U.S. friendship, although the two countries "do not see eye to eye" on all issues. The Israeli premier also proposes troop withdrawals in the Sinai for a state of non-belligerence with Egypt, including the end of Egyptian economic and diplomatic boycotts.

Sept. 18. Addressing the U.N. General Assembly, Ford says the oil-producing nations should define "a global policy on energy to meet the growing need and to do this without imposing unacceptable burdens on the international monetary and trade system."

Sept. 19. The U.N., heeding a request by Arab nations, slates a debate on the Palestine question.

Sept. 21. Meeting in Cairo, the Palestine Liberation Organization (PLO), Egypt and Syria recognize the PLO as the sole representative of the Palestinian people.

Sept. 23. Ford, speaking to the World Energy Conference in Detroit, Mich., says that exorbitant oil prices threaten worldwide depression and the breakdown of international order and safety.

Sept. 26. The shah of Iran turns down Ford's bid for lower oil prices.

Oct. 9. Israeli troops prevent at least 5,000 Israelis from establishing new settlements in the Israeli-occupied West Bank of the Jordan.

Oct. 12. About 8,000 Israelis in Jerusalem demonstrate in opposition to the surrender of any West Bank land to Jordan as Kissinger arrives on a new Middle East peace-seeking mission.

Oct. 13. After talks with King Faisal, Kissinger tells reporters he has assurances that Saudi Arabia will take "constructive steps" to lower the world price of oil.

Oct. 14. The U.N. General Assembly overwhelmingly passes a resolution inviting the Palestine Liberation Organization to take part in its debate on the Palestine question.

Oct. 15. Israel denounces the U.N. on the PLO vote.

Oct. 20. Israel says it will seek oil supply guarantees before any withdrawal from occupied Egyptian oil fields at Abu Rudeis in the Sinai.

Oct. 23. The U.N. extends its peacekeeping force in the Sinai for six months.

Oct. 28. The 20 Arab League heads of state in a summit meeting at Rabat, Morocco, recognize unanimously the PLO as the "sole legitimate representative of the Palestinian people on any liberated Palestinian territory." Hussein agrees to honor the PLO's claim to negotiate for the West Bank.

Nov. 4. King Hussein says Jordan will rewrite its constitution to exclude the West Bank from Jordan and that it is "totally inconceivable" that Jordan and a Palestinian state could form a federation.

Nov. 5. An estimated 100,000 persons demonstrate in New York City over the U.N. invitation to Palestine Liberation Organization leader Yasir Arafat.

Nov. 9. The Israeli pound is devalued by 43 per cent and other austerity measures are instituted to cope with economic woes brought on by the 1973 war.

Nov. 12. The United States and Algeria renew diplomatic ties severed during the 1967 war.

Nov. 13. Addressing the U.N. General Assembly, Arafat says the PLO goal is "one democratic [Palestinian] state where Christian, Jew and Moslem live in justice, equality and fraternity." In rebuttal, Israeli delegate Yosef Tekoah asserts the Arafat proposal would mean destruction of Israel and its replacement by an Arab state.

Nov. 20. UNESCO, by a 64-27 vote in Paris, adopts a resolution cutting off annual financial aid to Israel because of its "persistence in altering the historical features of Jerusalem."

Nov. 22. The U.N. General Assembly approves a resolution recognizing the right of the Palestinian people to independence and sovereignty and giving the PLO observer status at the U.N.

Dec. 1. Israel President Ephraim Katzir says Israel has the capacity to produce atomic weapons and will do so if needed.

Dec. 13. OPEC countries, meeting in Vienna, announce an oil price boost of 38 cents a barrel, effective Jan. 1, 1975, and to continue to Oct. 1, 1975.

Dec. 17. Israeli Defense Minister Shimon Peres charges that 3,000 Soviet soldiers are in Syria, manning ground-to-air missiles.

Dec. 30. Soviet leader Brezhnev calls off his January visit to Egypt, Syria and Iraq. No official reason is given. Speculation ranges from poor health to serious diplomatic differences between Cairo and Moscow.

Dec. 31. The *London Times* reports Libya has quietly ended its 14-month-old oil embargo against the United States.

1975

Jan. 1. Some 1,000 Egyptian industrial workers stage Cairo riots in protest against soaring prices and low wages.

Jan. 2. In a *Business Week* interview, Kissinger warns that the United States might use force in the Middle East "to prevent the strangulation of the industrialized world" by Arab oil producers. His remarks arouse angry world reaction.

Jan. 6. At the U.N., Lebanon charges Israel with 423 acts of aggression in the past month. These include border crossings made by Israel to wipe out guerrilla forces in South Lebanon.

Jan. 9. Saudi Arabia agrees to buy 60 American F-5 jets for $750-million.

Jan. 14. King Faisal begins a tour of Syria, Jordan and Egypt, pledging Saudi Arabia's oil wealth to the struggle against Israel.

Jan. 23. The Pentagon announces the Israeli purchase of 200 Lance missiles, to be armed with conventional warheads but capable of carrying nuclear ones. Israel also asks for $2-billion in U.S. military and economic aid.

Feb. 7. At the conclusion of a Paris meeting, the International Energy Agency of oil-consuming nations tentatively agrees on plans to continue the search for new fuel sources, to reduce their dependence on Arab oil and to eventually force down the price of petroleum.

Feb. 18. The shah of Iran says he will send additional oil to Israel if Israel cedes Abu Rudeis oil fields to Egypt in a general peace settlement. His offer comes after a meeting with Kissinger, who visited several Middle East countries seeking "a framework for new negotiations."

Feb. 18. Egypt confirms it is getting Soviet arms, including MIG-23 fighters, from the Soviet Union for the first time since the 1973 war, but Foreign Minister Ismail Fahmy says the weapons do not replace losses of that war and that Egypt will not return to the Geneva conference until they are replaced.

Feb. 21. The U.N. Commission on Human Rights passes resolutions condemning Israel for carrying out the "deliberate destruction" of Quenitra, a Syrian city in the Golan Heights, and for "desecrating" Moslem and Christian shrines.

Feb. 26. A U.S. Senate subcommittee publishes a list of 1,500 U.S. firms and organizations boycotted by Saudi Arabia because of links to Israel. In conjunction, the Anti-Defamation League of B'nai B'rith charges that two U.S. agencies and six private companies had violated civil rights laws by discriminating against Jews in order to do business with Arab states. President Ford calls such alleged discrimination "repugnant to American principles" and promises a probe.

March 5. Eighteen persons, including six non-Israeli tourists, are slain when eight Palestinian guerrillas seize a shorefront hotel in Tel Aviv. Israeli troops kill seven attackers and capture the other.

March 5. Iraq and Iran agree to end their long-standing dispute over frontiers, navigational claims and Iranian supply of the Kurdish rebellion in north Iraq.

March 7. Baghdad launches a major military offensive against the Kurds, who seek total autonomy.

March 8. In a new round of shuttle diplomacy, Kissinger seeks further disengagement in the Sinai.

March 22. Kissinger suspends his efforts to draw Israel and Egypt into new accords, calling the breakdown "a sad day for America." Obstacles include the fate of Israeli-held Abu Rudeis oil fields, Mitla and Gidi passes and an Egyptian pledge of non-belligerency.

March 22. The Kurdish rebellion collapses in Iraq. Rebel leader Mustafa al-Barzani flees into Iran.

March 24. After the breakdown of the Kissinger mission, Ford announces a total Middle East policy reassessment.

March 25. Saudi King Faisal is shot to death by his nephew, Prince Faisal Ibn Musaed. Crown Prince Khalid becomes new king. Prince Fahd is named heir apparent. The royal family says the assassin is deranged and acted alone.

April 13. Six days of fighting breaks out in Beirut between the rightist Christian Phalangist Party militia and Palestinian guerrillas, leaving 120 dead and 200 wounded.

May 5. The United States announces a $100-million sale of Hawk surface-to-air missiles to Jordan.

May 15. Criticized for his handling of the Lebanese riots in April, Premier Rashid al-Solh resigns.

May 17. Nine Lebanese children die from the explosion of a shell left over from an Israeli raid on a town.

May 19. The resumption of new Beirut clashes kills 130 persons.

May 19. Lebanese President Suleiman Franjieh appoints a military cabinet, which quits in three days, in an attempt to restore order in Beirut.

May 28. Rashid Karami is appointed Lebanese premier, promising an end to the bloody Christian-Moslem strife over the Palestinian refugee question.

June 2. At the end of a two-day conference with Egyptian President Sadat in Salzburg, Austria, President Ford says, "The United States will not tolerate stagnation in our efforts for a negotiated settlement."

June 2. During the Ford-Sadat parley, Israel orders partial withdrawal of its limited forces in the Sinai in response to reopening of the Suez Canal. Sadat cautiously hails the Israeli gesture as a step toward peace.

June 3. Syria agrees to release additional water from behind a dam on the Euphrates River, ending a bitter dispute with Iraq, which had claimed a severe water shortage.

June 5. The Suez Canal reopens after an eight-year closure to commercial shipping. President Sadat leads a ceremonial convoy of ships through the waterway to mark the opening.

June 5. Civil strife ends in Beirut with the dismantling of barricades and roadblocks.

June 11. Israeli Premier Rabin begins five days of talks in Washington with Ford and Kissinger. They agree on renewing efforts to negotiate another limited Israeli-Egyptian accord in the Sinai.

June 18. Prince Faisal Ibn Musaed is beheaded in Riyadh, Saudi Arabia, for the March 25 assassination of his uncle, King Faisal.

June 23. New fighting breaks out and continues for 10 days in Beirut between the right-wing Christian Phalangist Party and the left-wing Moslem Progressive Socialist supporters of the Palestinian guerrillas. At least 280 persons are killed and 700 are wounded.

June 29. A U.S. Army officer, Col. Ernest R. Morgan, is kidnaped in Beirut by radical Palestinian terrorists who threaten to kill him unless the United States pays a ransom.

July 1. Another truce is proclaimed in Beirut. Premier Rashid Karami, appointed May 28, forms a "salvation cabinet," which includes all major Moslem and Christian groups except the warring Socialists and Phalangists.

July 4. A Palestinian terrorist bomb explodes in Jerusalem's Zion Square, killing 14 Israelis and wounding 78.

July 7. Israel retaliates for the Jerusalem bombing by launching a sea and ground attack on suspected Palestinian bases in south Lebanon, killing 13 persons and destroying many buildings.

July 7. President Sadat releases 2,000 Egyptian prisoners convicted of trying to overthrow the late President Nasser, as a wave of anti-Nasser commentary runs through the Egyptian press.

July 8. Egypt, Israel and Secretary Kissinger deny world rumors that new Sinai accords are imminent.

July 11. Col. Morgan is released unharmed in Lebanon by his Palestinian abductors, who say he is freed because he is black and that blacks seldom obtain such high rank in the U.S. Army. The State Department says his release is secured without concessions to the kidnapers.

July 15. Egypt avows it will not renew the U.N. forces mandate in the Sinai unless progress toward peace is made, and accuses Israel of exploiting the state of "no war, no peace" to perpetuate its Sinai occupation.

July 20. In Cairo, King Khalid ends his first tour as Saudi Arabia's sovereign by loaning Egypt $600-million.

July 23. A senior Egyptian official denounces Israeli Premier Rabin's call for direct Egyptian-Israeli talks as secret U.S.-led negotiations for a new Sinai accord mount.

July 24. The U.N. Security Council approves a three-month extension of an emergency force in the Sinai after Egypt drops opposition to renewal of the mission.

July 25. Rabin terms Egypt's new disengagement proposal "substantially not acceptable," but better than the one in March, when Kissinger's shuttle diplomacy collapsed. In addressing an Arab-Socialist Union congress, Sadat says it does not matter whether or not a certain step in negotiations succeeds or fails, because "our armed forces are ready."

July 28. The Ford administration informs Congress it is deferring plans to sell Jordan 14 Hawk antiaircraft missile batteries. The action is taken after Congress, which has power to veto the sale, shows signs of balking over the number of Hawks in the arms package. Reacting angrily, King Hussein says Jordan will seek weapons elsewhere unless he can buy all 14 batteries.

Aug. 1. At a summit meeting in Kampala, Uganda, heads of member nations of the Organization of African Unity refuse to adopt an Arab-backed policy proposal calling for Israel's suspension from the United Nations.

Aug. 17. The Israeli cabinet endorses the Rabin stand on new Sinai negotiations, paving the way for new direct negotiations by Kissinger. Cairo also approves of new Kissinger mediation.

Aug. 21. Kissinger arrives in Israel to begin a new round of shuttle diplomacy. He is met by nationwide demonstrations by Israelis who dislike his brand of diplomacy and fear the projected agreement. At Tel Aviv airport, Kissinger says, "the gap in negotiations has been substantially narrowed by concessions on both sides."

Aug. 22. Reports indicate Israel wants written assurance for U.S.-Israeli joint consultations in the event of Soviet intervention in the Middle East.

Aug. 22. King Hussein concludes a 5-day visit to Syria and both countries announce a "supreme command" to coordinate foreign policy, information and military affairs.

Aug. 24. An Israeli official says agreement has been reached on setting up early warning installations, to be manned partly by American technicians, in the Sinai, thus clearing away one last major obstacle to new accords.

Aug. 25. Arab nations stop short of calling directly for Israeli expulsion from the U.N. at a ministerial conference of 108 nonaligned nations in Lima, Peru.

Aug. 28. A senior official with the Kissinger party in Jerusalem says implementation of a new Sinai agreement hinges on approval by Congress of the use of American civilian technicians to man Sinai monitoring posts.

Aug. 30. Egypt joins Israel in asking that Congress approve the stationing of American technicians.

Sept. 1. In separate ceremonies in Jerusalem and Alexandria, Israeli and Egyptian leaders initial the new Sinai pact. Israel yields to Egyptian demands that it withdraw from Sinai mountain passes and return Abu Rudeis oil fields to Egypt in return for modest Egyptian political concessions. Kissinger initials provisions for stationing of U.S. technicians in the Sinai. Ford asks that Congress approve the new U.S. Middle East role.

Sept. 4. Egypt and Israel sign the agreement at Geneva.

Sept. 5. Syria calls the Sinai pact "strange and shameful" while Zuhayr Muhsin of the PLO calls Sadat a "traitor and conspirator" for signing the accord.

Sept. 8. The U.S. announces talks with Israel on the sale of F-15s and other arms are resuming.

Sept. 11. Egypt closes the PLO's radio station, "Voice of Palestine," which broadcasts from Cairo.

Sept. 11. Israeli jets raid targets in southern Lebanon.

Sept. 16. Four Palestinians, for which the PLO denies responsibility, seize hostages at the Egyptian Embassy in Madrid threatening to blow up the embassy unless Egypt repudiates the Sinai accord.

Sept. 21. Israeli cabinet approves Israel's initiating the second Sinai agreement but delays formal signing until U.S. Congress approves sending U.S. technicians to Sinai.

Sept. 24. Israeli Foreign Minister Allon and Soviet Foreign Minister Gromyko hold 3-hour discussion at U.N.

Sept. 26. France tells U.N. it is willing to participate in military safeguards for a Middle East peace.

Sept. 26. The White House announces U.S. will now consider Egypt's request for limited types of arms.

Oct. 13. President Ford signs a congressional resolution authorizing U.S. technicians for Sinai.

Oct. 15. An Arab League Conference convenes in Cairo to discuss the fighting in Lebanon. Syria, Libya and the PLO boycott the meeting.

Oct. 22. Two USIA officials are kidnaped in Beirut.

Oct. 23. U.N. Security Council votes to extend the peace force in Sinai for one year by a vote of 16 to 0.

Oct. 26. President Sadat and President Ford meet in Washington with Ford emphasizing the U.S. will not "tolerate stagnation or stalemate."

Oct. 28. Sadat urges the U.S. to open a dialogue with the PLO while Egypt and the U.S. sign four economic and cultural exchange agreements.

Oct. 31. France authorizes the PLO to open an office in Paris while urging the PLO "to take a responsible and moderate course."

Nov. 3. PLO delegate Farouk Kaddoumi opens the U.N. General Assembly debate on the Palestinian issue.

Nov. 5. Sadat addresses a joint session of Congress.

Nov. 7. The U.S. says in the General Assembly that it will work for a settlement taking into account the legitimate rights of the Palestinians.

Nov. 8. Britain agrees to sell Egypt $2-billion in military equipment.

Nov. 10. U.N. General Assembly passes resolution defining Zionism as "a form of racism or racial discrimination" on a 72 to 35 vote with 32 abstentions and 3 absences. U.S. Ambassador Daniel Moynihan says, "The United States...does not acknowledge, it will not abide by, it will never acquiesce in this infamous act." Second resolution recognizes Palestinians' right to self-determination and to attend any U.N. Middle East negotiation.

Nov. 10. Israeli Knesset passes resolution rejecting the U.N. resolution on Zionism and indicating Israel will not participate in the Geneva talks if the PLO is ever invited.

Nov. 11. Kissinger states the U.S. will ignore the U.N. resolution on Zionism.

Nov. 12. State Department official Harold Saunders tells a committee of the Congress that "In many ways...the Palestinian dimension of the Arab-Israeli conflict is the heart of the conflict."

Nov. 13. A large bomb explodes in Jerusalem's Zion Square killing 6 persons and wounding 40.

Nov. 16. Syria's Foreign Minister says Syria will "bring down the Sinai agreement even if we have to shed blood for it."

Nov. 16. The Israeli Cabinet expresses considerable criticism of the U.S. State Department for the Saunders statement on the Palestinians.

Nov. 20. CIA Director Colby asserts to Congress that another Middle East war would probably result in 8,000 Israeli casualties.

Nov. 22. U.N. Secretary-General Waldheim arrives in Damascus to begin a Middle East tour.

Dec. 2. Israeli jets attack Palestinian refugee camps in northern and southern Lebanon killing 74 and wounding at least a hundred others.

Dec. 3. The Egyptian Minister of Petroleum announces Egypt will claim more than $2.1-billion for the oil Israel extracted from Sinai.

Dec. 4. U.N. Security Council votes 9 to 3 with 3 abstentions to invite the PLO to participate in debate about Israeli air attacks in Lebanon. PLO is granted speaking privileges of a member nation.

Dec. 5. U.N. General Assembly by 84 to 17 with 27 abstentions passes a resolution condemning Israel's occupation of Arab territories and calling upon all states to refrain from aiding Israel.

Dec. 17. U.N. General Assembly ends session after appointing a Committee on the Exercise of the Inalienable Rights of the Palestinian People.

Dec. 17. UNESCO votes by 36 to 22 with seven abstentions to insert reference to General Assembly resolution condemning Zionism as a form of racism into a draft policy declaration.

Dec. 22. Pope Paul VI appeals to Israel to "recognize the rights and legitimate aspirations" of the Palestinians.

Dec. 26. OPEC headquarters in Vienna are raided by 6 terrorists killing three and taking hostages including Saudi Petroleum Minister Yamani. PLO denounces the attack.

Dec. 26. Iran and Iraq sign a number of good neighbor agreements.

Dec. 31. Egypt reveals that its nonmilitary debt to foreign countries and international organizations is $7-billion.

1976

Jan. 7. Protesting the name "Arab Gulf" rather than Persian Gulf, the Shah reveals Iran's recall of ambassadors from a number of Arab states after word of plans for an "Arabian Gulf News Agency."

Jan. 12. The U.N. Security Council opens its Middle East debate by voting 11 to 1 with 3 abstentions to allow the PLO to participate with speaking rights of a member.

Jan. 12. The Israeli Council for Israeli-Palestinian Peace calls for negotiations with "a recognized and authoritative body of the Palestinian Arab people" which could lead to "establishment of a Palestinian Arab state" alongside Israel.

Jan. 13. Sadat says Egypt will attend a Geneva conference without the Palestinians but would then "fight for the Palestinians to join."

Jan. 28. Rabin addresses a joint meeting of the U.S. Congress ruling out any negotiations with the PLO.

Jan. 29. Rabin and Ford end talks in Washington reportedly with the understanding the U.S. would try to see if the Geneva conference can be convened without the PLO.

Feb. 13. The U.N. Commission on Human Rights votes 23 to 1 with 8 abstentions for a resolution accusing Israel of having committed "war crimes" in the occupied Arab territories. U.S. casts lone "no" vote.

Feb. 15. The Israeli cabinet approves appointment of Shlomo Avineri, a dovish critic of Rabin's policies toward the Palestinian Arabs, to be Director-General of the Foreign Ministry.

Feb. 18. A Syrian mediation team headed by Foreign Minister Kaddum continues efforts to find a solution to the Lebanese crisis.

Feb. 22. The final step in the previous September's Sinai accord is carried out with U.N. personnel turning over the final 89 square miles to Egyptian forces.

Feb. 25. The two American USIA officials kidnaped in Beirut 4 months earlier are released.

March 2. Ford administration informs key congressional leaders it wants to lift arms embargo to Egypt by selling 6 C-130 military transport planes.

March 7. In Cairo, U.S. Treasury Secretary William Simon praises Sadat for liberalizing Egypt's economy and pledges a total of $1.85-billion in U.S. aid for fiscal 1976 and 1977.

March 10. Demonstrations and strikes break out on the occupied West Bank and in East Jerusalem.

March 12. The 44-nation Islamic Conference at the U.N. adopts statement denouncing Israel's systematic policy "to change the status of Jerusalem and gradually obliterate the Moslem and Christian heritage in the Holy City."

March 13. Lebanese President Franjieh is presented with a petition signed by two-thirds of parliament asking him to resign, but he refuses.

March 14. The Shah of Iran warns the U.S. that Iran "can hurt you as badly if not more so than you can hurt us" if the Congress imposes an arms embargo on the Persian Gulf region.

March 14. Senior CIA officials estimate Israel has 10 to 20 nuclear weapons "ready and available for use."

March 19. Israel announces it will participate in Security Council debate on the West Bank disturbances even though the PLO will be represented.

March 23. U.S. Ambassador to the U.N., William Scranton, tells the Security Council the U.S. considers the presence of Israeli settlements in the occupied territories to be "an obstacle to the success of the negotiations for a just and final peace."

March 25. The Security Council in a 14 to 1 vote, the U.S. vetoing the action, deplores Israel's efforts to change the status of Jerusalem, calls on Israel to refrain from measures harming the inhabitants of the occupied territories, and calls for an end to Israeli settlements in the occupied territories.

March 29. West Bank rioting spreads into Israel itself.

March 30. Israeli Arabs hold general strike protesting government land expropriation scheme in Galilee area. Violent clashes with police take place in more than a dozen villages resulting in 6 Arabs killed and more than 70 persons wounded.

March 30. Former Ambassador L. Dean Brown is designated special envoy to Lebanon by the State Department.

April 4. While endorsing police action against Israeli Arabs in Galilee, Israeli cabinet decides to reexamine policy toward Arab citizens.

April 4. Kissinger pledges the U.S. "will never abandon Israel."

April 5. Israel says it is not a nuclear power and will not be the first to introduce nuclear weapons into the Middle East conflict.

April 12. Syrian army begins to cautiously advance into Lebanon.

April 13. Militants and PLO supporters sweep to victory in the voting on the West Bank for new mayors and municipal councilmen. New mayors are elected in 10 of 24 towns and 148 new councilmen are selected leaving only 43 incumbents.

April 18. About 30,000 Israelis march for 2 days through the West Bank under the leadership of Gush Emunim. Arab counter demonstrations in Nablus and Ramallah are broken up.

April 20. Prime Minister Rabin tours Jordan Valley settlements and assures the settlers in the 16 new villages that they are "here to stay for a long time."

April 21. Units of the Palestine Liberation Army (PLA) begin taking up positions in Beirut.

April 27. Arab League states at a meeting of Finance Ministers in Rabat agree to establish an Arab Monetary Fund.

April 28. The Soviet Union calls for resumption of the Geneva conference in two stages with Palestinian partition in each stage.

May 3. A bomb explodes near Zion Square in Jerusalem injuring 30.

May 3. An Arab march from Ramallah to Jerusalem, planned in response to the Gush Emunim march the month before, is prevented by Israeli authorities.

May 4. Former Secretary of Defense Schlesinger says President Ford is undermining American support for Israel by putting undue pressure on Israel to make concessions to the Arabs.

May 8. Several thousand Israelis participate in a demonstration sponsored by the Mapam Party for the ejection of 125 Gush Emunim settlers from Camp Kadum, an illegal settlement in the West Bank.

May 8. Elias Sarkis is elected President of Lebanon. Backers of Raymond Edde boycott the election protesting interference by Syria.

May 19. Israeli Arab protestors at Hebrew University are attacked by Jewish counter demonstrators led by Rabbi Meir Kahane of the Jewish Defense League.

May 19. The U.N. Committee on the Exercise of the Inalienable Rights of the Palestinian People adopts by consensus a set of recommendations stressing the Palestinians' "right to return" to their homeland.

May 23. The head of the largest Jewish women's organization in the U.S., Hadassah, criticizes American Jews "who have taken it upon themselves to publicly criticize certain policies of Israel at this critical juncture."

May 24. Israeli Prime Minister Rabin refuses a request by Israeli Arab leaders to cancel plans to take over Arab land in Galilee and turns down demands for a special inquiry into the Galilee riots of March.

May 25. A suitcase bomb explodes at Ben-Gurion International Airport outside Tel-Aviv.

May 26. The Security Council presents a majority statement deploring Israeli measures altering the demographic character of the occupied territories at the end of its debate on the Middle East. The U.S. disassociates itself from the statement.

May 30. A rally in Paris in support of Israel draws 100,000.

May 31. Egypt asks the Arab League to admit the PLO to full membership.

May 31. Syrian troops numbering 2,000 advance into Lebanon's Akkar Valley in the north.

June 1. An additional 3,000 Syrian troops advance into Lebanon.

June 3. About 300 students seize the Syrian embassy in Cairo to protest Syria's role in Lebanon.

June 7. Egypt announces the PLO will be allowed to resume broadcasting in Cairo.

June 8. Large numbers of additional Syrian forces enter Lebanon.

June 10. Member nations of the Arab League agree to send forces to Lebanon to "replace" the Syrian forces.

June 16. U.S. Ambassador Francis Meloy and Economic Counselor Robert Warring are shot to death on their way to a meeting with Sarkis.

June 18. King Hussein, while visiting Moscow to inquire about possible arms purchases, strongly backs Syrian intervention in Lebanon.

June 19. The U.S. strongly urges all remaining American citizens to leave Lebanon.

June 20. The U.S. evacuates from Lebanon by sea, with the help of the PLO, 263 American and foreign nationals.

June 21. About a thousand Syrian and Libyan troops, vanguard of the Arab League peacekeeping force, arrive in Lebanon.

The State Department confirms a U.S. message, through indirect channels, to the PLO leadership thanking them for their help in the sea evacuation of Western nationals from Beirut.

June 27. A jetliner on its way from Tel Aviv to Paris with 257 persons is hijacked in Athens and taken to Entebbe airport in Uganda the following day.

June 29. West Bank Arab merchants call for a strike to protest the new Israeli value-added tax and the Israeli Treasury decides to postpone imposition of the tax until August 1.

July 2. The seige of Tel al-Zaatar refugee camp near Beirut begins as Christian forces overrun the outer defenses.

July 4. Israeli commandos raid the hijacked airliner at Entebbe airport freeing the 103 remaining hostages.

July 14. The Security Council ends debate on the Israeli Entebbe raid without approving either a resolution condemning the raid or a resolution condemning the hijacking and terrorism.

July 16. The American embassy in Beirut announces closure of all consular services and urges all Americans to leave Lebanon.

July 27. U.S. evacuates 308 American and foreign nationals from Beirut, again with the PLO helping to provide security.

Aug. 28. The PLO announces a general conscription for all Palestinians between ages of 18 and 30.

Aug. 28. Terrorists in Iran kill 3 American employees of the American firm Rockwell International.

Sept. 6. The PLO is unanimously granted full voting membership in the Arab League.

Sept. 7. A report by the Israeli Interior Ministry's Representative in the Northern District of Israel, Yisrael Koenig, is published in the newspaper *Al Hamishmar*. The report predicts an Arab majority in Galilee by 1978 and recommends measures to minimize Arab influence there.

Sept. 15. Syria and Israel open their security fences in the Jawlan region and allow Druze villagers from Syria to visit with relatives from the Israeli occupied areas.

Sept. 17. Israeli Foreign Minister Allon publishes an article in *Foreign Affairs* advocating Israeli withdrawal from most occupied Arab territory and creation of a demilitarized joint Jordanian-Palestinian entity in the West Bank and Gaza Strip.

Sept. 17. President Sadat of Egypt receives over 99 per cent approval for a second 6-year term as President.

Sept. 18. Fighting in Lebanon intensifies on all fronts.

Sept. 23. Elias Sarkis is inaugurated as President of Lebanon before 67 members of the National Assembly.

Sept. 28. PLO leader Yasir Arafat sends an urgent appeal to all Arab heads of state asking for immediate intervention to prevent Syria from "liquidating the Palestinian resistance."

Oct. 1. A public council to advise the Ministerial Committee on Israeli Arabs is established with 54 Arabs and 46 Jews.

Oct. 6. France and Iran sign a nuclear cooperation agreement under which France will build two nuclear reactors for Iran.

Oct. 11. During the heat of the Presidential campaign President Ford agrees to sell Israel military equipment previously not offered for sale.

Oct. 14. Lebanon's U.N. representative tells the General Assembly that "the Palestinian revolution" and its supporters caused the civil war in Lebanon.

Oct. 16. President Sadat is sworn into office for a second six-year term.

Oct. 18. Arab leaders of Saudi Arabia, Kuwait, Syria, Egypt, Lebanon, and the PLO, meeting in Riyadh, sign a peace plan calling for a cease-fire and a 30,000-man peacekeeping force under the command of Lebanese President Sarkis.

Oct. 25. All members of the Arab League, meeting in Cairo, approve the Riyadh agreement except for Iraq and Libya.

Nov. 1. The Security Council decides, over U.S. opposition, to allow the PLO to participate in the debate on the occupied territories.

Nov. 2. The Israeli government raises prices of most basic staple foods and services by about 20 per cent in an effort to gradually abolish subsidies for essential foods.

Nov. 8. UNESCO's general conference votes by 70 to 0 with 14 abstentions to let each regional group select its own members, thus making it possible again for Israel to be included in the European group.

Nov. 8. An International Symposium on Zionism opens in Baghdad.

Nov. 11. The Security Council, in a consensus statement, deplores the establishment of Israeli settlements in occupied Arab territories and declares "invalid" the annexation of eastern Jerusalem by Israel.

Nov. 11. Sadat announces that the three political groupings that had participated in the election will now be called parties, though they will still come under the overall umbrella of the Arab Socialist Union (ASU).

Nov. 15. Syrian peacekeeping forces take up positions in both Christian and Moslem sections of Beirut.

Nov. 17. OPEC decides to increase aid to non-oil-producing counties by $800-million.

Nov. 18. The PLO formally registers with the Justice Department in a step towards opening an office in Washington.

Nov. 19. Beirut's airport reopens.

Nov. 22. In Israel, former Army Chief of Staff Yigal Yadin announces formation of a new party, the "Democratic Movement."

Nov. 22. UNESCO votes by 61 to 5 with 28 abstentions to condemn Israel's educational and cultural policies in the occupied Arab territories.

Israel is restored to full membership in UNESCO.

Nov. 23. The General Assembly of the U.N. votes 118 to 2 with 2 abstentions for a resolution calling on Israel to halt resettlement of Palestinian refugees in Gaza and to return all refugees to their camps.

Nov. 23. Israeli forces are deployed along the Lebanese border following reported movements of Syrian peacekeeping forces toward the border.

Nov. 24. U.N. General Assembly approves by 90 to 16 with 30 abstentions the report of the Committee on the Inalienable Rights of the Palestinian People proclaiming right of Palestinian Arab refugees to establish their own state and reclaim former properties in Israel.

Nov. 25. Iraq and Syria withdraw most forces from their common frontier.

Dec. 9. U.N. General Assembly votes by 122 to 2 with 8 abstentions for a resolution calling for reconvening the Geneva peace conference by March 1.

Dec. 15. A general strike is held by Arabs in the West Bank and Gaza to protest the value-added tax.

Dec. 15. Arab League special envoy Ghassan Tueni meets with U.S. Under Secretary of State for Political Affairs Philip Habib on U.S. economic aid to Lebanon.

Dec. 15. Beirut's port reopens.

Dec. 17. OPEC announces that 11 members will increase oil prices by 10 per cent on Jan. 1 and a further 5 per cent on July 1, 1977. Saudi Arabia and the UAE announce only a 5 per cent price rise for Jan. 1 thus setting up two prices within OPEC.

Dec. 19. Rabin ousts the National Religious Party from his coalition government 5 days after the NRP abstained on a vote of no-confidence in the Knesset.

Dec. 20. Rabin resigns and new elections are called.

Dec. 21. Egypt and Syria announce formation of a "united political leadership" and agree to study the possibility of future union.

1977

Jan. 2. Egypt permits the Abie Nathan "peace ship" from Israel to proceed through the Suez Canal.

Jan. 2. Israeli Council for Israeli-Palestinian Peace publishes what is said to be a joint statement with the PLO calling for Zionist and Palestinian states to coexist peacably.

Jan. 3. Press censorship begins in Lebanon.
Rabin forms temporary government until elections.

Jan. 5. Israeli Knesset votes to dissolve and calls election for May.

Jan. 11. French authorities free PLO terrorist Abu Daoud after a judicial hearing rejecting West German and Israeli extradition claims.

Jan. 15. Sadat and Hussein both call for a separate Palestinian delegation at the Geneva peace conference.

Jan. 18. Thousands of Egyptian workers demonstrate against price rises.

Jan. 19. Demonstrations and rioting continue in Egypt and Sadat cancels the price increases. At least 65 persons are reported killed in clashes with police.

Jan. 25. Lebanon lifts press censorship for outgoing dispatches but retains it for domestic publications.

Jan. 26. Egypt bans demonstrations and strikes.

Feb. 7. U.S. blocks sale of Israeli Kfir jets with American jet engines to Ecuador.

Feb. 10. In a referendum on Sadat's decree outlawing demonstrations and strikes, 99 per cent vote to approve, according to the government.

Feb. 12. About 400 students demonstrate in Cairo against the new law banning demonstrations.

Feb. 15. On his first international mission as Secretary of State, Cyrus R. Vance arrives in Israel on his first tour of the Middle East.

Feb. 15. The U.N. Human Rights Commission adopts a resolution accusing Israel of practicing "torture" and "pillaging of archeological and cultural property" in occupied Arab territories.

Feb. 20. Biweekly bus service begins between Haifa and points in southern Lebanon.

Feb. 22. By a vote of 1,445 to 1,404 the Israeli Labor Party selects Yitzhak Rabin over Shimon Peres as its candidate for prime minister.

Feb. 22. A delegation of Israeli Arabs—the first ever to visit an Arab country—returns after a 5-day visit to Amman.

Feb. 24. Jordan and the PLO agree, in their first talks since 1970, on forming some kind of "link."

Feb. 25. In exchange for a real peace agreement, Israeli Labor Party platform calls for return of some West Bank territory to Jordan.

Feb. 27. Israeli army halts an attempt by Gush Emunim to establish another illegal West Bank settlement.

Feb. 28. Carter administration comes out in opposition to secondary and tertiary aspects of the Arab boycott against Israel.

March 6. George Habash tells reporters that the Popular Front for the Liberation of Palestine (PFLP) and 3 other "rejectionist" groups will break from the PLO if the Palestine National Council decides to go to the Geneva conference or to recognize Israel.

March 7. Foreign Ministers from 60 African and Arab countries meet in Cairo for an unprecedented Afro-Arab summit conference. Saudi Arabia immediately pledges $1-billion in aid to black Africa.

March 7. President Carter welcomes Israeli Prime Minister Rabin to Washington and says Israel should have "defensible borders."

March 9. Yasir Arafat and King Hussein meet in Cairo publicly for the first time since "Black September" in 1970.

March 9. President Carter distinguishes between "defensible borders" (temporarily extending beyond Israel's sovereign frontiers) and "secure borders" (the eventual permanent and recognized borders).

March 9. Ambassadors from Egypt, Iran and Pakistan play a major role in negotiating the release of hostages being held by the Hanafi Muslims in the B'nai B'rith and two other buildings in Washington.

March 12. In a report to Congress on human rights conditions in 82 countries receiving U.S. aid, the State Department criticizes Israel's treatment of the Arabs in the occupied territories.

March 12. The Palestine National Council opens in Cairo and President Sadat pledges that Egypt "will not cede a single inch of Arab land."

March 15. The Israeli newspaper *Ha'aretz* reports that Prime Minister Rabin's wife has an illegal bank account in Washington.

March 16. At a Clinton, Massachusetts town meeting President Carter—the first American President ever to do so—endorses the idea of a Palestinian "homeland."

March 16. Leftist leader Kamal Jumblatt is assasinated near Beirut.

March 17. A PLO official shakes President Carter's hand at a reception following Carter's U.N. speech.

May 20. Prime Minister Rabin admits that he maintained an illegal Washington bank account with his wife.

May 20. The Palestine National Council concludes its nine-day 13th session in Cairo by adopting a 15-point political declaration by a vote of 194 to 13.

May 21. Samuel W. Lewis is revealed as President Carter's choice for Ambassador to Israel.

May 27. Arab League approves a six-month extension of the Arab peacekeeping force in Lebanon at the request of Lebanese President Sarkis.

May 30. Israel admits secretly holding two Germans and three Arabs who are accused of attempting to shoot down an El Al plane in Nairobi.

April 3. Kuwait concludes its first arms deal with the USSR.

April 4. President Sadat visits Washington and tells President Carter that the Palestinian question is the "core and crux" of the Arab-Israeli dispute.

April 5. About 15,000 Gush Emunim marchers march on the West Bank.

April 7. Arafat and Communist Party Secretary Brezhnev meet publicly in Moscow for the first time.

April 7. Prime Minister Rabin withdraws from the top spot on the Labor Party ticket with elections only 6 weeks away.

April 8. Sadat says relations with Israel could be normalized within 5 years, according to U.S. officials.

April 10. Shimon Peres is selected to head Labor Party ticket in Israeli election.

April 11. Prime Minister Rabin is fined $1,500 for his role in maintaining an illegal bank account in Washington.

April 13. While the fighting continues in southern Lebanon, the State Department reveals it is deeply involved in the diplomatic effort to stabilize southern Lebanon.

April 14. Cairo bars Libyan citizens from leaving Egypt in apparent retaliation for a similar restriction placed on Egyptians and others in Libya.

April 16. Egypt delivers to the Arab League a note accusing Libya of plotting against Sudan, seizing portions of Chad, and harboring "international criminals."

April 17. Mrs. Leah Rabin, wife of Prime Minister Rabin, pleads guilty to maintaining an illegal bank account and is fined $27,000.

April 17. A Library of Congress report urges the U.S. to seek a more secure airlift route to the Middle East.

April 22. Prime Minister Rabin begins an extended vacation and turns over the day-to-day affairs of the government to Shimon Peres.

April 25. King Hussein visits Washington for talks with President Carter.

April 27. Moscow accuses Egypt of attempting to provoke armed clashes between Egypt and Libya. Libya is reported planning to expel some of the 200,000 Egyptians working in Libya.

May 1. Egypt sends pilots to Zaire.

May 3. The White House announces a compromise between Jewish groups and business groups regarding anti-boycott legislation pending in the Congress.

May 5. For the first time, Syrian President Assad endorses the idea of demilitarized zones between Israel and the Arab states.

May 8. The U.S. State Department accuses Libya, Iraq, South Yemen, and Somalia of actively supporting terrorism.

May 9. President Carter and Syrian President Assad meet in Geneva to discuss Middle East peace prospects.

May 9. Saudi Crown Prince Fahd says the PLO would be likely to recognize Israel in the context of an overall peace settlement.

May 12. President Carter pledges "special treatment" for Israel in regard to arms request and co-production of advanced U.S. weaponry. Carter again calls for a Palestinian homeland and says "there's a chance that the Palestinians might make moves to recognize the right of Israel to exist."

May 12. Bahrain announces that after June 30 the U.S. base there will be downgraded to a "facility."

May 13. A serious fire in a major Saudi oil field is finally extinguished.

May 15. Israel reveals a new 56-ton tank with armor impenetrable to existing Arab shells.

May 17. Menahem Begin's right-wing Likud Party unexpectedly wins a plurality in the Israeli election.

May 18. Secretary of State Vance and Soviet Foreign Minister Gromyko begin several days of talks in Geneva about the Middle East and about nuclear arms limitations.

May 19. The leaders of Saudi Arabia, Syria and Egypt hold talks in Riyadh to "create a cohesive Arab position" in light of Begin's victory in Israel.

May 19. Begin calls for many new Jewish settlements in Israeli-occupied territories.

May 21. Talks in Baghdad between U.S. and Iraqi officials lead to speculation of resumed relations.

May 24. Saudi Arabian Crown Prince Fahd begins talks with Carter in Washington.

May 24. Sudan's President Nimeri asks for U.S. aid after expelling Soviet military personnel.

May 25. Moshe Dayan agrees to serve in a Begin government as Foreign Minister.

May 26. At a news conference, President Carter says the Palestinians should be compensated for their losses.

May 27. Lebanese rightist leaders declare the 1969 Cairo agreement invalid and call the Palestinian presence in Lebanon illegal.

May 27. Moshe Dayan resigns from the Labor Party due to the furor over his acceptance of the role of Foreign Minister in the government being formed by Menahem Begin.

May 30. Saudi Arabia declares that within one year all banks will be under majority control of Saudi nationals.

June 5. Businesses close in 3 West Bank towns to protest 10 years of Israeli occupation.

June 5. Acting Prime Minister Rabin tells the cabinet that Israel is strong enough to resist U.S. pressures to force Israel "to accept views inimical to our security."

June 7. President Katzair formally asks Menahem Begin to form Israel's next government.

June 10. The anti-boycott bill passes both Houses of Congress.

June 10. Menahem Begin's representative, Samuel Katz, meets with Carter's national security adviser, Zbigniew Brzezinski, preparing the way for Begin's expected visit in July.

June 10. The PLO signs an addition to the Geneva war convention that prohibits terrorism against civilians.

June 11. Egyptian Foreign Minister Fahmy and Soviet Foreign Minister Gromyko complete two days of talks in Moscow aimed at improving Egyptian-Soviet relations.

June 16. The U.S. delegation and others walk out of an African-sponsored U.N. gathering which excluded Israel.

June 17. Vice-President Walter Mondale delivers a major speech on the Middle East comprehensively outlining the Carter administration's views and emphasizing a three-point peace plan: return to approximately the 1967 borders, creation of a Palestinian homeland probably linked to Jordan, and establishment of complete peace and normal relations between the countries in the area.

June 19. The *London Sunday Times* in a major report charges that "Israeli interrogators routinely ill-treat and often torture Arab prisoners."

June 19. Two religious parties agree to join Begin's Likud Party in forming the next Israeli government.

June 21. The Labor Party wins a decisive victory in the Histadrut elections in Israel.

June 21. Menahem Begin officially becomes Prime Minister of Israel after winning a 63 to 53 vote of confidence in the new Knesset.

June 22. President Carter signs the bill aimed at limiting participation of American firms in the Arab boycott of Israel.

June 23. In his first major speech as Prime Minister, Begin announces that Israel will not "under any circumstances" relinquish the West Bank or allow the creation of a Palestinian state west of the Jordan River.

June 25. President Carter recommends the sale of an additional $115-million in arms to Israel.

June 26. Sixty-six Vietnamese refugees arrive in Israel and are granted asylum.

June 27. The State Department issues a statement on the Middle East warning that "no territories, including the West Bank, are automatically excluded from the items to be negotiated." Israel's government is infuriated by the timing as well as the substance of this statement, coming just weeks before Prime Minister Begin's scheduled visit to Washington.

June 29. The nine members of the European Economic Community issue a Middle East statement endorsing the idea of a Palestinian "homeland."

June 29. Nine top Democratic senators release a letter to President Carter giving him "strong support in the Senate for your efforts to help Israel and the Arab nations secure a genuine and lasting peace."

July 3. An extremist Moslem group in Cairo kidnaps a former cabinet minister who three days later is found murdered. The demand had been for the release of 30 imprisoned colleagues of the terrorists.

July 3. Saudi Arabia and the UAE raise their price for oil 5 per cent bringing them back in line with the other OPEC countries.

July 3. Israel "emphatically denies" with a detailed rebuttal the London *Sunday Times* torture story.

July 6. Carter holds an hour-long White House meeting with 53 prominent American Jewish leaders to reassure them of his commitment to Israel.

July 6. Carter again refuses to allow Israel to sell Ecuador 24 Kfir fighters.

July 10: King Hussein and President Sadat agree in Cairo to an "explicit link" between Jordan and the Palestinians.

July 12. The United States' 1977 contribution to the U.N. Relief and Work Agency's program of Palestinian refugee aid totals $48.7-million, after an additional pledge of $22-million.

July 12. Iran's and Saudi Arabia's leading oil officials, attending OPEC's first ministerial meeting this year, advocate, in view of the world economy, a freeze on crude oil prices during 1978.

July 12. President Carter notes his personal preference "that the Palestinian entity...should be tied in with Jordan and not independent."

July 13. It is reported that the PLO is seriously considering forming a government-in-exile.

July 13. A spokesman for the Palestinian "rejection front" threatens any Arab leader who signs a peace agreement with Israel with assassination.

July 13. President Sadat, speaking to American congressmen, expresses Egypt's willingness to establish diplomatic and trade relations with Israel within five years of signing a peace agreement.

July 15. U.S. congressmen report that PLO leader Yasir Arafat told them the PLO is prepared to settle for a West Bank-Gaza Strip state and to coexist with Israel.

July 16. Speaking live over Cairo radio, Sadat announces he is willing to accept Israel as a Middle East nation after a peace treaty is signed.

July 16. President Sadat reveals that Saudi Arabia has agreed to finance Egypt's military development for the next five years.

July 17. The new Israeli government announces a major austerity program reducing food subsidies and trimming the defense budget.

July 19. Egypt returns to Israel with full military honors 19 bodies of Israeli soldiers killed during the 1973 war.

July 20. Israeli Prime Minister Begin visits Washington and confers with President Carter, who after their meeting states, "I believe that we've laid the groundwork now...that will lead to the Geneva Conference in October."

July 20. At a news conference telecast live to Israel from Washington Begin reveals his "peace plan" which is loudly criticized by all Arab parties.

July 20. A Beirut newspaper reveals that William Scranton met in London in June with a high-ranking PLO official. Speculation is that Scranton was acting on behalf of the Carter administration.

July 21. Egypt and Libya have a major military clash on their common border.

July 22. Egyptian planes continue bombing Libya and Sadat vows to teach Qaddafi "a lesson he will never forget."

July 22. The PLO becomes the first non-state to have full membership in any U.N. body when it is accepted as a member of the Economic Commission for Western Asia of ECOSOC.

July 22. The Carter Administration announces approval of a $250-million arms package for Israel.

July 25. The home of the Executive Director of the American Israel Public Affairs Committee (AIPAC—Israel's most important lobbying body in Washington) is bombed. Morris Amitay and his family are not injured but the home is greatly damaged.

July 27. The Israeli government legalizes three formerly unapproved settlements in heavily-populated areas of the West Bank.

July 27. Secretary of State Vance and President Carter indicate their disapproval of the Israeli government's legalization of the West Bank settlements.

July 28. President Carter states at a press conference that "The major stumbling block" to reconvening the Geneva Conference "is the participation...by the Palestinian representative." He offers to discuss this matter with the PLO and to possibly advocate a Pales- tinian role at Geneva if the PLO will agree to recognize Israel and to negotiate on the basis of U.N. Resolutions 242 and 338.

July 31. Secretary of State Vance leaves for nearly a two-week trip to the Middle East in an attempt to narrow the differences between Israel and the Arabs and to find a way to proceed with diplomacy either at Geneva or through some other mechanism.

SELECTED BIBLIOGRAPHY ON THE MIDDLE EAST

Books and Reports

Abboushi, W. F., *The Angry Arabs* (Philadelphia, Westminster Press: 1974).

Abir, Mordechai, *Oil, Power and Politics* (London, Frank Cass: 1974).

Abu-Lughod, Ibrahim (ed.), *The Transformation of Palestine* (Evanston, Ill.; Northwestern University Press: 1971).

Adams, Michael, and Mayhew, Christopher, *Publish It Not...The Middle East Cover-Up* (New York, Longman: 1975).

Al-Marayati, Abid A., *The Middle East: Its Government and Politics* (Belmont, Calif.; Duxebury Press: 1972).

AlRoy, Gil Carl, *Behind the Middle East Conflict: The Real Impasse Between Arab and Jew* (New York, Capricorn: 1975).

AlRoy, Gil Carl, *The Kissinger Experience: American Policy in the Middle East* (New York, Horizon: 1975).

Anthony, John Duke, *Arab States of the Lower Gulf: People, Politics, Petroleum* (Washington, Middle East Institute: 1975).

Anthony, John Duke (ed.), *The Middle East: Oil, Politics, and Development* (Washington, American Enterprise Institute: 1975).

The Arabian Peninsula, Iran and the Gulf States: New Wealth, New Power (Washington, Middle East Institute: 1973).

Aruri, Naseer H. (ed.), *Middle East Crucible: Studies on the Arab-Israeli War of October 1973* (Wilmette, Ill.; Medina Univ. Press: 1975).

Becker, Abraham S., *The Economics and Politics of the Middle East* (New York, American Elsevier: 1975).

Begin, Menahem, *The Revolt: The Dramatic Inside Story of the Irgun* (Los Angeles, Nash: 1972).

Beling, Willard A., *The Middle East: Quest for an American Policy* (Albany, State University of New York Press: 1973).

Bell, J. Bowyer, *Terror Out of Zion* (New York, St. Martin's Press: 1977).

Bellow, Saul, *To Jerusalem and Back: A Personal Account* (New York, Avon: 1977).

Ben-Dak, Joseph, and Assousa, George, "Peace in the Near East: The Palestinian Imperative" (Muscatine, Iowa; Stanley Foundation: 1974).

Ben Ezer, Ehud, *Unease in Zion* (Jerusalem, Jerusalem Academic Press: 1974).

Bertelsen, Judy, "The Palestinian Arabs: A Non-State Nation Systems Analysis," Sage Professional Paper, Beverly Hills and London, 1976.

Blair, John M., *The Control of Oil* (New York, Pantheon: 1976).

Blandford, Linda, *Superwealth: The Secret Lives of the Oil Sheiks* (New York, Morrow: 1977).

Bill, James A., *The Middle East: Politics and Power* (Boston, Allyn and Bacon: 1974).

Bober, Arie, *The Other Israel: The Radical Case Against Zionism* (New York, Doubleday: 1972).

Bose, Tarun C., *The Superpowers in the Middle East* (New York, Asia Publishing House: 1972).

Bovis, Eugene H., *The Jerusalem Question* (Stanford, Calif.; Hoover Institute Press: 1971).

Brecher, Michael, *Decisions in Israel's Foreign Policy* (New Haven, Yale University Press: 1975).

Brecher, Michael, *The Foreign Policy System of Israel* (New Haven, Yale University Press: 1972).

Brookings Institution, "Toward Peace in the Middle East," Washington, December 1975.

Bruzonsky, Mark A., "A United States Guarantee for Israel?" Washington, Georgetown University Center for Strategic and International Studies, April 1976.

Bull, Vivian A., *The West Bank—Is It Viable?* (Lexington, Mass.; Lexington Books: 1975).

Bullock, John, *The Making of a War: The Middle East From 1967 to 1973* (London, Longman: 1974).

Carmichael, Joel, *Arabs Today* (Garden City, N.Y.; Anchor: 1977).

Cattan, Henry, *Palestine and International Law* (London, Longman: 1973).

Central Intelligence Agency, "The International Energy Situation: Outlook to 1985," Washington, April 1977.

Chill, Dan S., *The Arab Boycott of Israel* (New York, Praeger Special Studies in International Economics and Development: 1976).

Chubin, Shahram, *The Foreign Relations of Iran: A Developing State in a Zone of Great-Power Conflict* (Berkeley, University of California Press: 1975).

Churba, Joseph, *The Politics of Defeat: America's Decline in the Middle East* (New York, Cyrco Press: 1977).

Cline, Ray, *World Power Assessment* (Washington, Georgetown University Center for Strategic and International Studies: 1977).

Cohen, Aharon, *Israel and the Arab World* (Boston, Beacon: 1976).

Confino, Michael (ed.), *The U.S.S.R. and the Middle East* (Jerusalem, Israel Universities Press: 1973).

Cooley, John K., *Green March, Black September: The Story of the Palestinian Arabs* (London, Frank Cass: 1973).

Cooper, Charles A. (ed.), *Economic Development and Population Growth in the Middle East* (New York, American Elsevier: 1972).

Costello, Mary, "Foreign Policy After Kissinger," Washington, Editorial Research Reports; Vol. I, 1977; pp. 3-20.

Costello, Mary, "Arab Disunity," Washington, Editorial Research Reports, Vol. II, 1976, pp. 787-806.

Curtis, Michael (ed.), *Israel: Social Structure and Change* (New York, Transaction Books: 1973).

Curtis, Michael (ed.), *The Palestinians: People, History, Politics* (New York, Transaction Books: 1975).

Curtis, Michael (ed.), *People and Politics in the Middle East* (New York, Transaction Books: 1971).

Deans, Ralph C., "American Policy in the Middle East;" Washington; Editorial Research Reports; Vol. II, 1970; pp. 609-626.

Dobson, Christopher, *Black September: Its Short, Violent History* (New York, Macmillan: 1974).

Dowty, Alan, "The Role of Great Power Guarantees in International Peace Agreements," Jerusalem Papers on Peace Problems #3, Jerusalem, The Hebrew University, February 1974.

El-Asmar, Fouzi, *To Be an Arab in Israel* (London, Frances Pinter: 1975).

Eliav, Arie L., *Land of the Hart: Israelis, Arabs, the Territories and a Vision of the Future* (Philadelphia, Jewish Publication Society: 1974).

El Kodsy, Ahmad, and Lobel, Eli, *The Arab World and Israel* (New York, Monthly Review Press: 1970).

Elmessiri, Abdelwahab M., and Stevens, Richard P., *Israel and South Africa: The Progression of a Relationship* (New York, New World Press: 1976).

Elon, Amos, *The Israelis: Founders and Sons* (New York; Holt, Rinehart and Winston: 1971).

Elon, Amos, and Hassan, Sana, *Between Enemies: A Compassionate Dialogue between an Israeli and an Arab* (New York, Random House: 1974).

El-Rashidi, Galal, *The Arabs and the World of the Seventies* (New Delhi, Vikas: 1977).

Epp, Frank H., *The Palestinians: Portrait of a People in Conflict* (Scottdale, Pa.; Herald Press: 1976).

Erdman, Paul, *The Crash of '79* (New York, Simon and Schuster: 1977).

Evron, Yair, *The Middle East: Nations, Superpowers, and War* (New York, Praeger: 1973).

Fabian, Larry L., and Schiff, Ze'ev (eds.), *Israelis Speak* (Washington, Carnegie Endowment for International Peace: 1977).

Faddad, Mohammed I., *The Middle East in Transition: A Study of Jordan's Foreign Policy* (New York, Asia Publishing House: 1974).

Fenelon, K. G., *The United Arab Emirates: An Economic and Social Survey* (London, Longman: 1973).

First, Ruth, *Libya: The Elusive Revolution* (Baltimore, Penquin Books: 1974).

Friendlander, Saul, and Hussein, Mahmoud, *Arabs & Israelis: A Dialogue* (New York, Holmes & Meier: 1975).

Geyer, Georgie A., *The New 100 Years War* (Garden City, N.Y.; Doubleday: 1972).

Ghilan, Maxim, *How Israel Lost Its Soul* (London, Penguin: 1974).

Glassman, Jon D., *Arms for the Arabs: The Soviet Union and War in the Middle East* (Baltimore, Johns Hopkins University Press: 1976).

Golan, Galia, *Yom Kippur and After: The Soviet Union and the Middle East Crisis* (New York, Cambridge University Press: 1977).

Golan, Matti, *The Secret Conversations of Henry Kissinger* (New York, Quadrangle: 1976).

Gonen, Jay Y., *A Psychohistory of Zionism* (New York, Mason/Charter: 1975).

Haddad, George, *Revolution and Military Rule in the Middle East*, 3 volumes (New York, Speller: 1965-1973).

Halkin, Hillel, *Letters to an American Friend: A Zionist's Polemic* (Philadelphia, Jewish Publication Society: 1977).

Hamer, John, "Persian Gulf Oil:" Washington; Editorial Research Reports; Volume I, 1973; pp. 231-248.

Hamer, John, "World Arms Sales;" Washington; Editorial Research Reports; Vol. I, 1976; pp. 325-342.

Hammond, Paul Y. (ed.), *Political Dynamics in the Middle East* (New York, American Elsevier: 1971).

Handel, Michael I., *Israel's Political-Military Doctrine* (Cambridge, Harvard University Center for International Affairs: July 1973).

Harkabi, Yehoshafat, *Arab Strategies and Israel's Response* (New York, The Free Press: 1977).

Hatem, M. Abdel-Kader, *Information and the Arab Cause* (London, Longman: 1974).

Heikal, Mohamed, *The Road to Ramadan* (New York, Ballantine: 1975).

Heikal, Mohamed, *The Cairo Documents* (Garden City, N.Y., Doubleday: 1973).

Heradstveit, Daniel, *Arab and Israeli Elite Perceptions* (Atlantic Highlands, N.J.: 1973).

Hertzberg, Arthur, *The Zionist Mind* (New York, Atheneum: 1973).

Herzl, Theodore, *The Jewish State* (Tel Aviv, Rohald Press: 1956).

Hirschmann, Ira, *Red Star Over Bethlehem: Russia Drives for the Middle East* (New York, Simon and Schuster: 1971).

Hirschmann, Ira, *Questions and Answers About Arabs and Jews* (New York, Bantam: 1977).

Houthakker, Hendrik S., "The World Price of Oil," Washington, American Enterprise Institute: October 1976.

Howe, Irving, and Gershman, Carl, *Israel, the Arabs, the Middle East* (New York, Bantam: 1972).

Howe, Russell Warren, and Trott, Sarah Hays, *The Power Peddlers* (Garden City, N.Y.; Doubleday: 1977).

Hunter, Robert E., "The Energy 'Crisis' and U.S. Foreign Policy;" (New York, Foreign Policy Association: 1973).

Hurewitz, J. C., *The Arab-Israeli Dispute, and the Industrial World* (Boulder, Colo.; Westview: 1976).

Hurewitz, J. C., *The Struggle for Palestine* (New York, Shocken: 1976).

Hurewitz, J. C., *Middle East Dilemmas: The Background of United States Policy* (New York, Russell and Russell: 1973).

Hurewitz, J. C., *Middle East Politics: The Military Dimension* (New York, Octagon: 1974).

Hussain, Mehmood, *PLO* (Delhi, India; University Publishers: 1975).

Insight Team (London Sunday Times), *The Yom Kippur War* (Garden City, N.Y.; Doubleday: 1974).

Institute for Psychiatry and Foreign Affairs, Department of State, "The Psychological Impact of the Seventeen-Day War," March 29, 1974.

Isaac, Rael Jean, *Israel Divided: Ideological Politics in the Jewish State* (Baltimore, Johns Hopkins University Press: 1976).

Isaacs, Stephen D., *Jews and American Politics* (Garden City, N.Y.; Doubleday: 1974).

Ismael, Tareq Y., *Government and Politics of the Contemporary Middle East* (Homewood, Ill.; Dorsey Press: 1970).

Ismael, Tareq Y., *The Middle East in World Politics: A Study in International Relations* (Syracuse, N.Y.; Syracuse University Press: 1974).

Jabber, Fuad (ed.), *Israel and Nuclear Weapons* (Chester Springs, Pa.; Dufour: 1973).

Jiryis, Sabri, *The Arabs in Israel* (New York, Monthly Review Press: 1976).

Jureidini, Paul A., and Hazen, William E., *The Palestinian Movement in Politics* (Lexington, Mass.; Lexington Books: 1976).

Kalb, Marvin, and Kalb, Bernard, *Kissinger* (Boston, Little Brown: 1974).

Kaplan, Mordechai M., *A New Zionism* (New York, The Herzl Press: 1959).

Katz, Samuel, *Battleground: Fact and Fantasy in Palestine* (New York, Bantam: 1973).

Kennan, George F., *The Cloud of Danger: Current Realities of American Foreign Policy* (New York, Little, Brown: 1977).

Kerr, Malcolm H. (ed.), *The Elusive Peace in the Middle East* (Albany, State University of New York Press: 1975).

Kiernan, Thomas, *The Arabs: Their History, Aims and Challenge to the Industrial World* (Boston; Little, Brown: 1975).

Kiernan, Thomas, *Yasir Arafat* (London, Abacus: 1976).

Kimche, Jon, *There Could Have Been Peace* (New York, Dial Press: 1973).

Kimche, Jon, *The Second Arab Awakening* (New York; Holt, Rinehart, and Winston: 1974).

Klebanoff, Shoshana, *Middle East Oil and U.S. Foreign Policy: With Special Reference to the U.S. Energy Crisis* (New York, Praeger Special Studies in International Economics and Development: 1974).

Klug, Tony, "Middle East Impasse: The Only Way Out," (London, Fabian Society: 1977).

Kohler, Foy D., "The Soviet Union and the October 1973 Middle East War: The Implications for Detente;" Miami, Florida; University of Miami, Center for International Studies; 1974.

Koury, Enver M., *The Super Powers and the Balance of Power in the Arab World* (Beirut, Lebanon; Catholic Press: 1970).

Koury, Fred J., *The Arab-Israeli Dilemma* (New York, Knopf: 1976).

Laffin, John, *The Arab Mind Considered: A Need for Understanding* (New York, Taplinger: 1975).

Landau, Jacob M., *The Arabs in Israel* (London, Oxford: 1968).

Landis, Lincoln, *Politics and Oil: Moscow in the Middle East* (New York, Dunellen: 1973).

Laqueur, Walter (ed.), *The Arab-Israeli Reader* (New York, Bantam: 1976).

Laqueur, Walter, *A History of Zionism* (New York, Holt, Rinehart and Winston: 1972).

Laqueur, Walter, *Confrontation: The Middle East and World Politics* (New York, Quadrangle: 1974).

Laqueur, Walter, *The Struggle for the Middle East: The Soviet Union in the Mediterranean 1958-1968* (London, Routledge, Kegan Paul: 1970).

Lenczowski, George (ed.), "Political Elites in the Middle East;" Washington; American Enterprise Institute; 1973.

Lenczowski, George, "Soviet Advance in the Middle East;" Washington; American Enterprise Institute; 1973.

Lenczowski, George, "Middle East Oil in a Revolutionary Age;" Washington; American Enterprise Institute, March 1976.

Lewis, Bernard, *The Middle East and the West* (New York, Harper Torch Books: 1966).

Long, David E., *The Persian Gulf* (Boulder, Colo.; Westview: 1976).

Long, David E., "Saudi Arabia;" Washington; Washington Paper No. 39, Georgetown University Center for Strategic and International Studies; 1976.

Lucas, Noah, *The Modern History of Israel* (New York, Praeger: 1975).

Malone, Joseph J., *The Arab Lands of Western Asia* (Englewood Cliffs, N.J.; Prentice-Hall: 1973).

Mansfield, Peter (ed.), *The Middle East: A Political and Economic Survey* (New York, Oxford University Press: 1973).

Mansfield, Peter, *The Arab World: A Comprehensive History* (New York, Crowell: 1977).

Ma'oz, Moshe (ed.), *Palestinian Arab Politics* (Jerusalem, Jerusalem Academic Press: 1975).

Meir, Golda, *A Land of Our Own* (Marie Syrkin, ed.), Philadelphia, Jewish Publication Society, 1973.

Memmi, Albert, *Jews and Arabs* (Chicago, O'Hara: 1975).

Milson, Menahem (ed.), *Society and Political Structure in the Arab World* (New York, Humanities Press: 1973).

Moore, John N. (ed.), *The Arab-Israeli Conflict*, 3 volumes (Princeton, Princeton University Press: 1975).

Mosley, Leonard, *Power Plan: The Tumultuous World of Middle East Oil 1890-1973* (London, Weidenfeld and Nicolson: 1973).

Nakheleh, Emile A., *Arab-American Relations in the Persian Gulf* (Washington, American Enterprise Institute: 1975).

Norton, Augustus R., "Moscow and the Palestinians;" (Miami, Fla.; Center for Advanced International Studies, University of Miami: 1974).

Nutting, Anthony, *Nasser* (New York, E. P. Dutton: 1972).

O'Ballance, Edgar, *Arab Guerrilla Power: 1967-1972* (Hamden, Conn.; Shoe String Press: 1973).

O'Ballance, Edger, *The Third Arab-Israeli War* (Hamden, Conn.; Archon Books: 1972).

O'Ballance, Edgar, *The Electronic War in the Middle East 1968-1970* (Hamden, Conn.; Shoe String Press: 1974).

O'Neill, Bard E., *Revolutionary Warfare in the Middle East: The Israelis vs. The Fedayeen* (Boulder, Colo.; Paladin Press: 1974).

Pelkovits, N. A., "Security Guarantees in a Middle East Settlement," Sage Foreign Policy Papers No. 5, Beverly Hills and London, 1976.

Polk, William R., *The United States and the Arab World*, Third ed.; (Cambridge, Harvard University Press: 1975).

Pomerance, Michla, "American Guarantees to Israel and the Law of American Foreign Relations," Jerusalem Papers on Peace Problems No. 9, Jerusalem, The Hebrew University, December 1974.

Porath, Y., *The Emergence of the Palestinian-Arab National Movement, 1918-1929* (London, Frank Cass: 1974).

Pranger, Robert J., "American Policy for Peace in the Middle East, 1969-1971;" Washington; American Enterprise Institute; 1973.

Pranger, Robert J., and Tahtinen, Dale R.; "Nuclear Threat in the Middle East;" Washington; American Enterprise Institute; July 1975.

Pranger, Robert J., and Tahtinen, Dale R.; "Implications of the 1976 Arab-Israeli Military Status;" Washington; American Enterprise Institute; April 1976.

Pryce-Jones, David, *The Face of Defeat: Palestinian Refugees and Guerrillas* (New York; Holt, Rinehart and Winston: 1972).

Quandt, William B., *et. al. The Politics of Palestinian Nationalism* (Berkeley, University of California Press: 1973).

Rabinowicz, Oscar K., *Arnold Toynbee on Judaism and Zionism: A Critique* (London, W. H. Allen: 1974).

Rabinowitz, Ezekiel, *The Jews: Their Dream of Zion and the State Department* (New York, Vantage: 1973).

Ramazani, Rouhollah K., *The Persian Gulf: Iran's Rule* (1972).

Reich, Bernard, *The Quest for Peace* (New York, Transaction Books: 1977).

Reisman, Michael, *The Art of the Possible: Diplomatic Alternatives in the Middle East* (Princeton, Princeton University Press: 1970).

Roberts, Samuel, *Survival or Hegemony? The Foundations of Israel's Foreign Policy* (Baltimore, Johns Hopkins University Press: 1974).

Rodinson, Maxime, *Israel: A Colonial-Settler State?* (New York, Monad Press: 1973).

Rodinson, Maxime, *Israel and the Arabs* (Baltimore, Penguin: 1973).

Rosenfeld, Alvin, *The Plot to Destroy Israel* (New York, Putnam: 1977).

Rostow, Eugene V., *The Middle East: Critical Choices for the United States* (Boulder, Colo.; Westview: 1977).

Rubinstein, Alvin Z., *Red Star on the Nile: The Soviet-Egyptian Relationship Since the June War* (Princeton, Princeton University Press: 1977).

Sachar, Howard M., *A History of Israel: From the Rise of Zionism to Our Time* (New York, Knopf: 1976).

Safran, Nadav, *From War to War, 1948-1967* (New York, Pegasus: 1969).

Schiff, Ze'ev, and Rothstein, Raphael, *Fedayeen: Guerrillas Against Israel* (New York, David McKay: 1972).

Schleifer, Abdullah, *The Fall of Jerusalem* (New York, Monthly Review Press: 1972).

Schmidt, Dana Adams, *Armagedon in the Middle East* (New York, John Day: 1974).

Schurr, Sam H., *Middle East Oil and the Western World, Prospects and Problems* (New York, American Elsevier: 1971).

Schoenbrun, David, *The Israelis* (New York, Atheneum: 1973).

Segal, Ronald, *Whose Jerusalem? The Conflict of Israel* (London, Jonathan Cape: 1973).

Segal, Ronald, *Whose Jerusalem?* (New York, Bantam: 1974).

The Seventh Day: Soldiers Talk About the Six-Day War (London, Penguin: 1971).

Sheehan, Edward R. F., *The Arabs, Israelis, and Kissinger: A Secret History of American Diplomacy in the Middle East* (New York, Reader's Digest Press: 1976).

Sherman, Arnold, *When God Judged and Men Died* (New York, Bantam: 1973).

Sherbiny, Naiem A., and Tessler, Mark A. (eds.), *Arab Oil: Impact on the Arab Countries and Global Implications* (New York, Praeger Special Studies in International Business, Finance and Trade: 1976).

Shwadran, B., *Middle East, Oil and the Great Powers* (Jerusalem, Israel Universities Press: 1973).

Sid-Ahmed, Mohamed, *After the Guns Fall Silent* (New York, St. Martin's Press: 1977).

Sivard, Ruth Leger, *World Military and Social Expenditures* (Leesburg, Va.; WMSE Publications: 1977).

Slonim, Shlomo, "United States-Israel Relations, 1967-1973;" Jerusalem Papers on Peace Problems No. 8; Jerusalem, The Hebrew University, 1974.

Snetsinger, John, *Truman, the Jewish Vote and the Creation of Israel* (Stanford, Calif.; Hoover Institute Press: 1974).

Stephens, Robert, *Nasser: A Political Biography* (New York, Simon and Schuster: 1971).

Stoessinger, John G., *Henry Kissinger: The Anguish of Power* (New York, Norton: 1976).

Sykes, Christopher, *Crossroads to Israel 1917-1948* (Bloomington, Indiana University Press: 1973).

Tahtinen, Dale R., "Arms in the Persian Gulf;" Washington, American Enterprise Institute, 1974.

Tahtinen, Dale R., "The Arab-Israeli Military Balance Since 1973;" Washington, American Enterprise Institute, 1974.

Tahtinen, Dale R., "Arab-Israeli Military Status in 1976;" Washington, American Enterprise Institute, January 1976.

Talmon, J. L., *Israel Among the Nations* (New York, Macmillan: 1970).

Tekoah, Yosef, *In the Face of the Nations: Israel's Struggle for Peace* (New York, Simon & Schuster: 1976).

Teveth, Shabtai, *Moshe Dayan: The Soldier, the Man, the Legend* (Boston, Houghton Mifflin: 1973).

Tuma, Elias H., *Peacemaking and the Immoral War: Arabs and Jews in the Middle East* (New York, Harper: 1972).

Turki, Fawaz, *The Disinherited: Journal of a Palestinian Exile* (New York, Monthly Review Press: 1972).

Udovitch, A. L. (ed.), *The Middle East: Oil, Conflict & Hope* (Lexington, Mass.; Lexington Books: 1976).

Urofsky, Melvin I., *American Zionism from Herzl to the Holocaust* (Garden City, Anchor Press: 1976).

Van Arkadie, Brian, *Benefits and Burdens: A Report on the West Bank and Gaza Strip Economies Since 1967* (Washington, Carnegie Endowment for International Peace: 1977).

Vicker, Ray, *The Kingdom of Oil* (New York, Scribner: 1974).

Vital, David, *The Origins of Zionism* (Oxford, Clarendon Press: 1976).

Voss, Carl Hermann, *The Palestine Problem Today: Israel and Its Neighbors* (Boston, Beacon Press: 1953).

Wagner, Abraham R., *Crisis Decision-Making: Israel's Experience in 1967 and 1973* (New York, Praeger: 1974).

Waines, David, *The Unholy War: Israel & Palestine 1897-1971* (New York, Chateau Books: 1971).

Ward, Richard J. (ed.), *The Palestine State: A Rational Approach* (Port Washington, N.Y.; Kennikat Press: 1977).

Weizman, Ezer, *On Eagles' Wings* (New York, Macmillan: 1976).

Wells, Donald A., "Saudi Arabian Development Strategy;" Washington, American Enterprise Institute, September 1976.

Whetten, Lawrence L., *The Canal War: Four-Power Conflict in the Middle East* (Cambridge, MIT Press: 1974).

Wilson, Evan M., *Jerusalem: Key to Peace* (Washington, Middle East Institute: 1970).

World Energy Demands and the Middle East (Washington, Middle East Institute: 1972).

Articles

Adams, Michael; "Israel's Record in the Occupied Territories," *Journal of Palestine Studies*, Winter 1977.

Agha, Hussein J.; "What State for the Palestinians?" *Journal of Palestine Studies*, Autumn 1976.

Ajami, Fouad; "Middle East Ghosts," *Foreign Policy*, Spring 1974.

Allman, T. D., "Oppressor Israel," *New Times*, August 19, 1977.

Allon, Yigal; "Israel: The Case for Defensible Borders," *Foreign Affairs*, October 1976.

"The Arab-Israeli Wars from 1967 to 1973," *Commentary*, December 1973.

Aronson, Shlomo; "Nuclearization of the Middle East," *The Jerusalem Quarterly*, Winter 1977.

Avineri, Shlomo; "The Palestinians and Israel," *Commentary*, June 1970.

Avishai, Bernard, "A New Israel," *The New York Review*, June 23, 1977.

Ball, George: "The Coming War in the Middle East and How to Avert It," *Atlantic Monthly*, January 1975.

Ball, George; "Kissinger's Paper Peace: How Not to Handle the Middle East," *Atlantic Monthly*, February 1976.

Ball, George; "How to Save Israel in Spite of Herself," *Foreign Affairs*, April 1977.

Barberis, Mary A.; "The Arab-Israeli Battle on Capitol Hill," *Virginia Quarterly Review*, Spring 1975.

Bartov, Hanoch, "Israel after the War: Back to Normal," *Commentary*, March 1974.

Beim, David O.; "Rescuing the LDCs," *Foreign Affairs*, July 1977.

Bell, Coral; "The October Middle East War: A Case Study in Crisis Management During Detente," *International Affairs*, October 1974.

Berger, Elmer; "Memoirs of an Anti-Zionist Jew," *Journal of Palestine Studies*, Autumn 1975/Winter 1976.

Bill, J. A.; "Plasticity of Informal Politics," *Middle East Journal*, Spring 1975.

Breedham, Bruce; "The Gulf," *Economist, Survey Section,* May 1975.

Bruzonsky, Mark A.; "American Proposals for Peace in the Middle East," *International Problems* (Tel Aviv University), Fall 1975.

Bruzonsky, Mark A.; "The U.S., the PLO, and Israel," *Commonweal*, May 21, 1976.

Bruzonsky, Mark A.; "Washington: Uncertain, Waiting, But Getting Ready," *New Outlook* (Tel Aviv), Sept.-Oct. 1976.

Bruzonsky, Mark A.; "American Thinking About A Security Guarantee for Israel," *International Problems* (Tel Aviv University), Fall 1976.

Bruzonsky, Mark A.; "U.S. & Israel: The Coming Storm," *Worldview*, September 1976.

Bruzonsky, Mark A.; "What Washington Think-Tanks Think About the Middle East," *National Jewish Monthly*, December 1976.

Bruzonsky, Mark A.; "Carter and the Middle East," *The Nation*, December 11, 1976.

Bruzonsky, Mark A.; "Carter's 'Year of Decision' in the Middle East," *Dissent*, Winter 1977.

Bruzonsky, Mark A.; "Mr. Carter Grasps That Nettle," *The Nation*, April 23, 1977.

Bruzonsky, Mark A.; "We Want the Israelis To Understand Us" (an interview with Egyptian Ambassador Ashraf Ghorbal), *Worldview*, July-August 1977.

Bruzonsky, Mark A.; "Let Us Talk Face to Face" (an interview with Israeli Ambassador Simcha Dinitz), *Worldview*, July-August 1977.

Brzezinski, Zbigniew; "Unmanifest Destiny: Where Do We Go From Here?" *New York Magazine*, March 3, 1975.

Brzezinski, Zbigniew, *et. al.;* "Peace In An International Framework," *Foreign Policy*, Summer 1975.

Caradon, Lord; "The Shape of Peace in the Middle East (an interview)," *Journal of Palestine Studies*, Spring/Summer 1976.

Chenery, Hollis B.; "Restructuring the World Economy," *Foreign Affairs*, January 1975.

Cumming-Bruce, Nick, "U.S. and France Lead Scramble for Middle East Arms Orders," *Middle East Economic Digest*, February 7, 1975.

"Can OPEC Be Broken Up?" *Forbes*, February 15, 1977.

Diodatus, "Playing to the PLO, The New U.S. Policy in the Middle East," *The New Leader*, January 5, 1976.

Draper, Theodore; "United States and Israel: Tilt in the Middle East?" *Commentary*, April 1975.

Elon, Amos; "Yitzhak Rabin: After Father and Mother Figures, a Native Son Rules Israel," *New York Times Magazine*, May 4, 1975.

Enders, Thomas O.; "OPEC and the Industrial Countries: the Next Ten Years," *Foreign Affairs*, July 1975.

"Facing Facts in the Middle East: With An Analysis of Views of S. Hoffmann," *America*, May 10, 1975.

Feron, James; "The Israelis of New York," *The New York Times Magazine*, January 16, 1977.

Fulbright, William J.; "Getting Tough With Israel," *Washington Monthly*, February 1975.

Fulbright, William J.; "Beyond the Sinai Agreement," *Worldview*, November 1975.

Ghareeb, Edmund; "The American Media and the Palestine Problem," *Journal of Palestine Studies*, Autumn 1975/Winter 1976.

Goldmann, Nahum; "The Time for Peace," *Journal of Palestine Studies*, Winter 1975.

Goldmann, Nahum; "The Psychology of Middle East Peace," *Foreign Affairs*, October 1975.

Halkin, Hillel;"Driving Toward Jerusalem: A Sentimental Journey Through the West Bank," *Commentary*, January 1975.

Harbottle, Michael; "The October Middle East War: Lessons for UN Peacekeeping," *International Affairs*, October 1974.

Harkabi, Yehoshafat; "The Position of the Palestinians in the Israeli-Arab Conflict and Their National Covenant," *New York University Journal of International Law & Politics*, Spring 1970.

Harris, Louis; "Oil or Israel?" *The New York Times Magazine*, April 6, 1975.

Holden, David; "Which Arafat," *The New York Times Magazine*, March 23, 1975.

Hoffmann, Stanley; "New Policy for Israel," *Foreign Affairs*, April 1975.

Hoffmann, Stanley; "Tiptoeing Toward Peace in the Middle East," *Harvard Political Review*, Spring 1975.

Hudson, Michael; "The Lebanese Crisis," *Journal of Palestine Studies*, Spring/Summer 1976.

"Israel and Torture," *Sunday Times* (London), June 19, 1977.

Just, Ward; "Unease in Zion," *Atlantic*, June 1975.

Kennedy, Edward M.; "The Persian Gulf: Arms Race or Arms Control?" *Foreign Affairs*, October 1975.

Kipper, Judith, and Bruzonsky, Mark A.; "Washington and the PLO," *Middle East International* (London), February 1977.

Kollek, Teddy; "Jerusalem," *Foreign Affairs*, July 1977.

Kraar, Louis, "OPEC is Starting to Feel the Pressure," *Fortune*, May 1975.

Laqueur, Walter, "Israel After the War: Peace with Egypt?" *Commentary*, March 1974.

Laqueur, Walter, and Luttwak, Edward; "Kissinger and the Yom Kippur War," *Commentary*, September 1974.

Lelyveld, Joseph, "Katz in the Mountains," *The New York Times Magazine*, July 10, 1977.

Lewis, Bernard; "The Palestinians and the PLO: A Historical Approach," *Commentary*, January 1975.

Levy, Walter J.; "World Oil Cooperation or International Chaos," *Foreign Affairs*, July 1974.

Michaelson, Michael; "Peace Prospects Are Better Than Ever; An Interview With Egyptian President Sadat," *Parade*, February 6, 1977.

"The Middle East: 1975," *Current History*, February 1975.

"The Middle East: 1977," *Current History*, January 1977.

Merlin, Samuel; "Menahem Begin: Orator, Commander, Statesman," *National Jewish Monthly*, July-August 1977.

Middleton, Drew; "Who Lost the Yom Kippur War?" *Atlantic Monthly*, March 1974.

Moran, Theodore H.; "Why Oil Prices Go Up—The Future: OPEC Wants Them," *Foreign Policy*, Winter 1976-77.

Morgenthau, Hans J.; "An Intricate Web," *The New Leader*, December 24, 1973.

Morgenthau, Hans J.; "The Geopolitics of Israel's Survival," *The New Leader*, February 4, 1974.

Nissan, Mordechai; "PLO 'Moderates,'" *The Jerusalem Quarterly*, Fall 1976.

Oppenheim, V. H.; "Why Oil Prices Go Up—The Past: We Pushed Them," *Foreign Policy*, 1976-77.

Oren, Stephen; "The Struggle for Lebanon," *New Leader*, May 12, 1975.

Pearson, Anthony; "Mayday! Mayday!—The Attack on the Liberty," *Penthouse*, May 1976.

Pearson, Anthony; "Conspiracy of Silence," *Penthouse*, June 1976.

Peretz, Don; "The United States, the Arabs, and Israel: Peace Efforts of Kennedy, Johnson, and Nixon," *The Annals of the American Academy of Political and Social Science*, May 1972.

Peters, Joan; "A Conversation With Dayan," *Harper's*, November 1975.

Poole, James; "The Next Oil Crisis," *Atlas World Press Review*, July 1976.

Pranger, Robert J.; "Nuclear War Comes to the Mideast," *Worldview*, July-August 1977.

Quandt, William R.; "Kissinger and the Arab-Israeli Disengagement Negotiations," *Journal of International Affairs*, Spring 1975.

Roosa, Robert V.; "How Can the World Afford OPEC Oil?" *Foreign Affairs*, January 1976.

Rouleau, Eric; "The Palestinian Quest," *Foreign Affairs*, January 1975.

Rustow, Dankwart, "Who Won the Yom Kippur and Oil Wars," *Foreign Policy*, Winter 1974.

Rustow, Dankwart; "U.S.-Saudi Relations and the Oil Crises of the 1980s," *Foreign Affairs*, April 1977.

Safran, Nadav; "Engagement in the Middle East," *Foreign Affairs*, October 1974.

Safran, Nadav; "The War and the Future of the Arab-Israeli Conflict," *Foreign Affairs*, January 1974.

Salpeter, Eliahu; "Balancing Act in the Persian Gulf," *New Leader*, March 17, 1975.

Salpeter, Eliahu; "Dreams and Realities in the Shah's Iran," *New Leader*, March 3, 1975.

Schnall, David J.; "Native Anti-Zionism: Ideologies of Radical Dissent in Israel," *The Middle East Journal*, Spring 1977.

Sheehan, Edward R. F.; "A Proposal for a Palestinian State," *The New York Times Magazine*, January 30, 1977.

Shinn, Roger L.; "Nuclear War in the Mideast: Ethical Questions," *Worldview*, July-August 1977.

Simpson, Dwight J.; "Israel After Twenty-Five Years," *Current History*, January 1973.

Singer, Israel, and Bruzonsky, Mark A.; "Dependent Israel: The Two Options," *Worldview*, April 1976.

Smith, Terence; "Israel Journal; 1972-1976: Reflections on a Troubled People," *Saturday Review*, February 5, 1977.

Stevens, Georgiana G.; "1967-1977: America's Moment in the Middle East?" *Middle East Journal*, Winter 1977.

Stone, Christopher D., and McNamara, Jack; "How to Take on OPEC," *The New York Times Magazine*, December 12, 1975.

Szulc, Tad; "How Carter Fouled the Israeli Election," *New York Magazine*, June 6, 1977.

Tucker, Robert W.; "Oil: The Issue of American Intervention," *Commentary*, January 1975.

Tucker, Robert W.; "Israel and the United States: From Dependency to Nuclear Weapons," *Commentary*, November 1975.

Tueni, Ghassan, "After October: Military Conflict and Political Change in the Middle East," *Journal of Palestine Studies*, Summer 1974.

Turck, Nancy; "The Arab Boycott of Israel," *Foreign Affairs*, April 1977.

Turki, Fawaz; "The Palestinian Estranged," *Journal of Palestine Studies*, Autumn 1975/Winter 1976.

Ullman, Richard H.; "After Rabat: Middle East Risks and American Roles," *Foreign Affairs*, January 1975.

Ullman, Richard H.; "Alliance With Israel?" *Foreign Policy*, Summer 1975.

"United States Policy Toward the Middle East," *Middle East Review*, Spring 1977.

Veit, Lawrence A.; "Troubled World Economy," *Foreign Affairs*, January 1977.

Vital, David; "Israel After the War: The Need for Political Change," *Commentary*, March 1974.

Walters, Robert; "Big-Name Americans Who Work for Foreign Countries," *Parade*, June 20, 1976.

Wells, Benjamin; "Arab Power," *Washingtonian*, December 1975.

Williams, Maurice J.; "The Aid Programs of the OPEC Countries," *Foreign Affairs*, January 1976.

Reference

Statistical data and other factual information have been compiled from the following reference sources:

The CBS News Almanac 1977 (Maplewood, N.J.; Hammond Almanac: 1976).

Encyclopedia of Zionism and Israel (2 volumes), New York, Herzl Press, 1971.

Facts on File Publications:
  — *Egypt & Nasser* (3 volumes), 1973.
  — *Israel & the Arabs: Prelude to the Jewish State*, 1972.
  — *Israel & the Arabs: The June 1967 War*, 1968.
  —*Israel & the Arabs: The October 1973 War*, 1974.
  —*Palestinian Impasse: Arab Guerrillas & International Terror*, 1977.
  —*Political Terrorism*, 1975.
  —*U.S. & Soviet Policy in the Middle East 1945-1956*, 1972.
  —*U.S. & Soviet Policy in the Middle East 1957-1966*, 1974.

International Economic Report of the President, January 1977.

National Basic Intelligence Handbook, Central Intelligence Agency, January 1977.

Political Dictionary of the Middle East in the 20th Century (revised and updated edition), Yaacov Shimoni & Evyatar Levine (eds.), New York, Quadrangle, 1974.

Political Handbook of the World: 1977, Arthur S. Banks (ed.), New York, McGraw-Hill, 1977.

Congressional Documents

"The Middle East, 1971: The Need To Strengthen the Peace" (July 13, 14, 15, 27; August 3; Sept. 30; Oct. 5, 28), Hearings Before the House Committee on Foreign Affairs, 1971.

"Jerusalem: The Future of the Holy City For Three Monotheisms" (July 28), Hearing before the House Committee on Foreign Affairs, 1971.

"U.S. Interests In and Policy Toward the Persian Gulf" (Feb. 2; June 7; August 8, 15), Hearings before the House Committee on Foreign Affairs, 1972.

"Oil Negotiations, OPEC, and the Stability of Supply" (April 10; May 14, 16; July 11; Sept. 6), Hearings before the House Committee on Foreign Affairs, 1973.

"New Perspectives on the Persian Gulf" (June 6; July 17, 23, 24; Nov. 28), Hearings before the House Committee on Foreign Affairs, 1973.

"The Impact of the October Middle East War" (Oct. 3, 23; Nov. 29), Hearings before the House Committee on Foreign Affairs, 1973.

"U.S.-European Relations and the 1973 Middle East War" (Nov. 1, 1973; Feb. 19, 1974), Hearings before the House Committee on International Relations, 1974.

"Report of Special Study Mission to the Middle East" (Feb. 25), House Committee on Foreign Affairs, 1974.

"The International Petroleum Cartel, the Iranian Consortium and U.S. National Security" (Feb. 21), Senate Committee on Foreign Relations, 1974.

"Middle East Between War and Peace, November-December 1973" (March 5), Staff Report of Senate Committee on Foreign Relations, 1974.

"Problems of Protecting Civilians Under International Law in the Middle East Conflict" (April 4), Hearing before the House Committee on Foreign Affairs, 1974.

"The Middle East, 1974: New Hopes, New Challenges" (April 9, May 7, 14, 23; June 27), Hearings before the House Committee on Foreign Affairs, 1974.

"Old Problems—New Relationships, Report of a Study Mission to the Middle East and South Asia," House Committee on Foreign Affairs, May 1974.

"International Terrorism" (June 11, 18, 19, 24), Hearings before the Committee on Foreign Affairs, 1974.

"Resolution of Inquiry into Proposed Nuclear Agreements with Egypt and Israel" (July 9), House Committee on Foreign Affairs, 1974.

"Persian Gulf 1974: Money, Arms and Power" (July 30; Aug. 5, 7, 12), Hearings before the House Committee on International Relations, 1974.

"The Chronology of the Libyan Oil Negotiations," Senate Committee on Foreign Relations, 1974.

"Authorizing U.S. Funding for United Nations Middle East Peacekeeping Forces" (April 8), Hearing before the House Committee on International Relations, 1975.

"The Persian Gulf, 1975: The Continuing Debate on Arms Sales" (June 10, 18, 24; July 29), Hearings before the House Committee on International Relations, 1975.

"Oil Fields as Military Objectives: A Feasibility Study," Library of Congress Congressional Research Service for House Committee on International Relations, August 21, 1975.

"U.S. Missile Sale to Jordan" (July 15, 21), Hearings before the Senate Committee on Foreign Relations, 1975.

"Priorities for Peace in the Middle East" (July 23, 24), Hearings before the Senate Committee on Foreign Relations, 1975.

"The Palestinian Issue in Middle East Peace Efforts" (Sept. 30, Oct. 1, 8; Nov. 12), Hearings before the House Committee on International Relations, 1975.

"Early Warning System in Sinai" (Oct. 6, 7), Hearings before the Senate Committee on Foreign Relations, 1975.

"Select Chronology and Background Documents Relating to the Middle East," Senate Committee on Foreign Relations, 1975.

"U.S. Oil Companies and the Arab Embargo: The International Allocation of Constricted Supplies," Senate Committee on Foreign Relations, 1975.

"Middle East Peace Prospects" (May 19; June 7, 21, 30; July 19 and 26), Hearings before the Senate Committee on Foreign Relations, 1976.

"The Arab Boycott and American Business," Report of House Committee on Interstate and Foreign Commerce, September 1976.

"U.S. Arms Sales Policy" (Sept. 16, 21, 24), Hearings before the Senate Committee on Foreign Relations, 1976.

"Middle East Problems" (May 18, 20), Hearings before the Senate Committee on Foreign Relations, 1977.

"Recent Developments in the Middle East" (June 8), Hearings before the House Committee on International Relations, 1977.

"Economic and Military Aid Programs in Europe and the Middle East" (Feb. 22, 23; March 3, 7, 9, 14, 16, 21; April 21, 27), Hearings before the House Committee on International Relations, 1977.

Note: Most hearings were held before the appropriate subcommittee of the House or Senate Committee. All documents are available from the U.S. Government Printing Office.

INDEX